T0342093

Building Services Engineering

Building Services Engineering

Smart and Sustainable Design for Health and Wellbeing

Tarik Al-Shemmeri and Neil Packer

This edition first published 2021
© 2021 John Wiley & Sons Ltd

The right of Tarik Al-Shemmeri and Neil Packer to be identified as the authors of this work has been asserted in accordance with law.

Registered Offices
John Wiley & Sons, Inc., 111 River Street, Hoboken, NJ 07030, USA
John Wiley & Sons Ltd, The Atrium, Southern Gate, Chichester, West Sussex, PO19 8SQ, UK

Editorial Office
111 River Street, Hoboken, NJ 07030, USA

For details of our global editorial offices, customer services, and more information about Wiley products visit us at www.wiley.com.

Wiley also publishes its books in a variety of electronic formats and by print-on-demand. Some content that appears in standard print versions of this book may not be available in other formats.

Library of Congress Cataloging-in-Publication Data

Names: Al-Shemmeri, Tarik, author. | Packer, Neil, 1959- author.
Title: Building services engineering : smart and sustainable design for
    health and wellbeing / Tarik Al-Shemmeri and Neil Packer.
Description: First edition. | Hoboken, NJ : John Wiley & Sons, Inc., 2021.
    | Includes bibliographical references and index.
Identifiers: LCCN 2020055799 (print) | LCCN 2020055800 (ebook) | ISBN
    9781119722854 (hardback) | ISBN 9781119722984 (adobe pdf) | ISBN
    9781119722977 (epub) | ISBN 9781119722953 (obook)
Subjects: LCSH: Buildings–Environmental engineering. |
    Buildings–Mechanical equipment.
Classification: LCC TH6021 .A425 2021 (print) | LCC TH6021 (ebook) | DDC
    696–dc23
LC record available at https://lccn.loc.gov/2020055799

Cover Design: Wiley
Cover Image: Designed by Hannah Al-Shemmeri

Set in 9.5/12.5pt STIXTwoText by SPi Global, Chennai, India
Printed and bound by CPI Group (UK) Ltd, Croydon, CR0 4YY

10 9 8 7 6 5 4 3 2 1

Contents

Preface

It is often claimed that, in the developed world, people spend 90% of their entire lives indoors. Sometimes it is a requirement for employment or leisure activities, on other occasions it is a reaction to less than optimal external conditions.

Buildings are often packed with engineering services to try to maintain an acceptable working/living environment in the face of changes in the external environment and building usage/occupancy. This has resulted in a huge array of different devices for heating, cooling, ventilating, and lighting the internal environment.

And yet very few people seem to be entirely satisfied with the result leading to the recent study of health and well-being in the internal environment. Thermal comfort in buildings has been a study for academic research for some time. This may prove to be a subset of *emotional comfort*. Early work is beginning to show that this more holistic comfort parameter has a significant effect on productivity in the workplace.

Of course, total comfort can always be brought about with sufficient expenditure of energy and resources. This cannot, however, take place at excessive cost and/or emission production leading to an urgent need for the smart, sustainable design of building services. This does not require the abandonment of all our existing engineering knowledge. Instead, it entails the provision of flexible, resource-minimising, environment-optimising systems deploying, for example, connectivity, the use of renewable energy, and the introduction of some new technologies tailored to human needs and expectations.

This book describes existing building services design practices, contrasting these traditional methods with new developments in areas such as smart ventilation, smart glazing systems, smart batteries, smart lighting, smart soundproofing, smart sensors and meters, etc.

It introduces principles to the reader via a well-tried and tested method comprising 100+ fully worked examples followed up by 100+ tutorial questions.

United Kingdom
October 2020

Tarik Al-Shemmeri
Neil Packer

Structure of the Book

Chapter 1: Ambient Air

The ancient Greeks believed air to be one of four fundamental elements that made up their Universe. This belief prevailed for over 2000 years until the eighteenth century.

Swedish Chemist **Carl Wilhelm Scheele** proposed the existence of an atmospheric component known as 'fire air' (air necessary for combustion) essentially identifying gaseous oxygen in 1772. Unfortunately, he did not formerly publish his findings and formal credit goes to British scientist **Joseph Priestly** who did publish his own work in the same area in 1774. The gas was later given the name 'Oxygen' by French chemist **Antoine Lavoisier** who confirmed its elemental nature.

It is likely that Scheele was also the first to recognise the atmospheric component Nitrogen (which he termed 'foul air') but again did not publish and so credit is usually given to a number of his peers, e.g. British botanist **Daniel Rutherford**.

Nitrogen was named by French chemist **Jean-Antoine-Claude Chaptal** in 1790, and again it was considered to be an element by his compatriot Lavoisier.

Scottish chemist **Joseph Black** studied the reaction of quicklime (hydrated calcium oxide) with air in the 1750s. The proportion of the air that was used up in the reaction he called 'fixed air' but was, in fact, a minor but very important constituent of air, namely, carbon dioxide.

Another century elapsed before British physicists **Lord Rayleigh** and **Sir William Ramsey** identified the third largest constituent in atmospheric air, i.e. the inert gas argon.

Chapter 1 introduces the reader to the chemical, physical, and thermodynamic properties of ground level ambient air.

Chapter 2: The Thermodynamics of the Human Machine and Thermal Comfort

Given that humans expend much time, money, and ingenuity in the pursuit of the provision of an acceptable internal environment, it is surprising to find that the interrelationship between internal environmental factors was not seriously explored until the work of Danish engineer **Povl Ole Fanger** and American engineer **RG Nevins** in the middle of the twentieth century. Their work integrated biological factors with thermodynamics and heat transfer to enable the prediction of well-being.

This chapter is based on their work providing an analysis of the energy balance of the human machine and a quantitative method assessing of its thermal comfort.

Chapter 3: Ventilation

Examples of human powered efforts to deal with the aspects of providing fresh air date back over 2500 years and include Chinese hand fans and Punkah fans from South East Asia.

The need to supply external air to an internal space was also recognised some time ago.

Persian wind towers, traps, and scoops are part of traditional Arabian architecture and are thought to date back 2000 years.

Around the same time, Roman architect and engineer **Vitruvius Pollio** wrote his celebrated treatise on building structure and services *De Architectura* in which he made reference to the control of ventilation rate via the use of a high-level damper. Several Roman authors also describe the measures taken to ensure good ventilation in mines including the lighting of fires to induce updraft and the sinking of parallel shafts specifically for air flow purposes.

With the advent of practical steam power in the eighteenth century, reciprocating engines were used as power sources for air bellows and, later, rotary air movers.

In the nineteenth century, ventilation fans were supplying air to the USA capitol (1866), and by the early twentieth century, electricity had replaced steam as a source of fan motive power.

The fresh air quantity required to merely provide sufficient oxygen is small. Normal respiration rates for an adult are in the range 15–20 respirations per minute, and the amount of air required per inspiration is approximately 500 ml. As a result, fresh air requirements (usually referred to as *ventilation rates*) are based on the need to dilute other contaminants sourced from the occupants and/or business processes.

This chapter commences with an understanding of the need for a good quantity and quality of air and goes on to describe the design of the technology for its provision.

Chapter 4: Psychrometry and Air-Conditioning

As the nineteenth century turned in to the twentieth century, American engineer **Willis H. Carrier** was concerned with humidity control. Not only is he credited with producing the world's first 'air conditioner' but he also created a graphical and very practical representation of moist air properties – the Hygrometric or Psychrometric chart. Although property calculation is now commonly carried out by computational algorithm, for the design engineer the chart has not been bettered in terms of providing an instant understanding of mixture processes.

This chapter uses both charts and underlying equations to describe the behaviour of air–water vapour mixtures and the design and specification of technology to provide acceptable combinations.

Chapter 5: The Building Envelope

The necessity of a physical boundary for security and shelter drove the development of brick technology some 9000 years ago. The materials used were mud, clay, and straw. For the first few millennia, their production was solar power aided. This all changed with the development of kilns enabling innovation and product variation.

Until the sixteenth century, openings in walls of buildings were, for most people, covered by a range of materials such as cloth, paper, or wooden shutters. Renaissance architecture called for larger and more ornate window design and so glass technology production had to innovate to deliver. The production of glass, for jewellery, can be traced back some 6000 years. Glass for small window pane use is thought to be over 2000 years old. However, large pane production was not widespread until the nineteenth century with laminated glass available in the early twentieth century.

This chapter describes the interaction of a building's solid boundary with the changing external environment and the effect it has on its interior condition.

Chapter 6: Refrigeration and Heat Pumps

Two hundred years ago, blocks of ice from water bodies in New England, USA, were being harvested and sold to States in the south and beyond for use in drinks and ice cream making. The medical profession also saw a use for ice in fever reduction. The physical principles of cooling had been known by the mid-seventeenth century when Scottish medical scientist **William Cullen** used a lowered pressure to produce the evaporation of ether ($C_4H_{10}O$) that resulted in the cooling of local air. In early 1800s, British scientist and engineer **Michael Faraday** demonstrated the condensation of ammonia (NH_3) vapour by the application of pressure and a reduced temperature. An ether-based early ice-making machine was patented in the mid-1800s by Scottish typesetter and entrepreneur **James Harrison**. By the start of the twentieth century, however, the production of ice or a sensation of 'coolth' (opposite of warmth) using refrigeration plant based on the evaporation, compression, and condensation of a volatile fluid was widespread.

This chapter provides a thermodynamic analysis of cooling/heating refrigerant-based systems for tempering the indoor environment.

Chapter 7: Acoustic Factors

The nature of harmonious or pleasing sound was considered some 2500 years ago by Greek philosopher, **Pythagoras**. His investigations were related to the length of a vibrating string, the resulting sound, and its pleasurable nature or otherwise.

In the seventeenth century, British scientist **Robert Boyle** demonstrated with the use of a watch and an exhaust pump that sound waves cannot travel through a vacuum and require a medium to reach the ear.

The methodical study of sound science or acoustics expanded in the nineteenth century. In 1863 century, German physicist and physician **Hermann L.F. van Helmholtz** was bringing this old idea of the 'theory of music' back to life. His work ranged from the 'quality' of a tone, i.e. the presence of harmonics, to the physiology of the inner ear.

In 1877, British scientist **John William Strut** (Baron Rayleigh) produced his work 'The Theory of Sound' describing his hypotheses regarding the relationship between elasticity and vibration in both solid and fluid materials. He also went on to estimate the magnitude of the displacement by molecules experiencing the passage of a sound wave.

As the nineteenth century turned into the twentieth, Irish Engineer and Scientist **John Tyndall** was laying down the principles of sound transmission and amplification.

American Physicist **Wallace C.W. Sabine** was an early practitioner of the science of acoustics applied to the internal environment using his principles of reverberation (prolonged reflection), prediction, and modification to improve the quality of speech and music in auditoria.

This chapter describes the principles of sound waves, sound parameters, and their transmission and propagation.

Chapter 8: Visual Factors

The nature of light and sight is intimately linked with concepts such as colour, shade, hue, shadow, etc. as well as more emotive concepts such as good (presence of light) and evil (darkness). Its study has long preoccupied civilisations throughout history.

The Ancient Greeks, in the times of Aristotle and Plato, believed that objects were illuminated by 'emissions' from a viewer's eyes that became reflected back thus making the object visible.

Approximately 1000 years ago, the Persian scientist **Abu Ali al-Hasan** (*Alhazen*) wrote of his classical experiments demonstrating the principles of light refraction and reflection.

These themes were reviewed in the thirteenth century by British theologian and scientist **John Pecham (Peckham)**.

In the 1490s, Italian artist, sculptor, scientist, and engineer **Leonardo Da Vinci** expended a considerable amount of his energies studying the behaviour of light and the eye itself. He was well aware of the importance of the lens and his work provided coherent descriptions of subjects such as central and peripheral vision, depth perspective, and stereoscopic vision. He also employed his knowledge of water waves to hint at a possible analogue for the transmission of light. Using mounting empirical evidence on diffraction from experimenters such as Italian scientist **Francesco Grimaldi**, Dutch mathematician and theorist **Christiaan Huygens** is usually credited with the first firm statement of the light being a waveform in 1690.

This was soon followed by **Isaac Newton's** epic tome *Opticks* in 1704 describing his precise experiments in the reflection, refraction, and diffraction of light. In the early nineteenth century, British physicist and physician **Thomas Young** asserted that light waves were transverse in nature and furthermore went on to estimate the wavelengths of the *visible* spectrum.

This chapter provides the reader with an understanding of the basic laws of illumination, luminaire design, and the occurrence of unwanted light.

Chapter 9: Cleaning the Air

In 1851, Irish physicist and mathematician **George Gabriel Stokes** provided a model predicting the resulting velocity for a small particle falling under the effect of gravity. This simple understanding underpins much of Chapter 9 that looks at the nature of particulates in a fluid stream and describes the theory and operation of a range of pollution control devices to capture them and alleviate the problem.

This chapter provides standard models, performance parameters for particulate collection devices such as fabric filters, electrostatic precipitators, etc.

In addition to particulate generation, human activities have given rise to local atmospheric gas concentrations in excess of historical and geological ambient background levels. For example, Ozone (O_3), produced by the presence of electric fields was described by **Christian Friedrich Schönbein** in 1840. Its usefulness for disinfection purposes was soon exploited. However, it is a respiratory system irritant. Exposure to oxides of nitrogen (NO_x), produced by combustion, can also produce airway inflammation.

Chapter 9 also introduces the reader to technologies being deployed to alleviate local concentrations of these and other unwelcome gas/vapours.

Chapter 10: Solar Energy Applications

The sun is an enormous nuclear fusion reactor around which the earth orbits at a distance of approximately 150×10^9 m. The core of the sun is estimated to be at a temperature of some 15 million Kelvin, whilst its outer surface (photosphere) is at a relatively cool 5500 K.

Heat transfer theory tells us that the radiation power per m^2 of surface area of a body is the product of a constant times the body's temperature in Kelvin raised to the fourth power. Using this, some arithmetic will then show that the sun emits at a rate of around 70 MW/m^2!

Assuming the sun to have a radius of around 0.7×10^9 m, the total power output for the sun is of the order of 3.8×10^{18} MW. Thanks to an inverse square distance relationship, the resulting energy intensity at our solar orbit is found to be only about 1360 W/m^2. Distributed over the surface of the Earth's upper atmosphere, this is equivalent to an average incoming solar radiation level of 340 W/m^2.

Even after atmospheric absorption/reflection, surface irradiation levels are sufficient to provide ample opportunities for energy collection and storage using a range of thermal and electric technologies, as described in this chapter.

Chapter 11: Measurements and Monitoring

Devices to enable the evaluation of the state of prevailing atmospheric conditions are not new. Many attempts date back to the Renaissance period. In the mid-1400s, Italian philosopher and architect **Leon Battista Alberti** proposed using changes in the weight of a natural sponge to indicate the presence of atmospheric water vapour. Alberti also suggested a device for evaluating prevailing wind velocity. In the seventeenth century, temperature and pressure measurement caught up with the work of Italian Astronomer, physicist and engineer **Galileo Galilei**, and Italian physicist and mathematician **Evangelista Torricelli.**

To paraphrase nineteenth-century British physicist **William Thomson** (later *Lord Kelvin*), deep knowledge and understanding of a phenomenon cannot be gained without precise measurement. The output of a measurement transducer may be used to signal a control element or may be used for metering purposes. Today, accurate and accessible (by both supplier and customer) metering has become essential for utility cost and pollution control.

This chapter describes the working principles and application of a range of devices used to monitor and control the parameters and systems introduced in earlier chapters.

Chapter 12: Drivers, Standards, and Methodologies

To maintain a building's interior conditions at optimum level, in the face of a changing external environment and changing occupational use, requires energy. Until the start of the twenty-first century, fossil fuels consumption in all its forms was used, without much forethought, to supply this need. This practice has contributed to the unwelcome modification of the planet's atmospheric climatic conditions.

It is the nature of human beings to require a transparent, rules-based system to achieve a common, equitable goal, and thus, the planet and its countries now have carbon dioxide generation targets and limits that are being pursued with only a degree of success. As a major consumer of energy, the design of buildings must play its part in addressing this urgent goal.

However, it is not the aim of national building construction regulations and international standards to stifle design and innovation. No one wants an uninteresting or uninspiring built environment as the price paid for the recovery of the planet's climate. This is unnecessary and, in fact, well thought-out building regulations and standards can prompt designers to innovate and improve their schemes producing sustainable structures that minimise their environmental impact and at the same time provide internal spaces where people are happy to be.

This chapter provides an awareness of and examples of how sustainability now underpins the design, construction, and operation of the built environment.

Chapter 13: Emerging Technologies

Technology is the offspring of Science. Earlier chapters in this book have alluded to the fact that the scientific principles on which our current building technologies depend have, in some cases, been around for a few hundred years or even longer.

The phrase 'Emerging Technologies' is therefore a term worth consideration. Emerging does not necessarily equal new as a great deal of the ideas behind these systems are not new to science. Nevertheless, new materials and the use of microprocessors and the internet, in particular, are opening up new opportunities for the application of 'smart' technologies.

This chapter provides an awareness of how emerging technologies and strategies in all areas of the built environment can provide a better matching of energy supply and demand, lower impacts, lower costs, and an improvement in occupied space quality.

Chapter 14: Closing Remarks

Notation

A text having a broad range of topics will unavoidably require a large nomenclature. In general, parameters are introduced along with their units in the text.

However, there are a few cases where the reuse of a symbol in a different context has become necessary to maintain coherence with widely accepted practice.

The listings below highlight the reuse by indicating in brackets the chapter of its subsequent re-occurrence.

Occasionally, differentiation may require an extended subscript. For example, amb – ambient. Notations such as these are self-evident and will not be included in the list below.

English Symbols

A	area	m²
$\mathcal{A}$	total absorption	m²
B	coefficient of cubical expansion	K⁻¹
B	magnetic field strength (11)	T
C	conduction coefficient	W/K
C_p	specific heat capacity at constant pressure	kJ/kg K
C_v	specific heat capacity at constant volume	kJ/kg K
c	concentration	kg/m³
c	light velocity (10)	m/s
D	diameter, distance	m
DI_θ	directivity index	dB
d	diameter, distance	m
E	electric potential	Volts
EF	electric field	Volts/m
F	force	N
FSP	fan static pressure	Pa
FTP	fan total pressure	Pa
f	frequency	Hz

g	gravitational acceleration	m/s²
h	specific enthalpy	kJ/kg
H	height	m
h_c	convective heat transfer coefficient	W/m² K
h_r	radiative heat transfer coefficient	W/m² K
I	electrical current	Amp
I	sound intensity (7)	W/m²
I_E	illuminance	lumen/m² or lux
I_F	luminous flux	lumen
I_L	luminance	candela/m²
I_o	luminous intensity	candela
I_{irr}	solar irradiation	kWh/m²
i	angle of incidence	degrees
k	thermal conductivity	W/m K
k	absolute roughness (3)	m
L	length (3)	m
L	sound level (7)	dB
l	length	m
M	molar mass	kg/kmol, g/mol
$\dot{M}$	activity level or metabolic rate	W/m²
$m, \dot{m}$	mass, mass flow rate	kg, kg/s
N	rotational speed	rpm
N	number of air changes per hour (ACH)	h⁻¹
n	number of moles	kmol, mole
P	pressure	Pa, bar
P	perimeter (3)	m
Q, q	heat	kJ, kJ/kg
$\dot{Q}$	heat transfer	kW
$\dot{q}$	heat transfer	kW/m²
R	specific gas constant	kJ/kg K
R_{cl}	clothing resistance	m² K/W
R_{elec}	electrical resistance	Ω
R_o	universal gas constant	kJ/kmol K
R_{th}	thermal resistance	m² K/W
R_{vap}	vapour resistance	Ns/kg
r_{vap}	vapour resistivity	Ns/kg m
r	radius	m
r	angle of refraction, reflection (7)	degrees

S, s	entropy	kJ/K, kJ/kg K
SP	static pressure	Pa
T	temperature	°C, K
t, T	time, period (7)	s
U	overall heat transfer coefficient	W/m² K
u	specific internal energy	kJ/kg
$V, \dot{V}$	volume, volume flow rate	m³, m³/s
VP	velocity pressure	Pa
$\overline{V}$	velocity	m/s
v	specific volume	m³/kg
ν	wave number (8)	m⁻¹
W, w	work	kJ, kJ/kg
W	width (8)	m
$\dot{W}$	work transfer, power	W
x	thickness, distance	m
Y	admittance	W/m² K
y	day angle	degrees
Z	figure of merit	K⁻¹
z	height above a datum	m

Subscripts/Superscripts

A	absorber	i	incident (5)
a	air	i	inside (5)
B	buoyancy	isen	isentropic
b	building	L	latent
C	centrifugal	m	mean
C	condenser (6)	n	n-type
c	convection	o	initial state
c	contaminant (3)	o	outside (5)
c	cold side (6)	p	process
c	collection (9)	p	pump, p-type (6)
cl	clothing	p	projected (8)
cw	cold water	p	particle (9)
D	drag	pc	phase change
d	diameter	pn	perceived noise
da	dry air	R	room

db	dry bulb	Ref, R	refrigeration (6)
dp	dew point	R	resultant (9)
E	evaporator (6)	r	radiative
E	electrostatic (9)	r	radius (7)
e	evaporation	r	relative (9)
e	emission (9)	rev	reverberation
eq	equivalent continuous	S	sensible
exp	expander	S	supply (4)
F	fabric	s	surface
f	free stream, friction	s	solar (5)
f	saturated liquid (4)	sat	saturated
f	percentile (7)	T	total
fb	fibre	th	thermal
fg	liquid–gas condition	ts	terminal settling
G	gravity	w	wall
g	gas or vapour	w	water (4)
HP	heat pump	wb	wet bulb
h	hydraulic	wv	water vapour
h	hot side (6)		
i	component		

Greek Symbols

α	angle	degree
α	thermal diffusivity (5)	m^2/s
α	Seebeck coefficient (11)	V/K
α	temperature coefficient of resistance (11)	$^\circ$C^{-1}
β	coefficient of cubical expansion	K^{-1}
β	thermal effusivity (5)	W s$^{1/2}$/m^2 K
γ	angle	degree
δ	distance	m
δ	angle (5)	degree
ε_o	permittivity of free space	C/Vm
ϕ	relative humidity	%
ϕ	angle (8)	degree
η	efficiency	%
λ	wavelength (10)	m
λ	mean distance between collisions (9)	m

μ	dynamic viscosity	kg/ms
π	Peltier coefficient	V
ρ	density	kg/m^3
ρ	resistivity (6)	Ωm
σ	Stefan–Boltzmann constant	W/m^2 K^4
τ	time constant	hour
θ	angle	degree
ω	moisture content	kg$_{wv}$/kg$_{da}$
ω	hour angle (5)	degrees
ω	solid angle (8)	steradian
Ψ, ψ	thermodynamic property	various

Dimensionless Numbers

A	albedo
AM	air mass
α	absorptivity
BF	bypass factor
CF	contact factor
CF	configuration factor (5)
C_c	Cunningham correction factor
C_d	discharge coefficient
$\mathbb{C}_p$	pressure coefficient
γ	ratio of specific heat for a gas
DF	decrement factor
DF	daylight factor (10)
DF$_\theta$	directivity factor
DLOR	downward light output ratio
EA	energy absorptance
ER	energy reflectance
ET	energy transmittance
ε	emissivity
ε	dielectric constant (9)
F	shape factor
F_R	collector heat removal factor
FSE	fan static efficiency
FTE	fan total efficiency
f	friction factor
f_{cl}	clothing factor
f_f	fibre solid fraction
f_r	thermal response factor
ϕ	humidity

Gr	Grashof number
γ	ratio of specific heats
K	loss factor (3)
K	metre factor (11)
Kn	Knudsen number
LaMF	lamp maintenance factor
LLF	light loss factor
LOR	light output ratio
MF	maintenance factor
mf	mass fraction
Nu	Nusselt number
N	number of couples (6)
n	number of times air is changed (3)
n	expansion/compression index (6)
n	refractive index (8)
PMV	predicted mean vote
PPD	predicted percentage dissatisfied
Pr	Prandtl number
Re	Reynolds number
RI	room index
RMF	room maintenance factor
RRL	room ratio line
ρ	reflectivity
SF	solar factor
SHF	sensible heat factor
SHR	space height mounting ratio
Stk	Stokes number
St	Strouhal number
TSC	total shading coefficient
τ	transmissivity
UF	utilisation factor
ULOR	upward light output ratio
y	molar fraction
WI	well index

1

Ambient Air

1.1 Overview

The atmosphere is a mixture of gases and water separating the planet from space. Its maximum thickness is of the order of 8000 km. The mass of dry gas is of the order of 5.1×10^{18} kg, whilst the mass of water is about 1.3×10^{13} kg. Its interface with space is gradual, becoming indistinct at the outer edge. Its composition and density are, however, by no means constant throughout its thickness.

The ***troposphere*** is a dense turbulent layer closest to the earth comprising ~80% by mass of the earth's atmosphere. Its thickness varies from 8 km at the poles to 18 km at the equator. Seasonal variations in thickness can also occur.

An often-used average value is 11 km. It contains ~90% of the planet's atmospheric water content and is home to the world's weather systems. Its mass/energy interaction with the other components of the planet's environments and incoming solar radiation is central to life. We inhabit the very base level of this stratum.

Learning Outcomes

- To appreciate the importance of clean, fresh air to human health
- To be able to differentiate between fresh and respired air in terms of composition
- To be familiar with the basic thermodynamic properties of air
- To be able to evaluate air energy content

1.2 Why Ambient Air Is Important?

Humans cannot survive without air. On the spectrum of human needs, air is special. We are able to choose what to eat and drink, but this is not the case for the air we breathe.

Close to the planetary surface, the air comprises a mixture of oxygen and nitrogen in combination with a much smaller proportion of other elements. In the human body, the oxygen is absorbed by the blood stream in the lungs. Oxygen supports our life and oxidises or '*burns*' food to create energy for metabolic processes and heat for our bodies. This production of energy is termed respiration. Breathing is the means of obtaining the oxygen and

Building Services Engineering: Smart and Sustainable Design for Health and Wellbeing, First Edition.
Tarik Al-Shemmeri and Neil Packer.

the removal of gaseous by-products. The by-products of the process are carbon dioxide and water vapour.

The human body also needs nitrogen, but this is obtained by other means.

It is essential to maintain the necessary quality of atmospheric air in the internal environment, hence the need for impurity or pollution removal and the control of temperature and humidity.

Poor indoor air quality (IAQ) results in poor health. The human health effects of poor air quality are far reaching but principally affect the body's respiratory system and the cardio-vascular system. Individual reactions to low-quality air depend on the type of environment a person is exposed to, the degree of exposure, and the individual's health status and genetics.

Poor air quality sources can be divided into three broad categories, i.e. physical, chemical, and biological. In physical terms, the main problem is an inappropriate amount of heat and moisture. Particulates also play a major role, and sometimes noise and static electricity charge/discharge is added to this group. Chemical hazards include a wide range of compounds, e.g. carbon dioxide, carbon monoxide, volatile organic compounds (VOCs), etc. The biological category covers the presence of such hazards as bacteria, viruses, and moulds.

All of these unwanted contaminants can be minimised or even prevented if air quality is monitored and controlled.

To achieve this goal, familiarity with air composition and behaviour is essential.

1.3 Air Composition

The approximate composition of lower atmospheric 'clean' air is shown in Table 1.1. By volume, it largely comprises nitrogen ($\sim$78%) and oxygen ($\sim$21%) plus other gases present in much smaller volumes.

Oxygen is required by humans for cellular respiration. The human respiratory/pulmonary system comprises a series of dividing tubes and sacs (*alveoli*) supported and moved by bones and muscles. When a pressure slightly below atmospheric pressure is created in the system, then air is inhaled. When a pressure slightly greater than atmospheric pressure is created, air is exhaled. However, the system is never completely emptied and there is always a residual air volume. Gaseous exchange takes place at blood vessels in the alveoli. Oxygen is taken up by the blood and carried to the body's cells. At the same time, carbon dioxide which is a by-product of the respiration process is released from the blood to the alveolar air.

Adults at rest normally take between 12 and 20 breaths per minute. The average person at rest needs about 500 l of air per hour. This has an oxygen content of 130 g.

However, the absorption of oxygen does not have an efficiency of 100% and exhaled breath has an approximate oxygen content of 14–16%.

As detailed in Table 1.1, the carbon dioxide content of atmospheric air is approximately 0.04%. As a result of respiration, the CO_2 content of exhaled breath is closer to 4% by volume. Where intensive metabolic activity takes place (e.g. gymnasiums), the metabolic CO_2

Table 1.1 Approximate composition of "clean" dry air.

Component gas	Chemical symbol	Molar mass (kg/kmol)	Volume (%)	Concentration (ppm)
Nitrogen	N_2	28.01	78.09	780 900
Oxygen	O_2	32	20.94	209 400
Argon	Ar	39.95	0.93	9300
Carbon dioxide	CO_2	44.01	0.04	410
Neon	Ne	20.18	Trace	18.0
Helium	He	4	Trace	5.2
Methane	CH_4	16.04	Trace	1.2
Krypton	Kr	83.8	Trace	0.5
Hydrogen	H_2	2.02	Trace	0.5

Source: Adapted from Williams (2001).

production rate increases substantially and prevailing concentrations in the indoor environment may therefore be expected to rise considerably above the thresholds set for sedentary activities.

Carbon dioxide is also the product of commercial or domestic activities such as combustion for catering as well as other industrial processes which can contribute to local indoor concentrations.

Although Table 1.1 supplies data for dry air, it is not common for air to be entirely bereft of water vapour. Water in the ambient atmosphere is largely the result of contact with oceans, lakes, etc. Atmospheric air contents can range up to 0.03 kg of water vapour per kg of dry air. In a confined space, the relative proportion of water vapour existing is increased by perspiration and the respiration of occupants. Exhaled air is close to being saturated with water vapour and could have a water vapour content of 5–6% by volume at a temperature of 35 °C.

1.4 Gas Mixtures

1.4.1 Mixture Laws

Air is regarded as a *homogenous mixture* of *ideal* gases.

A *homogeneous mixture* comprises matter having a number of different compounds but still retaining uniform properties and composition throughout its bulk. Component compounds are visually indistinguishable but may be separated by physical means. Use of the term *ideal* assumes that the gas atoms are tiny spheres each having a volume that is insignificant relative to their containing boundary. Any collisions between these entities are assumed to be totally elastic and frictionless. Electrostatic forces are considered absent.

Consider a mixture comprising a number of ideal gases.

From a matter conservation perspective, the mixture's total mass (m_{mix}, kg) must be equal to the sum of the masses of its individual components, similarly, for its molar components (n_{mix}, kmol).

Table 1.2 Unit comparison of fundamental mixture laws.

Parameter	Mass basis	Molar basis
Totals	$m_{\text{mix}} = \sum_{i=1}^{j} m_i$	$n_{\text{mix}} = \sum_{i=1}^{j} n_i$
Fraction	$mf_i = \dfrac{m_i}{m_{\text{mix}}}$	$y_i = \dfrac{n_i}{n_{\text{mix}}}$

The contribution of any individual component (i) to the total can be described by either a mass fraction (mf_i) or a volumetric or molar fraction (y_i).

If the gas mixture comprises j components in total, then the above conservation considerations can be stated as given in Table 1.2.

Note that in both mass and molar cases, the fractional sum is equal to unity.

The relationship between the number of kilomoles of the air (n), its mass (m, kg), and molar mass (M, kg/kmol) is given by

$$m = nM \tag{1.1}$$

This relationship is extremely useful as it provides for a conversion from chemical to physical units. It can also be directly applied to a mixture, thus

$$m_{\text{mix}} = n_{\text{mix}} M_{\text{mix}} \tag{1.2}$$

The link between mass (kg) and volume (m^3) is provided by the density (kg/m^3)

$$m = \rho V \tag{1.3}$$

When expressed in flow rate terms, it is commonly described as the *Continuity equation*, thus

$$\dot{m} = \rho \dot{V} \tag{1.4}$$

At any cross section in a confined flow, the volumetric flow rate $\left(\dot{V}, m^3/s\right)$ is given by

$$\dot{V} = A\overline{V} \tag{1.5}$$

1.4.2 Dalton's Law

Pressure (P, N/m^2) is associated with the presence of matter and is defined most simply as the force (F, N) acting on a given area (A, m^2), i.e.

$$P = F/A \tag{1.6}$$

A more often used unit is the **Pascal** where 1 N/m^2 = 1 Pa; 1 Pa is a small amount of pressure and so multiples with distinctive names are in common use.

For example,

1 bar = 1×10^5 Pa

1 atm = 101 325 Pa

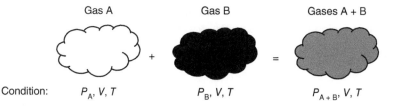

Gas A + Gas B = Gases A + B

Condition: P_A, V, T P_B, V, T P_{A+B}, V, T

Figure 1.1 Illustration of Dalton's law.

Pressure measurements using a perfect vacuum as a datum or baseline are termed **absolute pressures** (P_{abs}).

Measurements using ambient atmospheric pressure (P_{atm}) as a datum or zero are termed **gauge pressures** (P_{gauge}).

The contribution that component gases have on the physical behaviour of a mixture can be approximated by assuming that the gases are ideal and using *Dalton's law.*

Dalton's law of partial pressures states that the total pressure (P_{mix}) of a mixture of gases at a given temperature (T_{mix}) and volume (V_{mix}) is equal to the sum of the individual pressures (P_i) of each gas component if existing at the same temperature and volume, i.e.

$$P_{mix} = \sum P_i \quad \text{at } T_{mix}, V_{mix} \tag{1.7}$$

Consider two gases A and B at the same temperature and volume but at different pressures P_A and P_B, respectively. If these gases are now mixed and the mixture exists at the same temperature and volume as its components, then Dalton's law states that the resulting pressure of the mixture will be the sum of P_A and P_B as shown in Figure 1.1

The ratio of any individual pressure of a component gas to the total gas mixture pressure (P_i/P_{mix}) is called the **pressure fraction**. It can be shown that

$$\frac{P_i}{P_{mix}} = y_i \tag{1.8}$$

i.e. the pressure fraction of any gas component in a mixture of gases is equal to the molar fraction of the component gas in the mixture.

Rearranging Eq. (1.8), the product of molar fraction and mixture pressure ($y_i P_{mix}$) is called the **partial pressure** (P_i, Pa).

1.4.3 Gibbs–Dalton Law

The approach to mixture thermodynamic property evaluation is based on an extension to Dalton's law called the **Gibbs–Dalton law**. This assumes that each component gas has a thermodynamic property (e.g. enthalpy, specific heat capacity, specific gas constant) value equal to the value it would have if occupying the mixture volume alone at the same temperature.

Mass or molar properties of mixtures can be determined by the **addition** of component contributions (i).

Let ψ be any mass-*independent* thermodynamic property and $\overline{\psi}$ be its molar-independent counterpart, then mixture property values can be evaluated as shown:

$$\psi_{\text{mix}} = \sum_{i=1}^{j} mf_i \psi_i \tag{1.9}$$

$$\overline{\psi}_{\text{mix}} = \sum_{i=1}^{j} y_i \overline{\psi}_i \tag{1.10}$$

The above relationships are exact for ideal gases and approximate for real gases.

1.4.4 Ideal Gas Behaviour and the Equation of State

An ideal gas is governed by a relationship known as the perfect gas equation of state which relates the state pressure (P, Pa), volume (V, m^3), and temperature (T, K) of a fixed mass (m, kg) of a given gas:

$$\frac{PV}{T} = mR \tag{1.11}$$

For a gas, the specific gas constant (R, J/kg K) can be calculated from knowledge of the molar mass of the gas (M, kg/kmol), thus

$$R = R_o/M \tag{1.12}$$

where R_o is the Universal gas constant having a value of 8.314 kJ/kmol K.

The specific gas constant is a mass-independent quantity, and for a mixture of ideal gases, Eq. (1.9) can be employed:

$$R_{\text{mix}} = \sum_{i=1}^{j} mf_i R_i \tag{1.13}$$

Molar mass is (of course) a molar-independent quantity, and for a mixture of ideal gases, Eq. (1.10) can be employed, thus

$$M_{\text{mix}} = \sum_{i=1}^{j} y_i M_i \tag{1.14}$$

Equation (1.12) can then be modified and used thus

$$R_{\text{mix}} = R_o/M_{\text{mix}} \tag{1.15}$$

A commonly used value for the specific gas constant of dry air is 287 J/kg K.

If the mass and composition of a gas remain constant, the equation of state can be used to determine the behaviour of the air during a process, i.e. what happens to its temperature, volume, and pressure.

This is defined by a simple expression relating the initial (1) and final (2) states:

$$\frac{P_1 V_1}{T_1} = \frac{P_2 V_2}{T_2} \tag{1.16}$$

However, a number of change process paths, i.e. from initial to final condition, are possible modifying Eq. (1.16).

The most commonly encountered pathways are as follows:

Adiabatic process: A process characterised by the absence of heat transfer to or from the working fluid.

Isobaric process: A process characterised by the absence of change in the pressure of the working fluid.

Isochoric process: A process characterised by the absence of change in the volume of the working fluid.

Isothermal process: A process characterised by the absence of change in the temperature of the working fluid.

Polytropic process: A process in which pressure, temperature, and volume of the working fluid may all change.

1.5 Air Thermodynamic and Transport Properties

1.5.1 Gas Density

The density of an ideal gas can be estimated by combining Eqs. (1.3, 1.11):

$$\rho = m/V = P/RT \tag{1.17}$$

Water in its vapour form is regarded as a gas.

Thus, ambient air can be regarded as a mixture of gases *plus* water vapour. The presence of water vapour in atmospheric air renders the air *humid*.

For humid air, from Dalton's law, the total mixture air pressure is the sum of the component partial pressures:

$$P_{total} = P_{da} + P_{wv} \tag{1.18}$$

The density of humid air can then be calculated as the sum of the densities of the two gases, dry air and water vapour in proportion with their partial pressures:

$$\rho_{humid\ air} = \rho_{da} + \rho_{wv} = \frac{P_{da}}{R_{da}T} + \frac{P_{wv}}{R_{wv}T} \tag{1.19}$$

A commonly used value for the specific gas constant of water vapour (R_{wv}) is 461.5 J/kg K.

For an atmospheric pressure of 1 bar and a specific gas constant of 287 J/kg K, the approximate variation of dry (0% relative humidity or RH) air density (kg/m^3) over a range of -50 to $+50\,°C$ is shown in Figure 1.2.

Figure 1.2 also indicates the effect on density of air saturated with moisture (100% RH) over the range 0–50 °C. Saturating air with moisture reduces its density by up to 5%.

1.5.2 Dynamic Viscosity

The dynamic viscosity (μ, kg/ms) describes the resistance to flow in a liquid or a gas. It is sometimes described as its 'internal friction'. Its magnitude depends on molecular motion, cohesion, temperature, and pressure, and it has a significant effect on a range of important fluid flow phenomena.

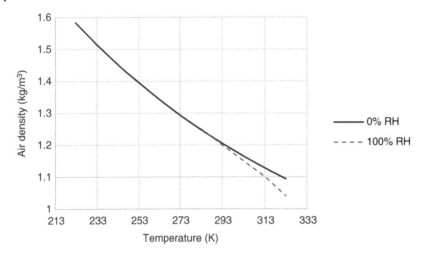

Figure 1.2 Variation of air density with temperature.

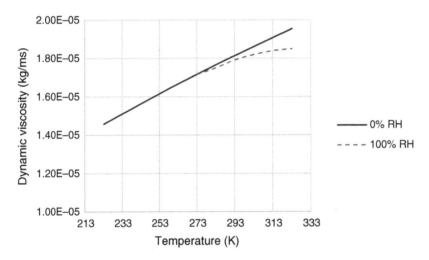

Figure 1.3 Variation of air viscosity with temperature.

The dynamic viscosity of dry air at 1 bar and 20 °C is approximately 1.8×10^{-5} kg/ms. The effect of temperature (T, K) on air dynamic viscosity (μ, kg/ms) at 1 bar over a range of -50 to $+50$ °C is shown in Figure 1.3 and can be estimated from

$$\mu = 1.458 \times 10^{-6} \left(\frac{T^{1.5}}{T + 110.4} \right) \tag{1.20}$$

Figure 1.3 also indicates that saturating the air with moisture (100% RH) over the temperature range 0–50 °C reduces its viscosity by up to 5%.

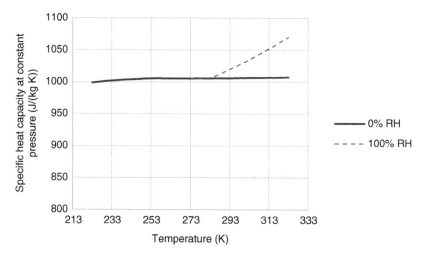

Figure 1.4 Variation of air specific heat capacity with temperature.

1.5.3 Specific Heat Capacity

Heat capacity is a property that describes the heat storage capabilities of a single phase (solid, liquid, or gas) substance. Note that this property is often quoted on a mass independent basis and is then termed the *specific heat capacity* (C: J/kg K). The heat capacity will determine the amount of energy required to change the temperature of 1 kg of substance by 1 °C.

As a result of being virtually incompressible, a single heat capacity parameter is usually sufficient for solids and liquids.

Gases are of course easily compressed, and hence, it is useful to further divide this property into two parameters depending on the prevailing conditions:

- Specific heat capacity *at constant volume* C_v
- Specific heat capacity *at constant pressure* C_p

For a given gas, these properties are related to specific gas constant (R):

$$R = C_p - C_v \tag{1.21}$$

The effect of temperature (T, K) on dry air thermal specific heat capacity at constant pressure at 1 bar over a range of −50 to +50 °C is shown in Figure 1.4.

A commonly used value of specific heat capacity at constant pressure for dry, atmospheric air is 1005 J/kg K.

The specific heat capacity of water vapour has a typical value of 1865 J/kg K over the range 0–50 °C, and the effect of moisture content (100% RH) can also be seen in Figure 1.4.

Like its constant pressure relation, the specific heat capacity at constant volume for air (C_v, J/kg K) varies little over the same temperature range. A commonly used value of specific heat capacity at constant volume for dry, atmospheric air is 718 J/kg K.

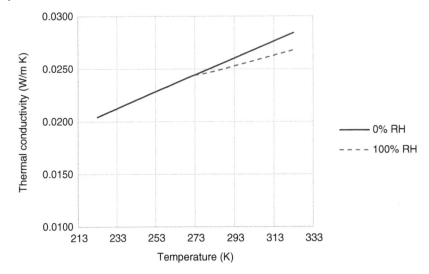

Figure 1.5 Variation of air thermal conductivity with temperature.

1.5.4 Thermal Conductivity

The thermal conductivity (k, W/m K) is a measure of the resistance to the flow of heat. Conduction of heat takes place as a result of energy transfer between adjacent atoms or molecules. The motion of the atoms or molecules is, in this case, regarded as vibrational, as opposed to translational, in nature. For liquids and gases in large-scale motion, heat transfer can be enhanced by other phenomena.

The thermal conductivity will determine the required thermal power $(\dot{Q}, \text{W})$ necessary to induce a 1 °C temperature reduction per unit linear length or thickness of material.

A typical value of thermal conductivity for dry, atmospheric air is 0.025 W/m K.

The effect of temperature (T, K) on dry air thermal conductivity at 1 bar over a range of -50 to $+50$ °C is shown in Figure 1.5.

For dry air (0% RH), the relationship can be approximated by

$$k = 8 \times 10^{-5}T + 0.0023 \tag{1.22}$$

Figure 1.5 also indicates that saturating the air with moisture (100% RH) over the temperature range 0–50 °C reduces its thermal conductivity.

Many building materials rely on encapsulated or integral air pockets for their insulating properties; hence, the resulting practical value will depend on the combination of solid and gaseous thermal conductivities.

1.5.5 Heat Transfer Coefficient

Heat transport in a fluid or between a fluid and a solid depends on heat conduction, energy storage, and fluid mixing motion. If the motion is instigated by a fan or blower, the heat transfer mechanism is termed *forced convection*.

If the motion is instigated as a result of a simple temperature difference, the heat transfer mechanism is termed natural, buoyancy-driven, or *free convection*.

In both cases, the resulting rate of heat transferred is a function of a parameter known as the heat transfer coefficient (h_c, W/m^2 K). Its value is dependent on system geometry, fluid transport properties, and flow modes, i.e. laminar/turbulent.

For a gas like air, free convection heat transfer coefficients are typically of the order of 5–25 W/m^2 K. For forced convection in air, this value could be increased by 10 times.

These values are quite small, however, compared with a light liquid, like water, where free convection heat transfer coefficients can be up to 1000 W/m^2 K and forced convection values up to 20 000 W/m^2 K. Water phase change (i.e. evaporating/condensing) values are even greater ranging up to 100 000 W/m^2 K.

1.5.6 Combinations of Properties

The properties described above are often used in *dimensionless* ratio combination to describe thermophysical and fluid behaviour. These ratios provide a means of expressing this behaviour in a useful empirical fashion.

The most common dimensionless groups are as follows:

1. Discharge coefficient $= \dfrac{\text{Actual flow rate}}{\text{Theoretical flow rate}}$

$$C_d = \frac{\dot{V}_{act}}{\dot{V}_{theo}} \tag{1.23}$$

The discharge coefficient is associated with flow openings, orifices, and nozzles. A value of zero indicates no flow and a value of unity indicates a theoretical maximum flow. Typical values for C_d are dependent of the prevailing flow regime through an opening. For a simple orifice, values between 0.5 and 0.7 are in common use.

2. Pressure coefficient $= \dfrac{\text{Static pressure}}{\text{Dynamic pressure}}$

$$C_p = \frac{P - P_o}{0.5\rho \overline{V}^2} \tag{1.24}$$

The pressure coefficient is defined as the difference between the mean local pressure (P) on a surface and that of the free stream (P_o) as fraction of the dynamic (or velocity-associated) pressure (Pa).

3. Nusselt number $= \dfrac{\text{Convective resistance}}{\text{Conductive resisance}}$

$$Nu = \frac{h_c l}{k} \tag{1.25}$$

The Nusselt number figures in both free and forced convection correlation equations. The parameter l is some representative length (m) of the system geometry.

For example, in a circular pipe or duct, this representative length would typically be its diameter.

4. Reynolds number $= \dfrac{\text{Inertia forces}}{\text{Viscous forces}}$

$$Re = \frac{\rho \overline{V} l}{\mu} \tag{1.26}$$

This parameter is ubiquitous in the study of fluid flow where its absolute value is used as a test to determine whether a flow is laminar (steady and undisturbed) or turbulent (unstable, perhaps with vortices).

The Reynolds number is also associated with forced convection correlation equations.

5. Grashof number $= \frac{\text{Buoyancy forces}}{\text{Viscous forces}}$

$$Gr = \frac{\rho^2 g \beta \Delta T l^3}{\mu^2} \tag{1.27}$$

The Grashof number is associated with free convection correlation equations, where ΔT is the temperature difference between a heated/cooled surface (T_s, K) and the fluid free stream (T_f, K) and β is the coefficient of cubical expansion (K^{-1}) given by

$$\beta = \frac{1}{(T_s + T_f)/2} \tag{1.28}$$

6. Prandtl number $= \frac{\text{Viscous transfer}}{\text{Heat transfer}}$

$$Pr = \frac{\mu C_p}{k} \tag{1.29}$$

The Prandtl number figures in both free and forced convection correlation equations. A typical value of Prandtl number for atmospheric air is 0.7.

7. Friction factor $= \frac{\text{Local (wall) shear stress}}{\text{Dynamic pressure}}$

$$f = \frac{\Delta P_w}{0.5 \rho \overline{V}^2} \tag{1.30}$$

This parameter is commonly used to estimate the frictional pressure losses associated with flow in conduits, in general, such as ducts and pipes. Values vary with flow regime (laminar/turbulent), flow rate, and conduit material.

Use of the above groups is not limited to air. They are in common usage for other fluids including water, steam, and refrigerants.

1.6 Important Energy Concepts

1.6.1 First Law of Thermodynamics

Laying aside considerations of mass–energy conversion, the principle of energy conservation as laid down in the First Law of Thermodynamics states that energy conversion from one form to another is permitted, but the sum of all energy forms remains constant.

In *flow* terms, the first law is commonly described by the following equation relating the initial (1) and final (2) energy states of a flowing fluid:

$$\dot{Q} - \dot{W} = \dot{m} \left[(h_2 - h_1) + \frac{\left(\overline{V}_2^2 - \overline{V}_1^2\right)}{2000} + g\frac{(z_2 - z_1)}{1000} \right] \tag{1.31}$$

Qualitatively:

Heat flow − Rate of thermodynamic work = Mass flow × (specific enthalpy change
+ kinetic energy change + potential energy change)

Since the equation utilises the mass flow rate, each term has the units of energy flow (Joules per second or Watts).

This equation is applicable to fluid flow in general and can be used with water, steam, and refrigerants as well as gases and can be employed in a reduced form depending on application.

The term **specific enthalpy** (*h*, kJ/kg) describes a bundle of energy forms including pressure-based and thermal (temperature and phase change-related) energy. The use of the term "specific" denotes the fact that the property is evaluated per unit mass of substance.

Applying Eqs. (1.9, 1.10) for a mixture of gases:

$$h_{mix} = \sum_{i=1}^{j} mf_i h_i \quad \text{and} \quad \overline{h}_{mix} = \sum_{i=1}^{j} y_i \overline{h}_i \tag{1.32}$$

Property changes in a mixture are evaluated by summing the individual component changes. For example,

$$\Delta h_{mix} = \sum_{i=1}^{j} \Delta h_i = \sum_{i=1}^{j} mf_i \Delta h_i = \sum_{i=1}^{j} y_i \Delta \overline{h}_i \tag{1.33}$$

A process realising no change in the enthalpy of the working fluid is termed *Isenthalpic*.

1.6.2 Thermal Energy

Knowledge of thermal energy content is important in both estimating building loads and the plant necessary to satisfy them.

In the absence of thermodynamic work and any changes in kinetic and potential energy, the rate of heat input/extract $(\dot{Q}, \text{kW})$ necessary to heat/cool a flow $(\dot{m}, \text{kg/s})$ of fluid is given by modifying Eq. (1.31), thus

$$\dot{Q} = \dot{m}\left(h_2 - h_1\right) \tag{1.34}$$

To produce positive design values in all cases, the following convention is adopted for calculation purposes:

Heat added to a fluid flow is positive, i.e. $+\dot{Q}$.
Heat extracted from a flow is negative, i.e. $(-)\dot{Q}$.

For *ideal gases*, enthalpy is proportional to specific heat capacity at constant pressure; thus, for a gas mixture, heat flow can be evaluated by

$$\dot{Q}_{gas} = \dot{m}_{gas} C_p \Delta T \tag{1.35}$$

where ΔT is the difference in gas temperature between final (2) and initial (1) states and C_p (kJ/kg K) is the specific heat capacity at constant pressure for the gas mixture.

The water component of air requires further consideration as, in maintaining the internal environment, evaporation, condensation, and perhaps the freezing of water will be of significance.

For the *water* component of the air, the enthalpy change associated with phase change (or *latent heat*) will be important. It may be regarded as the amount of energy required for the complete conversion of 1 kg of substance from one phase to another. In thermodynamics, the temperature at which phase change occurs is termed the (energy) *saturation temperature*.

For a phase change-associated mass flow of water, the amount of heat flow is given by

$$\dot{Q}_{pc} = \dot{m}_{water} h_{pc} \tag{1.36}$$

where h_{pc} (kJ/kg) is the specific latent heat value.

Again, values of latent heat vary with pressure and temperature. Typical values of the latent heat of vaporisation/condensation and the melting/freezing of water at ambient temperatures are in the area of 2450 kJ/kg (20 °C) and 335 kJ/kg (0 °C), respectively.

More accurate tables of values and charts are available in many design handbooks that account for both sensible and latent heat components in enthalpy terms.

1.6.3 Rate of Thermodynamic Work

This term $\left(\dot{W}\right)$ in Eq. (1.31) accounts for mechanical or electrical energy transfers that take place in devices such as fans, compressors, and pumps or even simple electrical heaters.

In general, thermodynamic terms, the work (J) associated with a fluid changing its state through a process (1) to (2) is given by

$$W = \int_1^2 PdV \tag{1.37}$$

The calculation of the work integral will depend on the actual process pathway undergone by the fluid.

1.6.4 Nonthermal Energy

A reduced and modified First Law of Thermodynamics is commonly used for ductwork design, the specification of air movement devices and conduits in general.

Nonthermal energy is normally considered to have three forms:

- Pressure energy associated with the presence of matter on a microscopic scale
- Kinetic energy associated with velocity on a large scale
- Potential or gravitational energy associated with position (relative to a datum)

Hence, for any given set of conditions:

$$\text{Pressure energy} + \text{Kinetic energy} + \text{Potential energy} = \text{Constant}$$

Using the SI unit of energy (Joule) as a basis, this can be written as

$$PV + \frac{1}{2}m\overline{V}^2 + mgz = \text{Constant} \tag{1.38}$$

This equation can be expressed in some other different ways. In the field of fluid mechanics when applied to inviscid, incompressible conditions, it is commonly known as the *Bernoulli equation*.

Here, each term is expressed using *head* (m) as its unit:

$$\text{Pressure head} + \text{Kinetic head} + \text{Potential head} = \text{Constant}$$

i.e.

$$\frac{P}{\rho g} + \frac{\overline{V}^2}{2g} + z = \text{Constant} \tag{1.39}$$

(The use of the term '*head*' alludes to the height of a column of fluid exerting a pressure on its base.) The constant is termed the *total head*.

In indoor climate studies, the potential term is often ignored and the energy relationship is expressed in *pressure* (Pa) terms, i.e.

Static pressure + Velocity pressure = Constant

i.e.

$$P + \rho\frac{\overline{V}^2}{2} = \text{Constant} \tag{1.40}$$

The constant here is termed the *stagnation or total pressure*.

Taking atmospheric air to have a density of $1.2\,\text{kg/m}^3$, the velocity pressure term is commonly expressed as

Velocity pressure $= 0.6\overline{V}^2$ (1.41)

1.6.5 Non-flow Conditions

The First Law of Thermodynamics caters for *non-flow* conditions with the following expression:

$$Q - W = m\left(u_2 - u_1\right) \tag{1.42}$$

where u (kJ/kg) is the **specific internal energy** which is a term that accounts for the sum of sensible, latent, chemical bond, and nuclear bond energy.

Being applicable to a non-flow condition, the units of each term in Eq. (1.42) are expressed in Joules.

For ideal *gases*, internal energy is proportional to specific heat capacity at constant volume; thus, for dry air, the heat content can be evaluated by

$$Q = m_{\text{gas}}C_v\Delta T \tag{1.43}$$

Note that there is a relationship between specific enthalpy and specific internal energy:

$$h = u + PV \tag{1.44}$$

1.6.6 Entropy

Enthalpy is concerned with the *quantity* of energy, and the First Law of Thermodynamics states that the quantity of energy is conserved in a process.

Entropy $(S, \text{kJ/K})$ is a property that is concerned with energy *quality* or *usefulness*. This is the business of the Second Law of Thermodynamics. Entropy is a nonconservative property, and, in general, engineering processes result in an increase in *global* entropy or a reduction in energy usefulness. Entropy increases as a result of the presence of *irreversibilities*, i.e. nonrecoverable energy transfers like, for example, friction.

Thus, a gas compression over a given pressure range will result in a temperature higher than that suggested by simple gas law theory.

To compare the ideal constant entropy case (termed *reversible adiabatic* or *isentropic*) with a real process, the term *isentropic efficiency* (η_{isen}) is often employed.

For example, if the ideal or isentropic final temperature in a compression process is designated ($T_{2,\,isen}$), then the isentropic efficiency for the process would be defined as

$$\eta_{isen} = \frac{T_{2,isen} - T_1}{T_{2,actual} - T_1} \tag{1.45}$$

This can also be described in enthalpy terms:

$$\eta_{isen} = \frac{h_{2,isen} - h_1}{h_{2,actual} - h_1} \tag{1.46}$$

Again, this approach can also be applied to other fluids such as steam and refrigerants.

1.7 Worked Examples

Worked Example 1.1   A Mole to Mass Conversion
A sample of air contains 0.8 kmol of nitrogen (N_2).
   If the molar mass of nitrogen is 28 kg/kmol, determine the mass (kg) of the nitrogen.

Solution
Given: $n_{N_2} = 0.8$ kmol, $M_{N_2} = 28$ kg/kmol
   Find: m_{N_2}

$$m_{mix} = n_{mix} M_{mix}$$
$$= 0.8 \times 28$$
$$= 22.4 \text{ kg}$$

Worked Example 1.2   The Molar Mass of Air
Ignoring minor components, it has been determined that the composition of air can be approximated as follows in Table E1.2:
   Using the above, determine its mixture molar mass (kg/kmol).

Solution
Given: $M_{N_2} = 28$ kg/kmol, $M_{O_2} = 32$ kg/kmol, $y_{N_2} = 0.79$, $y_{O_2} = 0.21$
   Find: M_{air}

$$M_{mix} = \sum_{i=1}^{j} y_i M_i$$
$$= \left(y_{N_2} \times M_{N_2} \right) + \left(y_{O_2} \times M_{O_2} \right)$$
$$= (0.79 \times 28) + (0.21 \times 32)$$
$$= 28.84 \text{ kg/kmol}$$

Table E1.2 Data for Worked Example 1.2.

Component gas	% by volume	Molar mass (kg/kmol)
Nitrogen (N_2)	79	28
Oxygen (O_2)	21	32

Worked Example 1.3 Calculation of Partial Pressure

Consider atmospheric air at a pressure of 101 325 Pa to consist of 79% nitrogen (N_2) and 21% oxygen (O_2) by volume.

Determine the partial pressures (Pa) for each gas in the mixture.

Solution

Given: $y_{N_2} = 0.79$, $y_{O_2} = 0.21$, $P_{mix} = 101\ 325$ Pa

Find: P_{N_2}, P_{O_2}

From Dalton's law, the volumetric or molar fractions are equal to the pressure fractions, hence

$$\frac{P_{N_2}}{P_{mix}} = y_{N_2}$$

$$\frac{P_{N_2}}{101\ 325} = 0.79$$

$$P_{N_2} = 80\ 046.75 \text{ Pa}$$

Since

$$P_{mix} = \sum P_i \quad \text{at } T_{mix}, V_{mix}$$

$$P_{mix} = P_{N_2} + P_{O_2}$$

$$101\ 325 = 80\ 046.75 + P_{O_2}$$

$$P_{O_2} = 21\ 278.25 \text{ Pa}$$

Worked Example 1.4 Density of Dry Air

An office of dimensions 20 m × 10 m × 3 m is kept at a temperature of 18 °C and absolute pressure of 100 kPa. See Figure E1.4. Determine:

(a) The mass of air in the room
(b) The density of air

Take the specific gas constant for air as 287 J/kg K.

Solution

Given: $P = 100 \times 10^3$ Pa, $V = 20 \times 10 \times 3 = 600$ m^3, $T = 18 + 273 = 291$ K, $R_{air} = 287$ J/kg K

Find: m, ρ

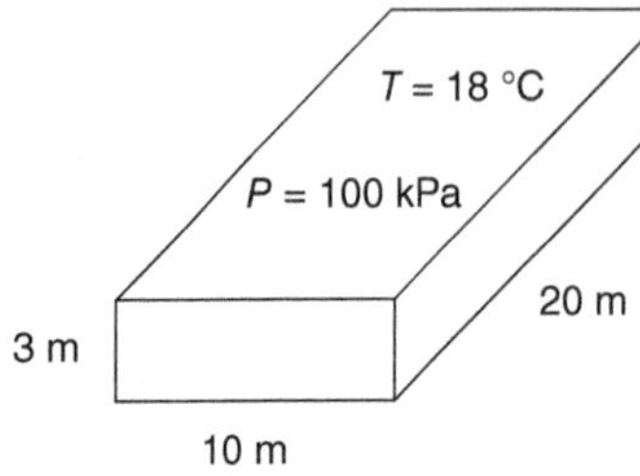

Figure E1.4 Office dimensions for Worked Example 1.4.

(a) Using the equation of state:

$$\frac{PV}{T} = mR$$

$$\frac{100\,000 \times 600}{291} = m \times 287$$

$$m = 718 \text{ kg}$$

(b) By definition, density is the ratio of mass to volume:

$$\rho = \frac{m}{V} = \frac{718}{600} = 1.197 \text{ kg/m}^3$$

Worked Example 1.5 Density of Moist Air

A sample of moist air has a total pressure of 100 000 Pa and a temperature of 27 °C.

Determine the density (kg/m³) of the dry portion of the air and, assuming a water vapour partial pressure of 3.6 kPa, the effect of humidity on the value of the mixture density.

Take $R_{da} = 287$ J/kg K and $R_{wv} = 461.5$ J/kg K.

Solution

Given: $P_{total} = 100\,000$ Pa, $T = 27 + 273 = 300$ K, $R_{da} = 287$ J/kg K, $R_{wv} = 461.5$ J/kg K, $P_{wv} = 3.6 \times 10^3$ Pa

Find: ρ_{da}, $\rho_{moist\ air}$

$$P_{da} = P_{total} - P_{wv}$$

$$= 100\,000 - 3600$$

$$= 96\,400 \text{ Pa}$$

Rearranging the Ideal gas law:

$$\frac{m}{V} = \rho = \frac{P_{da}}{R_{da}T} = \frac{96\,400}{287 \times 300} = 1.12 \text{ kg/m}^3$$

Using Eq. (1.19):

$$\rho_{moist\ air} = \frac{P_{da}}{R_{da}T} + \frac{P_{wv}}{R_{wv}T}$$

$$= \left(\frac{96\,400}{287 \times 300}\right) + \left(\frac{3600}{461.5 \times 300}\right)$$

$$= 1.12 + 0.026 = 1.146 \text{ kg/m}^3$$

Worked Example 1.6   Thermophysical Properties of Air

Estimate the dynamic viscosity (kg/ms) and thermal conductivity (W/m K) for a sample of dry air at 22 °C.

Solution

Given: $T = 22 + 273 = 295\,\text{K}$

Find: μ, k

$$\mu = 1.458 \times 10^{-6} \left(\frac{T^{1.5}}{T + 110.4} \right)$$

$$= 1.458 \times 10^{-6} \left(\frac{295^{1.5}}{295 + 110.4} \right)$$

$$= 1.82 \times 10^{-5}\,\text{kg/ms}$$

$$k = 8 \times 10^{-5}T + 0.0026$$

$$= 8 \times 10^{-5}\,(295) + 0.0026$$

$$= 0.0262\,\text{W/m K}$$

Worked Example 1.7   A Heated Air Supply

A workspace of dimensions $(10\,\text{m} \times 6\,\text{m} \times 5\,\text{m})$ is to be kept at 20 °C when the outside air temperature is 0 °C. See Figure E1.7.

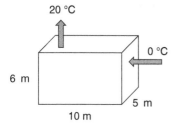

20 °C

0 °C

6 m

5 m

10 m

Figure E1.7   Work space dimensions for Worked Example 1.7.

If the air supply to the space is changed once *every hour*, determine:

(a) The mass flow rate (kg/s) of air to room.

(b) The heat required (kW) to maintain the room at its desired temperature.

Assume the density of air to be $1.2\,\text{kg/m}^3$ and $C_p = 1.0\,\text{kJ/kg K}$.

Solution

Given: $V = 10 \times 6 \times 5 = 300\,\text{m}^3$, $\Delta T = 20 - 0 = 20\,\text{K}$, $\rho = 1.2\,\text{kg/m}^3$, $C_p = 1.0\,\text{kJ/kg K}$, 1 air change per hour, i.e. (1/3600) air changes per second

Find: $\dot{m}, \dot{Q}$

(a) Volume flow rate of air through the room (m³/s) = Room volume × number of times air is replaced per second

$$\dot{V} = V \times (1/3600)$$

$$= 300 \times 0.000\,27$$

$$= 0.083 \text{ m}^3/\text{s}$$

$$\dot{m} = \rho \dot{V}$$

$$= 1.2 \times 0.083$$

$$= 0.1 \text{ kg/s}$$

(b) The hear (kW) required to maintain the room at 20 °C

$$+\dot{Q} = \dot{m}C_p\Delta T$$

$$= 0.1 \times 1.0 \times 20$$

$$= 2.0 \text{ kW}$$

Worked Example 1.8   An Extract System

How much heat will be carried away by an extractor fan rated at 2 m³/min from a sports hall kept at constant pressure of 1×10^5 Pa and a temperature of 20 °C while the temperature of air outside is 0 °C?

For air assume: $C_p = 1.005 \text{ kJ/kg K}$ and $R = 287 \text{ J/kg K}$.

Solution

Given: $\dot{V} = 2 \text{ m}^3/\text{min} = 2/60 \text{ m}^3/\text{sec}, \quad \Delta T = 20 - 0 = 20 \text{ K}, \quad P = 1 \times 10^5 \text{ Pa},$
$C_p = 1.005 \text{ kJ/kg K}, R = 287 \text{ J/kg K}$

Find: $\dot{Q}$

Calculate the density of air through the fan using the equation of state

$$\frac{m}{V} = \rho = \frac{P_{air}}{RT} = \frac{100\,000}{287 \times 293} = 1.189 \text{ kg/m}^3$$

$$\dot{m} = \rho \dot{V}$$

$$= 1.189 \times (2/60)$$

$$= 0.0396 \text{ kg/s}$$

Heat extracted by fan will be equal to heat added by heating external air to inside temperature.

$$(+)\dot{Q} = \dot{m}C_p\Delta T$$

$$= 0.0396 \times 1005 \times 20$$

$$= 796 \text{ W}$$

Worked Example 1.9   Latent Heat Condensation

A flow of moist air has a total mass flow rate of 0.25 kg/s.

The water vapour component comprises 1.5% by mass of the flow.

Assuming the water component is cooled to its saturation temperature, determine the rate of heat transfer (kW) required for the complete condensation of the water vapour.

Take the phase change enthalpy associated with the water condensation to be 2480 kJ/kg.

Solution

Given: $\dot{m}_{total} = 0.25 \text{ kg/s}, \quad h_{pc(cond)} = 2480 \text{ kJ/kg}$

Find: $\dot{Q}_{cond}$

$$\dot{m}_{wv} = 0.015 \times \dot{m}_{total}$$
$$= 0.015 \times 0.25$$
$$= 3.75 \times 10^{-3} \text{ kg/s}$$

$$\dot{Q}_{cond} = \dot{m}_{wv} h_{pc(cond)}$$
$$= 3.75 \times 10^{-3} \times 2480$$
$$= 9.3 \text{ kW}$$

Worked Example 1.10   Nonthermal Energy

A flow of air has a velocity of 3.5 m/s.

Express its nonthermal energy content as a kinetic head (m) and as a velocity pressure (Pa).

When flowing over a building, a pressure of 100 005 Pa resulted on the exterior surface. Using the data supplied, determine the pressure coefficient on the surface.

Data

Ambient air density: 1.2 kg/m³, Ambient atmospheric pressure: 100 000 Pa.

Solution

Given: $\overline{V} = 3.5$ m/s, $P = 100\,005$ Pa, $\rho = 1.2$ kg/m³, $P_o = 100\,000$ Pa

Find: Kinetic head (m), velocity pressure (Pa)

Kinetic head:

$$\frac{\overline{V}^2}{2g} = \frac{3.5^2}{19.62} = 0.624 \text{ m}$$

$$\text{Velocity pressure} = \rho\frac{\overline{V}^2}{2} = 1.2 \times \frac{3.5^2}{2} = 7.35 \text{ Pa}$$

$$C_p = \frac{P - P_o}{0.5\rho\overline{V}^2}$$
$$= \frac{100\,005 - 100\,000}{0.5 \times 1.2 \times 3.5^2} = 5/7.35$$
$$= +0.68$$

1.8   Tutorial Problems

1.1   A sample of oxygen (O_2) has a mass of 20 kg.

If the molar mass of oxygen is 32 kg/kmol, determine the number of moles of oxygen.

[Answer: 625 mol]

1.2 Ignoring minor components, it has been determined that the composition of air can be approximated as follows in Table P1.2.

Using the above, determine its mixture specific gas constant (J/kg K).

Table P1.2 Data for tutorial Problem 1.2.

Component gas	% by mass	Specific gas constant (J/kg K)
Nitrogen (N_2)	75.5	296.8
Oxygen (O_2)	23.2	259.8
Argon (Ar)	1.3	208.1

[Answer: 287.06 J/kg K]

1.3 Carbon dioxide comprises 0.04% of atmospheric air by volume.

If air pressure is taken as 101 325 Pa, determine the partial pressure contribution of the carbon dioxide.

[Answer: 4053 Pa]

1.4 Estimate the dynamic viscosity (kg/ms) and thermal conductivity (W/m K) for a sample of dry air at 25 °C.

[Answers: 1.84×10^{-5} kg/ms, 0.0264 W/m K]

1.5 Use the Gibbs–Dalton law and the approximate data supplied in Table P1.5 to estimate the enthalpy (kJ/kg) of air at 27 °C.

If the tabulated value for air is 300.19 kJ/kg, determine the % error associated with the approximation.

Table P1.5 Data for tutorial Problem 1.5.

Component gas	Mass fraction (%)	Enthalpy (kJ/kg) at 27 °C
Nitrogen (N_2)	75.5	311.5
Oxygen (O_2)	23.3	273

[Answers: 298.79 kJ/kg, 0.47%]

1.6 A room is kept at a temperature of 10 °C and absolute pressure of 100 kPa. Determine:

- The density of the air if water vapour is ignored.
- The density allowing for humidity assuming the partial pressure of water vapour at this temperature to be 1200 Pa.

Assume $R_{da} = 287$ J/kg K, $R_{wv} = 461.5$ J/kg K.

[Answers: 1.231 kg/m³, 1.226 kg/m³]

1.7 An occupied space of dimensions $4\,\text{m} \times 4\,\text{m} \times 3\,\text{m}$ is to be kept at $20\,°\text{C}$ when the outside air temperature is $10\,°\text{C}$. The air in the space is changed once every hour. Determine:

(a) The mass flow rate (kg/s) of air assuming ambient pressure to be $90\,\text{kPa}$.

(b) The heat required (kW) to maintain the gym at $20\,°\text{C}$.

Assume $C_{p,\,\text{air}} = 1.005\,\text{kJ/kg K}$, $R_{\text{air}} = 287\,\text{J/kg K}$.

[Answers: $0.0143\,\text{kg/s}$, $143\,\text{W}$]

1.8 A swimming pool ventilation system heats $60\,\text{l/s}$ of air. The pool air is kept at constant pressure ($100\,\text{kPa}$) and a temperature of $25\,°\text{C}$ while the temperature of supplied external air is $-5\,°\text{C}$.

Determine the output (kW) of the ventilation heater.

[Answer: $2.115\,\text{kW}$]

1.9 A flow of moist air has been cooled until its water vapour is on the point of condensation. It is then passed over a heat exchanger with a capacity to extract latent heat at a rate of $5\,\text{kW}$. The total air flow rate is $0.1\,\text{kg/s}$.

If the latent heat of condensation for the water is $2470\,\text{kJ/kg}$, determine the maximum mass flow rate of water vapour (kg/s) in the air that can be completely condensed by the heat exchanger.

Express this as a percentage of the total incoming air flow.

[Answers: $2 \times 10^{-3}\,\text{kg/s}$, 2%]

1.10 A flow of air has a velocity pressure of $3.75\,\text{Pa}$.

Determine its velocity (m/s).

For air assume: $\rho = 1.2\,\text{kg/m}^3$.

[Answer: $2.5\,\text{m/s}$]

2

The Thermodynamics of the Human Machine and Thermal Comfort

2.1 Overview

As a machine, the human body operates safely over a small range of functioning temperature.

It is essential that it is able to exchange heat efficiently and effectively with its surroundings for good health.

The normal (deep core) temperature of the human body is 37.6 °C.

A rise of less than 3 °C will result in **heat stroke** which, without treatment, is fatal.

Less severe symptoms resulting from an overheated environment include **heat cramps, heat exhaustion** (the result of heavy perspiration), and **heat syncope** (pooling of blood in dilated vessels).

At the other end of the operational range, a reduction of deep body core temperature to approximately 35 °C is regarded as **hypothermic**.

Optimum (i.e. *comfortable*) humidity for humans is in the range 30–70%.

However, operating at the extremes of this range is not recommended.

Additionally:

High humidity, i.e. >65% (and low temperature ~16 °C), is believed to be conducive to the growth of moulds and fungi resulting in an increased incidence of respiratory problems (asthma, rhinitis).

Low humidity can cause drying of the upper respiratory tract reducing the flow of mucous resulting in micro-fissures which act as points of entry for infection.

Subjectively, low humidity is also associated with increased aromatic perception.

The effect of ambient temperature and moisture content is modified by human factors such as level of clothing and activity. Consequently, thermal comfort will be a nonlinear combination of air parameters and human behaviour.

Learning Outcomes

- To be familiar with the energy balance of the human body and calculations involving body heat dissipation.

Building Services Engineering: Smart and Sustainable Design for Health and Wellbeing, First Edition. Tarik Al-Shemmeri and Neil Packer.

- To understand, from a thermodynamic viewpoint, the analysis of the different mechanisms of heat loss from the human body.
- To be able to use a quantitative method of assessing thermal comfort based on International Standards.

2.2 Thermal Comfort of Human Beings

Thermal comfort is defined as that state of mind in which satisfaction is expressed with the thermal environment.

The mechanisms underpinning human thermal comfort are governed by the basic modes of heat and mass transfer. Using these fundamentals, it is possible to establish a heat balance equation for a given person in a given activity, clothing, and environment. However, the heat balance in itself is not sufficient to establish thermal comfort. In the wide range of environmental conditions where a heat balance may be achieved, there is only a narrow range for mean skin temperature and sweat loss where thermal comfort is established. The mean skin temperature and sweat loss will vary with the activity of individuals and is generally coupled with six external parameters.

- Air temperature T_{db}
- Mean radiant temperature $\overline{T}_r$
- Relative air velocity $\overline{V}_a$
- Humidity ϕ
- Activity level $\dot{M}$
- Clothing thermal resistance or index R_{cl}

A combination of these six parameters resulting in a thermally neutral sensation is said to provide a *comfort condition* for the majority of individuals.

2.3 Energy Balance of the Human Body

The thermoregulatory system of the human body maintains a constant temperature when there is no net heat storage. The chemical energy supplied by the body may be equated to the heat and work output, as shown in Figure 2.1. The metabolic rate $\dot{M}$ is the total energy released by the oxidation processes in the human body per unit time. The total heat flow rate $\dot{Q}_T$ dissipated by the body may be considered as the sum of the latent heat flow rate $\dot{Q}_L$ due to mass exchange effects such as sweating and respiration and the sensible heat flow rate $\dot{Q}_S$ due to heat transfer from the skin, including that through any clothing to the surroundings. The third energy element involved in the balance is the physical work output, $\dot{W}$.

An energy balance on the human body then yields:

Rate of Energy Intake ($\dot{M}$) = Rate of Energy Expenditure (Heat $\dot{Q}_T$ and Work $\dot{W}$)

or rearranging:

$$\dot{M} - \dot{W} - \dot{Q}_L - \dot{Q}_S = 0 \tag{2.1}$$

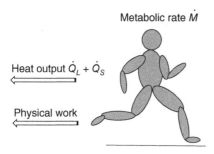

Metabolic rate $\dot{M}$

Heat output $\dot{Q}_L + \dot{Q}_S$

Physical work

Figure 2.1   Energy balance for the human body.

Table 2.1   Metabolic rates for some common activities.

Activity	Metabolic rate	
	W/m²	met
Sleeping	46	0.8
Seated, relaxed	58	1.0
Seated, light work, standing at rest	70	1.2
Standing, light activity	93	1.6
Walking (2 km/h)	110	1.9
Standing, medium activity	116	2.0
Domestic work	145–170	2.5–2.9
Sporting activities	232–550	4.0–9.5

Source: Adapted from Engineering ToolBox (2004b).

2.4   Metabolism (Ṁ) and Physical Work (Ẇ)

Energy is released in the body by oxidation at a rate equivalent to the needs of the body's function. The *Metabolic Equivalent of Task* (met) is a physiological measure of the energy consumed during a physical activity. In SI terms, 1 met = 58.2 W/m². Met values vary from 0.8 when sleeping to 9.5 for running (see Table 2.1). Metabolic rate ($\dot{M}$)can vary between individuals, and as a result, some people have slightly higher core body temperature than others. Compared to adult males, generally older people and children have a slower metabolic rate (~75%), while the metabolic rates of women are ~85%.

For any person, the metabolic rate varies on a regular daily cycle.

The efficiency (η_{met}) of human activity, defined as the ratio of physical work to the metabolic rate ($\dot{W}/\dot{M}$), is usually below 20% and 0% when motionless.

Expressing Eq. (2.1) in terms of body surface area (A) and rearranging gives

$$\dot{M} - \dot{W} = A(\dot{q}_S + \dot{q}_L) \tag{2.2}$$

where $\dot{q}_S$ is the 'sensible or dry' heat flow rate per unit area of skin due to convection, radiation, and conduction and $\dot{q}_L$ is the 'latent or wet' heat flow rate per unit area of skin due to sweating, evaporation, and respiratory effects. See Figure 2.2.

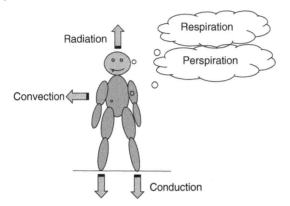

Figure 2.2 Heat flow mechanisms from the human body.

Note that the above values in Table 2.1 are expressed in W/m² of skin surface and should therefore be multiplied by the mean body area of 1.77 m² to obtain typical overall flow rates (W).

2.4.1 Latent Heat Loss

The latent heat transfer $\dot{q}_L$ (W/m²), which is accompanied by mass exchange effects, may be subdivided into two parts:

- The heat flux due to evaporation of sweat or perspiration on the skin surface, $\dot{q}_e$
- The respiration heat loss due to exhaling warm/inhaling cooler air, $\dot{q}_{res}$

i.e.

$$\dot{q}_L = \dot{q}_e + \dot{q}_{res} \tag{2.3}$$

2.4.1.1 Heat Loss by Perspiration

The amount of heat loss due to water diffusion and sweat evaporation $\dot{q}_e$ (W/m²) takes place all the time and is not controlled by the thermoregulatory system; it can be calculated by

$$\dot{q}_e = 3.05 \times 10^{-3}(256T_{skin} - 3373 - P_{wv}) + 0.42((\dot{M}/A - \dot{W}/A) - 58.15) \tag{2.4}$$

Equation (2.4) indicates that heat transfer by diffusion is strongly dependent on two variables:

(a) The skin temperature, which can be approximated by

$$T_{skin} = 35.7 - 0.0275(\dot{M}/A - \dot{W}/A) \tag{2.5}$$

(b) The water vapour pressure P_{wv} (Pa), which is a function of the relative humidity and temperature of the ambient air, given by

$$P_{wv} = \phi \times 0.6105e^{\frac{17.27T_{db}}{237.3+T_{db}}} \tag{2.6}$$

where ϕ is the relative humidity (% or fraction) and T_{db} is the dry bulb temperature (°C).

2.4.1.2 Heat Loss by Respiration

Aside from extreme ambient conditions, air necessary for metabolic processes is taken from surroundings which are normally at a temperature lower than the body temperature ($\sim$37 °C). The exhaled air temperature is in the region of 34 °C. This heat exchange is due to the difference in temperature of the air leaving the lungs and that entering. Since this air is moist, the equation to calculate the heat loss ($\dot{q}_{res}$, W/m^2) is made up of two terms having both sensible and latent components as shown in the following expression:

$$\dot{q}_{res} = 0.0014(\dot{M}/A)(34 - T_{db}) + 1.72 \times 10^{-5}(\dot{M}/A)(5867 - P_{wv}) \tag{2.7}$$

For typical indoor activity and an ambient temperature of 20 °C, the heat loss by respiration is of the order of 5 W/m^2 maximum and hence it is often neglected.

2.4.2 Sensible Heat Loss

The term $\dot{q}_S$ in Eq. (2.2) refers to the heat flux through the clothing by conduction and then to the surroundings from the outer surface of clothing by convection and radiation.

2.4.2.1 Heat Loss by Conduction

Conduction through the clothing ($\dot{q}_s$, W/m^2) is expressed as

$$\dot{q}_s = \left(\frac{A_{cl}}{A}\right)\left(\frac{k}{x}\right)(T_{skin} - T_{cl}) \tag{2.8}$$

The ratio (A_{cl}/A) is termed the *area factor* for the clothes which corrects for the difference between total body area and clothed body area.

The term (Ax/kA_{cl}) is the total thermal resistance of the clothing and has units of m^2 K/W. Thus, it is usually expressed as a *clothing index or clothing resistance* (R_{cl}, units: clo).

In SI terms, 1 clo = 0.155 m^2 K/W. By way of illustration, a typical summer lightweight business suit has a value of 1 clo, whereas an individual dressed for cold conditions with quilted garments would have an insulation level of around 2.5 clo.

The heat transfer by conduction through clothing can then be expressed as

$$\dot{q}_s = \frac{(T_{skin} - T_{cl})}{0.155R_{cl}} \tag{2.9}$$

The clothes temperature, T_{cl}, is calculated by the following expression:

$$
\begin{aligned}
T_{cl} = {}& 35.7 - 0.028(\dot{M}/A - \dot{W}/A) - 0.155R_{cl}(\dot{M}/A - \dot{W}/A) \\
& -0.003\,05(5733 - 6.99(\dot{M}/A - \dot{W}/A)) - P_{wv} \\
& -0.42((\dot{M}/A - \dot{W}/A) - 58.15) - 0.000\,017(\dot{M}/A)(5867 - P_{wv}) \\
& -0.0014(\dot{M}/A)(34 - T_{db})
\end{aligned}
\tag{2.10}
$$

Some clothing indices are shown in Table 2.2.

An overall insulation – or *clo* – value can be calculated by simply taking the *clo* value for each individual garment worn by a person and adding them together.

Table 2.2 Clothing indices.

Clothing		Insulation	
		clo	m² K/W
Nude	None	0	0
Underwear	Upper body	0.01–0.14	0.0002–0.022
	Lower body	0.02–0.1	0.003–0.016
Shirts	Short sleeves	0.09	0.029
	Long sleeves	0.25	0.039
Trousers	Briefs	0.04	0.006
	Shorts	0.06	0.009
	Lightweight trousers	0.20	0.031
	Normal trousers	0.25	0.039
Coveralls	Daily wear, belted	0.49	0.076
	Work	0.50	0.078
Jacket	Vest	0.13	0.020
	Light summer jacket	0.25	0.039
	Jacket	0.35	0.054

Source: Adapted from Engineering ToolBox (2004a).

2.4.2.2 Heat Loss by Convection

Convection from the bare skin or through the clothes ($\dot{q}_c$, W/m²) is given by

$$\dot{q}_c = h_o f_{cl}(T_{cl} - T_{db}) \tag{2.11}$$

where the clothing factor (f_{cl}) is expressed as

$$f_{cl} = 1.0 + 0.2R_{cl} \quad \text{for } R_{cl} < 0.5\,\text{clo}$$

and

$$f_{cl} = 1.05 + 0.1R_{cl} \quad \text{for } R_{cl} > 0.5\,\text{clo}$$

The convection heat transfer coefficient (h_c) can be estimated with the following correlations:

$$h_c = 2.38(T_{cl} - T_{db})^{0.25} \quad \text{for free convection (no air movement)}$$

and

$$h_c = 12.1(\overline{V}_a)^{0.5} \quad \text{for forced convection (with air movement)}$$

where $\overline{V}_a$ is the relative air velocity in m/s.

2.4.2.3 Heat Loss by Radiation

Radiation from a body to its surroundings ($\dot{q}_r$, W/m²) is given by

$$\dot{q}_r = h_r f_{cl}(T - T_{db}) \tag{2.12}$$

where h_r is the radiation heat transfer coefficient (W/m² K).

This is constant having a value of 3.90 W/m² K for a typical human body.

2.5 Optimum Comfort Temperature

The human body is equipped with a natural mechanism to adjust to all weather conditions. Indoor air-conditioned spaces are usually provided with automatic controls to maintain a preferred air temperature and perhaps humidity. This desired value may vary, depending on location, the nature of the activity, and the characteristics of the occupants.

The following expression is used to calculate the comfort temperature (T_{com}, °C) in an occupied space:

$$T_{com} = 33.5 - 3R_{cl} - (\dot{M}/A)[0.08 + 0.05R_{cl}] \tag{2.13}$$

The Chartered Institution of Building Services Engineers (CIBSE) in the United Kingdom recommend the following comfort criteria for general office occupancy:

- Winter dry resultant temperature: 21–23 °C (for clothing level 0.85 clo; activity 1.2 met)
- Summer dry resultant temperature: 22–24 °C (for clothing level 0.7 clo; activity 1.2 met)

Under conditions considerably different from the comfort temperature, the body reacts to preserve the internal temperature by an involuntary process known as *vasomotor regulation*. The tissues under the skin contain a network of blood vessels, which expand or contract under the command of the nervous system. In a warm environment, a copious flow of blood is allowed through these vessels, bringing the skin temperature closer to 35 °C. In contrast, when the environment is cooler, the veins are constricted, conserving the body heat by increasing the distance between the skin surface and the blood as demonstrated in Figure 2.3.

2.6 Estimation of Thermal Comfort

Thermal comfort involves creating a condition such that the maximum number of people in a given space would recommend no adjustment to the temperature or relative humidity. Since it is impossible to please everybody all the time, an index has been developed to express an overall feeling of comfort in that environment.

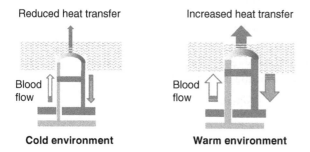

Figure 2.3 Body thermal regulation mechanisms.

Table 2.3 Range of comfort factors applicable to PMV.

Comfort parameter	Units	Symbol	Typical range
Metabolic rate	W/m^2	$\dot{M}$	46–232
Clothing resistance	m^2 °C/W	R_{cl}	0–0.31
Dry bulb temperature	°C	T_{db}	10–30
Radiant temperature	°C	T_r	10–40
Air velocity	m/s	$\overline{V}_a$	1–10
Vapour pressure	Pa	P_{wv}	0–2700

The criterion of thermal comfort is assessed by two indicators which are interrelated:

PMV-INDEX (predicted mean vote): This index gives a subjective thermal reaction of a large group of subjects according to a psychophysical scale. It predicts the mean value of the thermal votes of a large group of persons exposed to the same environment. The PMV value is specified for seven distinct sensations: hot (+3), warm (+2), slightly warm (+1), neutral (0), slightly cool (−1), cool (−2), and cold (−3).
The PMV is given by

$$
\begin{aligned}
\text{PMV} =&(0.303e^{-0.036M} + 0.028)\left\{(\dot{M} - \dot{W}) - 3.05 \times 10^{-3}[5733 - 6.99(\dot{M} - \dot{W}) - P_{wv}]\right. \\
&- 0.42 \times 0.0014(\dot{M}/A)(34 - T_{db}) - 3.96 \times 10^{-8}f_{cl}[(T_{cl} + 273)^4 - (T_r + 273)^4] \\
&\left.- f_{cl}h_c (T_{cl} - T_{db})\right\}
\end{aligned}
$$

(2.14)

Typical input parameters are given in Table 2.3.

PPD-INDEX (predicted percent dissatisfied): Individual votes on the PMV index show scatter and it is preferable in practice to predict the number of people likely to feel uncomfortably warm or cool, since these are the people most likely to complain about the environment. The PPD gives a quantitative prediction of the number of thermally dissatisfied persons. The PPD is given by

$$
\text{PPD} = 100 - 95e^{-(0.033\,53\text{PMV}^4 + 0.2179\text{PMV}^2)}
$$

(2.15)

The relationship between the PMV and PPD is shown in Figure 2.4.
Note that even when the PMV is zero, there will be 5% dissatisfied persons!
The PMV–PPD Index is an international standard for evaluating moderate thermal environments ISO (DIS 7730, 2005). The following limits are recommended:

$$-0.5 < \text{PMV} < 0.5 \text{ corresponding to PPD} < 10\%.$$

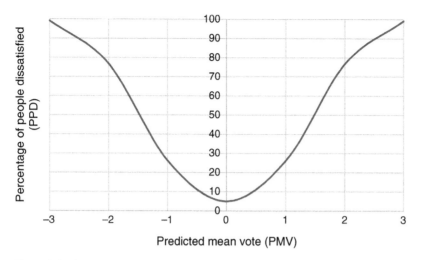

Figure 2.4 Relationship between PPD and PMV.

2.7 Worked Examples

Worked Example 2.1 Metabolic Rate

(a) Determine the metabolic rate, muscle power output, and the heat loss for an office worker (1.2 met, Efficiency $\eta_{met} = 15\%$).
(b) Compare these results with those determined for his state when he returns home and is relaxing in front of the television (0.8 met, Efficiency $\eta_{met} = 5\%$).

Assume the surface area of the human body to be $A = 1.77\,\mathrm{m^2}$.

Solution

Given: $A = 1.77\,\mathrm{m^2}$, (a) 1.2 met, $\eta_{met} = 15\% = 0.15$; (b) 0.8 met, $\eta_{met} = 5\% = 0.05$

Find: $\dot{M}, \dot{W}, Q$

(a) At 1 met $= 58.2\,\mathrm{W/m^2}$ then 1.2 met $\sim 70\,\mathrm{W/m^2}$

$$\dot{M} = 70 \times A$$
$$= 70 \times 1.77$$
$$= 123.9\,\mathrm{W}$$

$$\eta_{met} = \dot{W}/\dot{M}$$
$$0.15 = W/123.9$$
$$\dot{W} = 18.58\,\mathrm{W}$$

The total heat loss is given by

$$\dot{Q} = \dot{M} - \dot{W} = 123.9 - 18.58 = 105.32 \text{ W}$$

(b) At 1 met = 58.2 W/m² then 0.8 met ~ 46 W/m²

$$\dot{M} = 46 \times A$$

$$= 46 \times 1.77$$

$$= 81.42 \text{ W}$$

$$\eta_{met} = \dot{W}/\dot{M}$$

$$0.05 = W/81.42$$

$$\dot{W} = 4.07 \text{ W}$$

The total heat loss is given by

$$\dot{Q} = \dot{M} - \dot{W} = 81.42 - 4.07 = 77.35 \text{ W}$$

Worked Example 2.2 Latent Heat Loss

Determine the total latent heat loss (W) for a person engaged in a sedentary activity $(\dot{M}/A) = 58$ W/m² in a space where the air temperature is 20 °C and the relative humidity is 50%. Assume a skin temperature of 34 °C, an activity efficiency of 10%, and the surface area of the human body to be 1.77 m².

Solution

Given: $(\dot{M}/A) = 58$ W/m², $T_{db} = 20°C$, $\phi = 50\% = 0.5$, $T_s = 34°C$, $\eta_{met} = 0.1$, $A = 1.77$ m²

Find: $\dot{Q}_L$

$$\eta_{met} = (\dot{W}/A)/(\dot{M}/A)$$

$$0.1 = (\dot{W}/A)/58$$

$$\dot{W}/A = 5.8 \text{ W/m}^2$$

The partial pressure of the water vapour is given by

$$P_{wv} = \phi \times 0.6105 e^{\frac{17.27 T_{db}}{237.3 + T_{db}}}$$

$$= 0.5 \times 0.6105 \, e^{\frac{17.27 \times 20}{237.3 + 20}}$$

$$= 1168 \text{ Pa}$$

The heat loss associated with perspiration is given by

$$\dot{q}_e = 3.05 \times 10^{-3}(256 T_s - 3373 - P_{wv}) + 0.42(\dot{M}/A - \dot{W}/A - 58.15)$$

$$= 3.05 \times 10^{-3} \times (256 \times 34 - 3373 - 1168) + 0.42 \times (58 - 5.8 - 58.15)$$

$$= 10.19 \text{ W/m}^2$$

The heat loss associated with respiration is given by

$$\dot{q}_{res} = 0.0014(\dot{M}/A)(34 - T_{db}) + 1.72 \times 10^{-5}(\dot{M}/A)(5867 - P_{wv})$$

$$= 0.0014 \times 58 \times (34 - 20) + 1.72 \times 10^{-5} \times 58 \times (5867 - 1168)$$
$$= 5.824 \, \text{W/m}^2$$

The total latent heat loss (W) associated is given by

$$\dot{Q}_L = A(\dot{q}_e + \dot{q}_{res})$$
$$= 1.77(10.19 + 5.824)$$
$$= 28.35 \, \text{W}$$

Worked Example 2.3   Sensible Heat Loss

Determine the sensible heat loss from the body of an office worker wearing a typical summer lightweight business suit having a value of 0.25 clo.

Assume typical body temperature of 34 °C and the clothes temperature to be 28 °C. Take $A = 1.77 \, \text{m}^2$

Solution

Given: $R_{cl} = 0.25$ clo, $T_{skin} = 34 \degree$ C, $T_{cl} = 28 \degree$ C, $A = 1.77 \, \text{m}^2$

Find: $\dot{Q}_S$

$$\dot{q}_S = \frac{(T_{skin} - T_{cl})}{0.155 R_{cl}}$$
$$= \frac{(34 - 28)}{0.155 \times 0.25}$$
$$= 154.8 \, \text{W/m}^2$$

$$\dot{Q}_S = \dot{q}_S A$$
$$= 154.8 \times 1.77$$
$$= 274 \, \text{W}$$

Worked Example 2.4   Effect of Activity on Convective Heat Release

Determine the heat release (W) by convection from a human body for two situations:

(a) When jogging at an average speed of 5 m/s with a clothing index of 0.06 clo.
(b) When sitting in a draught-free room with a clothing index of 1.0 clo.

Assume body temperature to be 34 °C and air temperature 24 °C for both cases. Assume the surface area of the human body to be $A = 1.77 \, \text{m}^2$.

Solution

Given: $T_{cl} = 34 \degree$ C, $T_{db} = 24 \degree$ C, $A = 1.77 \, \text{m}^2$

(a) $\overline{V}_a = 5 \, \text{m/s}$, $R_{cl} = 0.06$ clo
(b) $R_{cl} = 1.0$ clo

Find: $\dot{Q}_c$

(a) Jogging at 5 m/s, i.e. with forced convection
The convective heat transfer coefficient is given by

$$h_c = 12.1(\overline{V}_a)^{0.5}$$
$$= 12.1(5)^{0.5}$$
$$= 27 \, \text{W/m}^2 \, \text{K}$$

Clothing resistance $R_{cl} < 0.5$ clo, therefore

$$f_{cl} = 1.0 + 0.2R_{cl}$$
$$= 1.0 + 0.2 \times 0.06$$
$$= 1.012$$

$$\dot{Q}_c = h_c f_{cl}(T_{cl} - T_{db}) \times A$$
$$= 27 \times 1.012 \times (34 - 24) \times 1.77$$
$$= 484 \, \text{W}$$

(b) Sitting still, i.e. with free convection
The convective heat transfer coefficient is given by

$$h_c = 2.38(T_{cl} - T_{db})^{0.25}$$
$$= 2.38(34 - 24)^{0.25}$$
$$= 4.23 \, \text{W/m}^2 \, \text{K}$$

Clothing resistance $R_{cl} > 0.5$ clo, therefore

$$f_{cl} = 1.05 + 0.1R_{cl}$$
$$= 1.05 + 0.1 \times 0.06$$
$$= 1.15$$

$$\dot{Q}_c = h_c f_{cl}(T_{cl} - T_{db}) \times A$$
$$= 4.23 \times 1.15 \times (34 - 24) \times 1.77$$
$$= 86 \, \text{W}$$

Worked Example 2.5 Effect of Clothing on Radiant Heat Release
Determine the radiant heat transfer from a young child (assume area of body to be $A = 1.0 \, \text{m}^2$ and body temperature 34 °C) in a building where the dry bulb temperature of the air is kept constant at 24 °C, for two types of clothing:

(a) Clothing resistance $= 0$ clo,
(b) Clothing resistance $= 1.0$ clo.

Solution

Given: $h_r = 3.9\,\text{W/m}^2\,\text{K}$, $T_{cl} = 34\,^\circ\text{C}$, $T_{db} = 24\,^\circ\text{C}$, $A = 1.0\,\text{m}^2$

(a) $R_{cl} = 0\,\text{clo}$
(b) $R_{cl} = 1.0\,\text{clo}$

Find: $\dot{Q}_r$

(a) Clothing index $R_{cl} < 0.5\,\text{clo}$, therefore

$$f_{cl} = 1.0 + 0.2R_{cl}$$
$$= 1.0 + 0.2 \times 0$$
$$= 1.0$$

$$\dot{Q}_r = h_r f_{cl}(T_{cl} - T_{db}) \times A$$
$$= 3.9 \times 1.0 \times (34 - 24) \times 1.0$$
$$= 39\,\text{W}$$

(b) Clothing index $R_{cl} > 0.5\,\text{clo}$, therefore

$$f_{cl} = 1.05 + 0.1R_{cl}$$
$$= 1.05 + 0.1 \times 1.0$$
$$= 1.15$$

$$\dot{Q}_r = h_r f_{cl}(T_{cl} - T_{db}) \times A$$
$$= 3.9 \times 1.15 \times (34 - 24) \times 1.0$$
$$= 44.85\,\text{W}$$

Worked Example 2.6 Evaluation of Comfort Temperature

Determine the comfort temperature for personnel in a factory (light activity, $\dot{M}/A = 70\,\text{W/m}^2$) under the following conditions:

(a) In winter condition with $R_{cl} = 1.5$.
(b) In summer time with $R_{cl} = 0.5$.

Discuss the limitations of this method.

Solution

Given: $\dot{M}/A = 70\,\text{W/m}^2$

(a) $R_{cl} = 1.5\,\text{clo}$
(b) $R_{cl} = 0.5\,\text{clo}$

Find: T_{com}

(a)

$$T_{com} = 33.5 - 3R_{cl} - (\dot{M}/A)[0.08 + 0.05R_{cl}]$$
$$= 33.5 - (3 \times 1.5) - (70)[0.08 + (0.05 \times 1.5)]$$
$$= 18.21\,^\circ\text{C}$$

(b)
$$T_{com} = 33.5 - 3R_{cl} - (\dot{M}/A)[0.08 + 0.05R_{cl}]$$
$$= 33.5 - (3 \times 0.5) - (70)[0.08 + (0.05 \times 0.5)]$$
$$= 24.7°C$$

This method does not allow for any variation in air velocity.

However, results using the ISO method ($\overline{V}_a < 0.1$ m/s) give 18 and 24 °C, respectively.

Worked Example 2.7   Effect of Ambient Conditions on Heat Loss Associated with Respiration

Investigate the rate of heat dissipation by respiration with the change of ambient dry bulb temperature T_{db} varying between 0 and 40 °C for office work (metabolic rate of 93 W/m²). Assume a relative humidity of 50%.

Solution

Given: $T_{db} = 0$–40 °C, $\phi = 50\% = 0.5$, $\dot{M}/A = 93$ W/m²

Find: $\dot{q}_{res}$

At 10 °C, the partial pressure of water vapour is given by

$$p_{wv} = \phi \times 0.6105 e^{\frac{17.27T_{db}}{237.3+T_{db}}}$$
$$= 0.5 \times 0.6105 e^{\frac{17.27 \times 10}{237.3+10}}$$
$$= 614 \text{ Pa}$$

The heat loss associated with respiration is given by

$$\dot{q}_{res} = 0.0014(\dot{M}/A)(34 - T_{db}) + 1.72 \times 10^{-5}(\dot{M}/A)(5867 - p_{wv})$$
$$= 0.0014 \times 93 \times (34 - 10) + 1.72 \times 10^{-5} \times 93 \times (5867 - 614)$$
$$= 11.528 \text{ W/m}^2$$

Repeating the above calculations for 0, 20, 30, and 40 °C gives Table E2.7.

Table E2.7   Results for Worked Example 2.7.

T_{db} (°C)	0	10	20	30	40
p_{wv} (Pa)	305.250	613.680	1168.566	2120.491	3685.995
$\dot{q}_{res}$ (W/m²)	13.323	11.52	9.338	6.514	2.70

Worked Example 2.8   Estimation of PMV

Determine the predicted mean vote (PMV) for the set of conditions supplied in Table E2.8.

Solution

Given: $(\dot{M}/A) = 116$ W/m², $\eta_{met} = 0.1$, $R_{cl} = 1.0$ m²°C/W, $T_{db} = 16°C$, $T_r = 18°C$, $T_{cl} = 34°C$, $\overline{V}_a = 0.1$ m/s, $P_s = 1404$ Pa.

Find: PMV

$$(\dot{W}/A) = \eta_{met}(\dot{M}/A) = 0.1 \times 116 = 11.6 \text{ W/m}^2$$

Table E2.8 Results for Worked Example 2.8.

Comfort parameter	Value
Metabolic rate	116 W/m^2
Body work efficiency	10%
Clothing resistance	1 clo
Dry bulb temperature	16 °C
Radiant temperature	18 °C
Body/clothes temperature	34 °C
Air velocity	0.1 m/s
Vapour pressure	1404 Pa

$$h_c = 12.1(\overline{V}_a)^{0.5} = 12.1(0.1)^{0.5} = 3.826 \text{ W/m}^2$$

$$f_{cl} = 1.0 + 0.2R_{cl} = 1.0 + (0.2 \times 1) = 1.2$$

$$
\begin{aligned}
\text{PMV} = {} & (0.303e^{-0.036\dot{M}} + 0.028)\{(\dot{M}/A - \dot{W}/A) \\
& - 3.05 \times 10^{-3}[5733 - 6.99(\dot{M}/A - \dot{W}/A) - P_{wv}] \\
& - 0.42 \times 0.0014\dot{M}(34 - T_{db}) - 3.96 \times 10^{-8}f_{cl}[(T_{cl} + 273)^4 - (T_r + 273)^4] \\
& - f_{cl}h_c\,(T_{cl} - T_{db})\} \\
= {} & (0.303e^{-0.036} + 0.028)\,\{(116 - 11.6) - 3.05 \times 10^{-3} \\
& [5733 - 6.99(116 - 11.6) - 1404] \\
& - 0.42 \times 0.0014 \times 104.4(34 - 16) - 3.96 \times 10^{-8} \times 1.2 \times \\
& [(34 + 273)^4 - (18 + 273)^4] \\
& - 1.2 \times 3.826\,(34 - 16)\} \\
= {} & +0.596
\end{aligned}
$$

Worked Example 2.9 Estimation of PPD

Calculate the predicted percent dissatisfied (PPD) for a predicted mean vote (PMV) of +1.5. Compare with the value indicated on PMV–PPD chart. Comment on the result.

Solution

Given: PMV

 Find: PPD

(a) Using Eq. (2.15)

$$
\begin{aligned}
\text{PPD} &= 100 - 95e^{-(0.033\,53 \times \text{PMV}^4 + 0.2179 \times \text{PMV}^2)} \\
&= 100 - 95e^{-(0.033\,53 \times 1.5^4 + 0.2179 \times 1.5^2)} \\
&= 50.9
\end{aligned}
$$

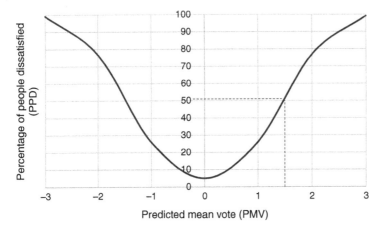

Figure E2.9 PPD–PMV plot for Worked Example 2.9.

(b) PMV–PPD chart:
At PMV = +1.5, draw a line vertically upward until it intersects the PPD curve, then read value from y-axis ~50%. See Figure E2.9.
This value is unacceptable as the ISO stipulates a maximum value of ±5% to keep the majority of occupants happy and satisfied but is similar to that provided by the equation.

Worked Example 2.10   Thermal Sensation
In a hospital ward, the air temperature T_a is controlled by an air thermostat set at 18 °C. The activity level and clothing of the occupants of the ward are given in Table E2.10a. Determine the likely thermal sensation of the occupants and comments on the suitability of the environment.

Table E2.10a   Data for Worked Example 2.10

Occupant	Activity (M) (W/m²)	Clothing (R_{cl}) (clo)
Staff	116	0.75
Patients sitting	70	1.0
Patients in bed	58	1.25

Solution
Given: $\dot{M}/A$, R_{cl}
  Find: T_{com}
  Using the comfort equation:

$$T_{com} = 33.5 - 3R_{cl} - (\dot{M}/A)(0.08 + 0.05R_{cl})$$

Given: For staff : $\dot{M}/A = 116\ \text{W/m}^2$, $R_{cl} = 0.75\ \text{clo}$

$$T_{com} = 33.5 - 3R_{cl} - (\dot{M}/A)[0.08 + 0.05R_{cl}]$$

$$= 33.5 - (3 \times 0.75) - (70)[0.08 + (0.05 \times 0.75)]$$

$$= 17.62°C$$

Repeat calculation for patients.

The three sets of occupants will feel differently, depending on the value of M and R, for each, hence if the building temperature is set to 18 °C, some will feel warm, others cold. See Table E2.10b.

Table E2.10b   Results for Worked Example 2.10.

Occupant	Comfort temperature (°C)	Sensation
Staff	17.6	Slightly warm
Patients sitting	19.2	Slightly cool
Patients in bed	21.5	Cool

2.8   Tutorial Problems

Where applicable in the following, assume the surface area of the human body A to be 1.77 m^2.

2.1   A person working in an office (1.0 met) has a body work efficiency $\eta_{met} = 15\%$. Determine his heat loss. Compare this result with that determined for his state when he returns home to recline on his sun bed (0.8 met, Efficiency $\eta_{met} = 5\%$).
[Answers: 87.3 W, 78.3 W]

2.2   Determine the total latent heat loss for a person engaged in office work ($\dot{M} = 58$ W/m^2) in an environment where the air temperature is 23 °C and the relative humidity is 50%.
Assume a skin temperature of 34 °C and a physical efficiency of 10%.
[Answers: 9.48 W/m^2, 5.34 W/m^2, 26 W]

2.3   Determine the sensible heat loss from the body of a worker in a manufacturing facility wearing coveralls having a value of 0.5 clo.
Assume typical body temperature of 34 °C and the clothes temperature to be 28 °C.
[Answer: 137 W]

2.4   Determine the heat release by convection from a runner for two situations:
(a) When the average air velocity is 7 m/s.
(b) When sitting in a draught-free room.
Assume clothing resistance of 0.5 clo, a body temperature of 34 °C, and air temperature of 24 °C.
[Answers: 623 W, 82 W]

2.5 Determine the radiant heat transfer from the human body for two situations:
(a) Clothing resistance $= 0\,\text{clo}$, a body temperature of $34\,°\text{C}$, and an air dry bulb temperature of $20\,°\text{C}$.
(b) As above but with a clothing resistance $= 1.0\,\text{clo}$.
[Answers: 96 W, 111 W]

2.6 Determine the comfort temperature for personnel in a factory (activity level: 2 met) under the following conditions:
(a) In winter conditions with $R_{cl} = 1.0$.
(b) In summer time with $R_{cl} = 0.25$.
[Answers: $15.42\,°\text{C}$, $22.02\,°\text{C}$]

2.7 Compare the rate of heat dissipation by respiration for two activity levels:
(a) Sedentary work, 1 met.
(b) Industrial work, 3 met.
Assume for both cases, the ambient dry bulb temperature T_{db} to be $20\,°\text{C}$, relative humidity of 50%, and the barometric pressure to be 1 bar.
[Answers: for 1 met: $10.3\,\text{W/m}^2$, for 3 met: $30.9\,\text{W/m}^2$]

2.8 Determine the predicted mean vote (PMV) for the circumstances described in Table P2.8.

Table P2.8 Data for tutorial Problem 2.8.

Comfort parameter	Value
Metabolic rate	$70\,\text{W/m}^2$
Body work efficiency	10%
Clothing resistance	$1\,\text{m}^2\,°\text{C/W}$
Dry bulb temperature	$20\,°\text{C}$
Radiant temperature	$22\,°\text{C}$
Body/clothes temperature	$34\,°\text{C}$
Air velocity	$0.1\,\text{m/s}$
Vapour pressure	1404 Pa

[Answer: −0.034]

2.9 Using a suitable correlation, determine the predicted percentage dissatisfied (PPD) for a predicted mean vote (PMV) of 0.75.
[Answer: 16.8%]

2.10 In a nursing home, the air temperature T_a is controlled by an air thermostat set at 22 °C. The activity level and clothing of the occupants of the ward are as supplied in Table P2.10. Determine the likely thermal comfort of the occupants and comment on the suitability of the environment.

Table P2.10 Data for tutorial Problem 2.10.

Occupant	Activity (W/m²)	Clothing (clo)
Staff	116	1.0
Patients	58	1.5

[Answers: 15.4 °C, 20 °C]

3

Ventilation

3.1 Overview

The act of breathing exchanges carbon dioxide for oxygen, reducing its ambient concentration in any confined space. High carbon dioxide concentrations are physiologically detrimental to humans making '*fresh*' air replacement necessary. The act of replacement is termed ventilation.

The function of a ventilation system is to provide and maintain a good quantity of high-quality air in the occupied space. This has taken on an extra dimension with the advent of the Covid-19 virus (2020).

Furthermore, ventilation dilutes odours, provides a sense of cooling, and removes other airborne solid and gaseous contaminants resulting from human activities.

European Standards define air quality and specify that, to achieve high air quality, a minimum outdoor air ventilation flow rate of 15 l/s for each occupant present in a space is required. The UK Building Regulations (Part F 2010) stipulates a minimum ventilation rate for office premises of 10 l/s per person. Rates in excess of these values may be required for areas with specialist process applications.

This chapter describes both natural and mechanically induced flows through a building as well as the design of flow conduits necessary to achieve suitable distribution of fresh air to occupants.

Learning Outcomes

- To understand the need for ventilation and control pollutant concentrations in the internal environment.
- To introduce the principles of providing fresh external air to a space by the use of natural phenomena.
- To introduce the principles of providing fresh external air to a space by mechanical means, i.e. fans.
- To understand the calculations involved in the design of air conduits or ducts.

Building Services Engineering: Smart and Sustainable Design for Health and Wellbeing, First Edition.
Tarik Al-Shemmeri and Neil Packer.

3.2 Concentrations, Contaminants, and the Decay Equation

3.2.1 Concentrations

The purpose of ventilating a space is to maintain desirable internal conditions limiting potential contaminants to acceptable concentrations.

The term concentration applies to the amount of substance contained or existing **within another substance**. Note that the SI units are the same as that of density (kg/m^3); however, in the case of concentration, the numerator and denominator refer to two different materials.

There are many ways of expressing the amount or concentration of any component within the whole. Some expressions of concentration are unit-free, whilst others have mixed units. Multiples and submultiples are in common use.

Some of the most common examples of concentration expression are given below:

(a) **Simple fractions**

Fractions are commonly mass-, volume-, or molar-based.

In general, the fraction of any component of interest in a combination of substances is given by:

$$\text{Simple fraction} = \frac{\text{Amount of component}}{\text{Amount of combination}} \tag{3.1}$$

Two aspects are notable: first, the units of the numerator and denominator are the same and free of multiples or submultiples, and second, the summation of all the component fractions making up the whole must be equal to unity.

(b) **Percentage fractions (%)**

Percentage fractions are simple fractions multiplied by 100, and the summation of all the component fractions making up the whole must therefore be equal to 100%.

(c) **'Parts per' fractions**

Another unit-free device for expressing very small concentrations that splits the whole up into a large number of parts.

Examples:

1 pph – 1 part per hundred or more commonly denoted by %
1 ppm – 1 part per million
1 ppb – 1 part per billion or 1 part per 1000 million.
1 ppt – 1 part per trillion or 1 part per 1 000 000 million

To avoid confusion, it must be made clear as to the basis, e.g. mass or volume, of these types of fractions and this is usually indicated by adding a postscript, '*w*' for mass and '*v*' for volume, to the term.

A comparison between simple fractions, percentage, and parts per million (ppm) is given in Table 3.1.

For illustration purposes, consider carbon dioxide. External air has a typical carbon dioxide concentration of around 410 ppmv. However, interior carbon dioxide concentrations in the internal environment can exceed this as a result of human exhalation. At rest, a typical rate of carbon dioxide generation is 5×10^{-6} m^3/s person. This can increase 20 times when undergoing physical exertion.

Table 3.1 Table of fractional concentration unit comparison.

Concentration unit equivalents		
Simple fraction	**Percentage (%)**	**ppm**
1	100	1 000 000
0.1	10	100 000
0.01	1	10 000
0.001	0.1	1 000
0.0001	0.01	100

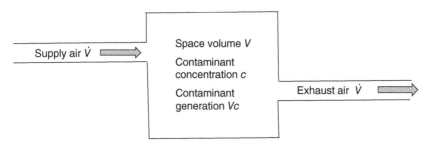

Figure 3.1 Basis of decay equation.

An increase in ambient CO_2 level from a normal value of around 0.04–2% increases the depth of respiration by 30%. When the concentration of CO_2 exceeds 3%, there is conscious respiratory discomfort. An unlimited increase would allow carbon dioxide to accumulate in the blood and other tissues.

According to the UK Health and Safety Executive (HSE) the recommended eight hour average workplace exposure limit for CO_2 is 0.5%.

3.2.2 The Decay Equation

The effect that ventilating a space has on its prevailing concentration conditions is described by the *Decay equation*. Consider a space having a volume V (m³) and a flow rate of ventilating air entering (and leaving) the space of $\dot{V}$ (m³/s) applied for a time t (seconds). See Figure 3.1.

It can be shown that:

- For a room with an initial contaminant concentration (c_o, ppm), *zero* contaminant in the ventilation air, and *zero* contaminant generation (V_c, m³/s) within the space, the variation of contaminant (c, ppm) in the space with time is given by:

$$c = c_o e^{-n} \qquad (3.2)$$

where n (no units) is the number of times the cubical content of the air is changed, given by:

$$n = \dot{V}t/V \qquad (3.3)$$

- For a room with *zero* initial contaminant concentration, a contaminant concentration in the ventilation air (c_a, ppm), and *zero* contaminant generation within the space:

$$c = c_a(1 - e^{-n}) \qquad (3.4)$$

- For a room with *zero* initial contaminant concentration, a contaminant concentration in the ventilation air (c_a, ppm), and contaminant generation within the space of V_c (m³/s person):

$$c = ([V_c \times 10^6 / \dot{V}] + c_a)(1 - e^{-n}) \qquad (3.5)$$

- For a room with an initial contaminant concentration (c_o, ppm), a contaminant concentration in the ventilation air (c_a, ppm), and contaminant generation within the space of V_c (m³/s person) and $\dot{V}$ (m³/s person) is the rate of fresh air supply:

$$c = ([V_c \times 10^6 / \dot{V}] + c_a)(1 - e^{-n}) + c_o e^{-n} \qquad (3.6)$$

When the time period under study is 3600 seconds, n usually transforms to its uppercase version and is termed the *number of air changes per hour* (N, h⁻¹).

Values of design air change rates vary with building application, thermal loads, and the presence of contaminants. A simple dwelling might have an air change rate of 1 h⁻¹. Buildings having a high occupancy density could, typically, have air change rates of 5–15 h⁻¹, and specialist applications like commercial kitchens could require air changes of up to 40 h⁻¹.

3.3 Natural Ventilation

Natural ventilation systems use ambient fluid density and pressure differences to induce a flow of air through a space.

3.3.1 Stack Effect Ventilation

Warm air rises inside a space due to convective buoyancy and exits the building at any high-level opening. Replacement with colder, ambient air is induced into the space at any lower level openings. The effect is enhanced in tall buildings.

Consider the building in Figure 3.2 with a low-level opening at position 1 and a high-level opening at position 2.

From fluid-static law considerations, the pressures at the inlet (1) are given by:

$$P_{o@1} = P_{o@2} + \rho_o g H \qquad (3.7)$$

$$P_{i@1} = P_{i@2} + \rho_i g H \qquad (3.8)$$

Subtracting, the pressure difference across the inlet:

$$P_{o@1} - P_{i@1} = \Delta \rho g H - (P_{i@2} - P_{o@2}) \qquad (3.9)$$

Applying the *Continuity equation* between inlet (1) and outlet (2), we have:

$$\rho_o \dot{V}_1 = \rho_i \dot{V}_2 \qquad (3.10)$$

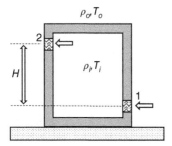

Figure 3.2 Stack ventilation principles.

Using the *Bernoulli equation* and assuming the inlet and outlet can be regarded as simple orifices:

$$\rho_o C_d A_1 \left[\frac{2(P_{o@1} - P_{i@1})}{\rho_o} \right]^{0.5} = \rho_i C_d A_2 \left[\frac{2(P_{i@2} - P_{o@2})}{\rho_i} \right]^{0.5} \tag{3.11}$$

Therefore,

$$P_{i@2} - P_{o@2} = (P_{o@1} - P_{i@1}) \frac{\rho_o (A_1)^2}{\rho_i (A_2)^2} \tag{3.12}$$

Substituting into (3.9):

$$P_{o@1} - P_{i@1} = gH\Delta\rho - (P_{o@1} - P_{i@1}) \frac{\rho_o (A_1)^2}{\rho_i (A_2)^2} \tag{3.13}$$

i.e.

$$P_{o@1} - P_{i@1} = \frac{gH\Delta\rho}{1 + \frac{\rho_o (A_1)^2}{\rho_i (A_2)^2}} \tag{3.14}$$

and

$$\dot{V}_{stack} = C_d A_1 \left[\frac{2(P_{o@1} - P_{i@1})}{\rho_o} \right]^{0.5} \tag{3.15}$$

Substituting for Eq. (3.14) into (3.15):

$$\dot{V}_{stack} = C_d A_1 \left[\frac{2gH\Delta\rho}{\rho_o \left\{ 1 + \frac{\rho_o A_1^2}{\rho_i A_2^2} \right\}} \right]^{0.5} = C_d \left[\frac{2gH\Delta\rho}{\rho_o \left\{ \frac{1}{A_1^2} + \frac{\rho_o}{\rho_i A_2^2} \right\}} \right]^{0.5} \tag{3.16}$$

Let the mean system density $\rho_m = (\rho_o + \rho_i)/2$
And mean system temperature T_m (K) $= (T_o + T_{in})/2$
Giving:

$$\dot{V}_{stack} = C_d A_{stack} \left[\frac{2gH\Delta T}{T_m} \right]^{0.5} \tag{3.17}$$

where

$$\frac{1}{A_{stack}^2} = \frac{1}{A_1^2} + \frac{1}{A_2^2} \tag{3.18}$$

For buildings with a series of openings, ΣA_1 and ΣA_2 can be used to replace A_1 and A_2, respectively.

3.3.2 Wind Effect Ventilation

Contact with a building modifies the energy distribution in the wind. The air flow may be accelerated, decelerated, or brought to a standstill. These changes in kinetic energy will result in pressure energy changes relative to the undisturbed, free stream flow.

For example, on the windward side of a building, wind brought to a standstill has its kinetic energy converted to pressure energy. If the resulting pressure is in excess of that inside a building, an induced air flow will occur through any openings in the structure. On the leeward side of a building, the modified air flow might result in a reduced pressure relative to the building interior resulting in a flow out of the structure.

Pressure coefficients for buildings are dependent on building form factors such as

- Building section geometry
- Roof angle
- Orientation with respect to wind direction.
- Surrounding environment, i.e. open/urban

Values for simple shapes can found in handbooks. Values for complex shapes can be determined empirically through physical and/or computational modelling.

Consider the building in Figure 3.3 with a windward opening at position 1 and a leeward opening at position 2.

Applying the continuity equation to a small ventilation opening:

$$\dot{V} = C_d A \overline{V} \tag{3.19}$$

The pressure difference across the opening is given by:

$$\Delta P = \frac{\rho_o \overline{V}^2}{2} \tag{3.20}$$

Hence:

$$\dot{V} = C_d A \overline{V} = C_d \left(\sum A \right) \left(\frac{2 \Delta P}{\rho_o} \right)^{0.5} \tag{3.21}$$

where ΣA is the sum of all opening areas on a surface.

At inlet:

$$\dot{V}_1 = C_d \left(\sum A_1 \right) \left(\frac{2\{P_1 - P_i\}}{\rho_1} \right)^{0.5} \tag{3.22}$$

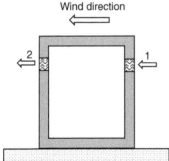

Figure 3.3 Wind ventilation principles.

And at outlet:

$$V_2 = C_d \left(\sum A_2 \right) \left(\frac{2\{P_i - P_2\}}{\rho_2} \right)^{0.5} \tag{3.23}$$

Let

$$\dot{V}_1 = \dot{V}_2 = \dot{V}_{\text{wind effect}} \quad \text{and} \quad \rho_1 = \rho_2 = \rho_0$$

Therefore,

$$P_1 - P_2 = \frac{\rho_0 \dot{V}_{\text{wind effect}}^2}{2C_d^2} \left(\frac{1}{\left(\sum A_1\right)^2} + \frac{1}{\left(\sum A_2\right)^2} \right) \tag{3.24}$$

Considering pressure coefficients:

$$C_{p1} = \frac{P_1 - P_0}{0.5\rho_0 \overline{V}_{\text{wind}}^2} \tag{3.25}$$

and

$$C_{p2} = \frac{P_2 - P_0}{0.5\rho_0 \overline{V}_{\text{wind}}^2} \tag{3.26}$$

Therefore,

$$P_1 - P_2 = 0.5(C_{p1} - C_{p2})\rho_0 \overline{V}_{\text{wind}}^2 = \frac{\Delta C_p \rho_0 \overline{V}_{\text{wind}}^2}{2} \tag{3.27}$$

Substituting:

$$\frac{\Delta C_p \rho_0 \overline{V}_{\text{wind}}^2}{2} = \frac{\rho_0 \dot{V}_{\text{wind effect}}^2}{2C_d^2 \left(\sum A_{\text{wind}}\right)^2} \tag{3.28}$$

where

$$\frac{1}{\left(\sum A_{\text{wind}}\right)^2} = \frac{1}{\left(\sum A_1\right)^2} + \frac{1}{\left(\sum A_2\right)^2} \tag{3.29}$$

And so

$$\dot{V}_{\text{wind effect}} = C_d \left(\sum A_{\text{wind}} \right) \overline{V}_{\text{wind}} (\Delta C_p)^{0.5} \tag{3.30}$$

Again, a typical value for C_d is 0.6.

3.3.3 Combined Wind and Stack Effect Ventilation

Under some circumstances, wind (pressure) and stack (density) effects can re-enforce each other.

From basic orifice theory, volumetric flow rate is proportional to the square root of pressure difference. The combined wind and stack effect is therefore calculated from:

$$\dot{V} = (\dot{V}_{\text{wind effect}}^2 + \dot{V}_{\text{stack effect}}^2)^{0.5} \tag{3.31}$$

For buildings with both potential inlet and outlet openings on two faces, the situation is more complex as under some circumstances wind and stack effect flows may be in opposition. See Figure 3.4.

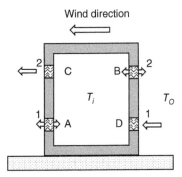

Figure 3.4 Combined wind and stack ventilation principles.

Empirical data indicate that for the above arrangement when:

$$\dot{V}_{\text{wind effect}} > \dot{V}_{\text{stack effect}} \quad \text{then } \dot{V} = \dot{V}_{\text{wind effect}}$$

and

$$\dot{V}_{\text{wind effect}} < \dot{V}_{\text{stack effect}} \quad \text{then } \dot{V} = \dot{V}_{\text{stack effect}}$$

and

$$\dot{V} = \dot{V}_{\text{wind effect}} \quad \text{when } \overline{V}_{\text{wind}} A_{\text{wind effect}} (\mathbb{C}_p)^{0.5} > 0.26 A_{\text{stack effect}} (H \Delta T)^{0.5} \tag{3.32}$$

$$\dot{V} = \dot{V}_{\text{stack effect}} \quad \text{when } \overline{V}_{\text{wind}} A_{\text{wind effect}} (\mathbb{C}_p)^{0.5} < 0.26 A_{\text{stack effect}} (H \Delta T)^{0.5} \tag{3.33}$$

Care is required in quantifying the flow areas, for example in Figure 3.4:

$$A_{\text{wind effect}} \text{ is based on } \sum A_1 = A_B + A_D \text{ and } \sum A_2 = A_A + A_C \tag{3.34}$$

$$A_{\text{stack effect}} \text{ is based on } \sum A_1 = A_A + A_D \text{ and } \sum A_2 = A_B + A_C \tag{3.35}$$

3.3.4 Infiltration

The above principles are used to design non-mechanical ventilation systems.

However, depending on the 'air-tightness' of a building, the same processes will induce an uncontrolled air flow through a space via cracks and imperfections in the building envelope. This flow is termed infiltration and, in 'leaky' buildings, can typically result in rates of between 0.5 and 2 air changes per hour.

3.4 Mechanical Ventilation

Forced or mechanical ventilation relies on using a fan-powered system; this type is generally preferred for bigger buildings such as theatres and sports halls, where occupancy is dense, or heat and odour need to be removed from the building. There are three ways in which forced ventilation is used:

(a) **Supply Systems**

In this system, the air is supplied by a fan. This arrangement lends itself to a combination with a heating/cooling system. Usually, the supply system introduces the air at the lower part of the building and relies on doors, window openings, vents, and other leakage paths through the building structure for exit.

(b) **Extract Systems**

In this kind of system, the air is exhausted by fans at roof or high level and makeup air enters naturally at ground level, either through the building structure or correctly positioned air inlet louvres.

(c) **Supply and Extract Systems**

A combination of the above, one supply air fan and one exhaust air fan ventilate the building through ducted systems. In this type of system, the added energy requirement for the second fan may be offset by the introduction of a heat exchanger, so that the incoming air is heated by the warmer 'used' air.

Typically the supply air flows to bedrooms and living rooms, while the exhaust air is drawn from the kitchen, bathroom, and utility room. Heat is transferred from the warm exhaust air to the cold outdoor air in the heat exchanger. The energy saving could be 50–60% compared to a system in which heat is not recovered. See Figure 3.5a–c.

3.5 Fan Types and Selection

There are four main types of fans used in air conditioning systems. They are perhaps most easily recognised with reference to their impeller and the geometry of the air flow through it. See Figure 3.6a–d.

(a) **Propeller Fans**

This type is usually used at free openings in walls or windows and for other types of low-pressure applications. It is not usually employed for ducted ventilation systems. *Shaft power to fluid power* efficiencies are low <40%.

(b) **Axial-Flow Fans**

This type of fan is designed for mounting inside a duct system and is suitable for moving air in complete systems of ductwork. Flow is parallel to the fan axis. This type of fan is common in low-pressure applications. Efficiencies of up to 85% are possible for complex forms having inlet guide vanes.

(c) **Centrifugal Fans**

The impeller typically comprises two rings connected by a number of blades with the whole rotating in a volute casing. The inlet of the fan is at 90° to the outlet and like the axial-flow fan is suitable for moving air in complete systems of ductwork. This type of fan is more common for higher pressure applications. Efficiencies of 70% are common with complex aerodynamically designed forms attaining up to 90%.

(d) **Crossflow Fans**

Commonly used in supply point applications, e.g. fan coil units. Discharge direction is narrow with air entering and exiting at 90° to the fan's axis of rotation. Total efficiencies are low, typically under 50%.

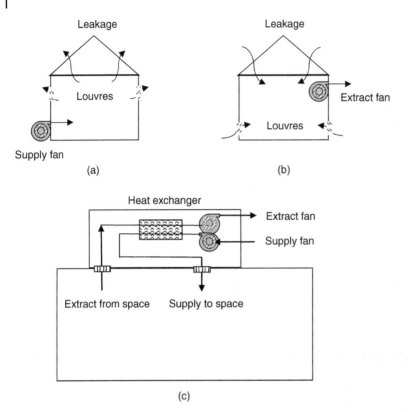

Figure 3.5 Mechanical ventilation systems. (a) Supply ventilation, (b) extract ventilation, and (c) supply and extract ventilation with heat recovery.

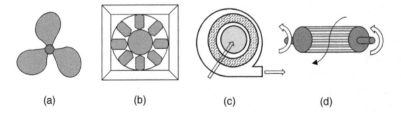

Figure 3.6 Types of ventilation fans. (a) Propeller, (b) axial, (c) centrifugal, and (d) crossflow.

3.5.1 Selection of Fans

In order to select a fan for a given duty, reference should be made to fan performance graphs supplied by the manufacturer.

A notional example is shown in Figure 3.7. The principle axes are pressure rise (Pa) or head rise (m) across the fan against delivery or flow rate through the fan (m³/s). Specifications often come with supplementary axes indicating the fan power requirement (W) and efficiency (%) vs. flow rate.

For both pressure and flow, multiples and submultiples are also in use, e.g. kPa, l/s.

Some consideration should be given to fan pressure.

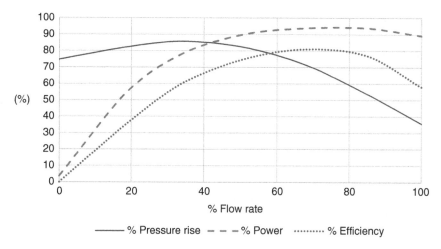

Figure 3.7 Example of fan performance data.

Let atmospheric pressure be the datum in terms of zero static pressure and velocity. Velocity pressure at fan inlet and outlet will therefore both have positive values.

However, static pressure at fan inlet must be negative relative to datum and that at fan outlet must be positive to facilitate flow and overcome system resistance.

The term fan total pressure (FTP) is defined as the rise in total pressure across the fan:

$$\text{FTP} = (\text{Static pressure} + \text{Velocity pressure})_{\text{fan out}}$$

$$- (\text{Static pressure} + \text{Velocity pressure})_{\text{fan in}}$$

$$= (P + \rho \overline{V}^2 / 2)_{\text{fan out}} - (P + \rho \overline{V}^2 / 2)_{\text{fan in}} \tag{3.36}$$

The term fan static pressure (FSP) is defined as the static pressure at fan outlet minus the total pressure at fan inlet:

$$\text{FSP} = (\text{Static pressure})_{\text{fan outlet}} - (\text{Total pressure})_{\text{fan in}}$$

$$= (P)_{\text{fan out}} - (P + \rho \overline{V}^2 / 2)_{\text{fan in}} \tag{3.37}$$

Alternatively, the FSP can be expressed as the FTP minus the velocity pressure at fan outlet

$$\text{FSP} = \text{FTP} - (\text{Velocity pressure})_{\text{fan outlet}} \tag{3.38}$$

to yield the same result.

FSP is often favoured by fan manufacturers in their performance specifications.

A fan should be chosen to match the application. This is usually done by calculating the system duty, then choosing a fan with the rated power higher than the calculated one in order to allow for a drop in efficiency with age.

A number of efficiency expressions are in use:

$$\text{Fan Total Efficiency (FTE)} = \text{Fluid power (based on FTP)/Shaft power} \tag{3.39}$$

$$\text{Fan Static Efficiency (FSE)} = \text{Fluid power (based on FSP)/Shaft power} \tag{3.40}$$

$$\text{Overall Efficiency} = \text{Fluid power/Electrical power input} \tag{3.41}$$

3.6 Duct Sizing and Fan Matching

3.6.1 Duct Pressure Losses

A duct is simply the means by which air is confined on the destination path to the point of application. The necessary volume flow rate in a duct is dependent on the requirement of its application and is usually outside the control of the designer.

The duct will respond to air movement with friction manifesting itself as a static pressure reduction and so, unless wind or buoyancy forces are sufficient, a pressure increase is required to encourage the movement of air along a duct.

This is supplied by a fan or blower.

The total ductwork system pressure resistance will be decided by the choice of duct size, duct type, and flow velocity and is given by:

$$\Delta P_{sys} = \Delta P_{friction} + \Delta P_{fittings}$$

$$= \left[\left(f \frac{L}{D_h} \right) \left(\rho_i \frac{\overline{V}^2}{2} \right) \right] + \left[\sum K \left(\rho_i \frac{\overline{V}^2}{2} \right) \right] \tag{3.42}$$

Some of these terms require a little more explanation.

- $\Delta P_{friction}$

  This evaluates the pressure drop along straight length (L, m) of duct having air moving at a chosen velocity ($\overline{V}$, m/s). Clearly, higher velocities will result in a higher pressure resistance, but acceptable velocities are often based, in practice, on minimising noise.

  Not all conduits for fluid transfer have a simple circular cross section. For example, air ducts may be rectangular, triangular, or spiral circular in section. In order to accommodate this, the circular *hydraulic diameter* (D_h, m) is used in the calculation of duct friction where:

$$D_h = 4 \times \text{cross sectional area} /' \text{wetted}' \text{perimeter}$$

$$= 4A/P \tag{3.43}$$

The (Darcy) *friction factor* (f) is a dimensionless parameter whose value is dependent on the prevailing flow regime and the duct material. To evaluate the friction factor, it is first necessary to make an estimate of the prevailing Reynolds number (Re) where:

$$Re = \frac{\rho \overline{V} D_h}{\mu} \tag{3.44}$$

The Reynolds number in this context is a predictor of flow behaviour.

If the resulting Reynolds number has a calculated value of <2000, then the flow is regarded as *laminar* and for any conduit material:

$$f = \frac{64}{Re} \tag{3.45}$$

If the duct can be assumed to be 'smooth' (e.g. plastic) and the resulting Reynolds number has a value of ≥4000, then the flow is regarded as *turbulent,* i.e. dominated by eddies and vortices and:

$$f = 1 / \left(1.8 \log \frac{Re}{6.9} \right)^2 \tag{3.46}$$

Iterative and complex explicit approximate correlations for non-smooth or 'rough' pipes with turbulent flow are available but are beyond the scope of this text.

Alternatively, friction factors can be estimated by using an extremely useful dimensionless chart known as the Moody diagram which is applicable to all flow regimes and duct materials. See Figure 3.8. The chart plots friction factor values on the y-axis against Reynolds number on the x-axis.

To use the chart requires knowledge of the absolute roughness (k, m) of the duct material. Clean, non-corroded steel-based ducts are likely to have an absolute roughness of the order of 0.1 mm, while plastic ducts would typically have a lower absolute roughness of <0.01 mm.

With these data, the relative roughness $(k/D_h$, no units) is determined and along with the prevailing Reynolds number, the friction factor can be estimated as shown in Figure 3.9. Note that curves describing both laminar and smooth/turbulent conditions are also available on the chart.

Air ducts are commonly made from steel. Sheet steel can be used to form both rectangular and circular cross sections. Seamed strip steel is used to form spiral ducting. Circular cross section is common with flattened circular cross sections used in confined spaces. Materials such as aluminium, copper, stainless steel, and PVC are used in specialist applications.

- $\Delta P_{\text{fittings}}$

The minor (**head** or **pressure**) loss coefficient (K) is a measure of the flow resistance of duct fittings such as bends, junctions, filters, heat exchangers, final outlet grilles, etc. In general, duct direction changes and transitions are a function of geometric change ratio, velocity change ratio, flow rate change ratio, and/or Reynolds number.

Some indicative values can be found in Table 3.2 where the subscripts u, d, and b denote upstream, downstream, and branch locations, respectively.

Loss coefficients for specialist in-duct items are usually supplied by the manufacturer. If the loss associated with a fitting is to be expressed as a head loss $(\Delta h_{\text{fitting}}, \mathrm{m})$, then:

$$\Delta h_{\text{fitting}} = K\overline{V}^2/2g \tag{3.47}$$

Traditionally, ductwork fitting-associated losses are often expressed in terms of velocity pressure. Using a value of air density as 1.2 kg/m^3, a fitting pressure loss $(\Delta P_{\text{fitting}}, \mathrm{Pa})$ can be calculated thus:

$$\Delta P_{\text{fitting}} = 0.6K\overline{V}^2 \tag{3.48}$$

Loss factors are associated with the velocity pressure at a particular location in the fitting, i.e. entry, exit, or branch. This must be indicated in the design data.

3.6.2 Selecting Duct Sizes

Three methods are in common use:

- Velocity method

All ducts (m) are sized by limiting the air velocity (m/s) to a value that does not result in unacceptable noise (dB) associated with air turbulence. Some guidance is given in Table 3.3.

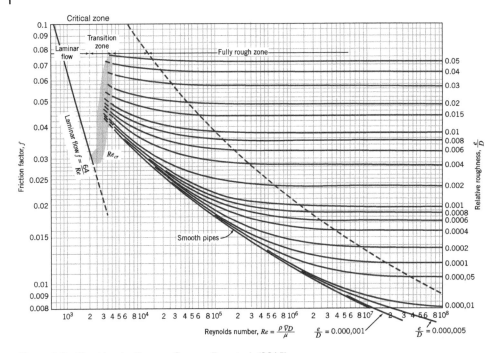

Figure 3.8 The Moody diagram. Source: Fox et al. (2015).

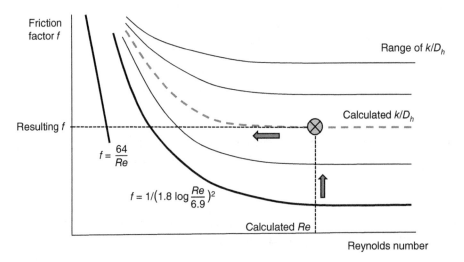

Figure 3.9 Use of the Moody diagram.

Noise in ventilation systems can be exacerbated by poor duct detail design and low-efficiency mechanical/electrical fan drives.

- Equal pressure drop method

  This is a method suitable for controlling excessive system pressure drop and fan energy consumption.

Table 3.2 Minor loss coefficients for some common fittings.

Component or fitting	Minor loss coefficient K
Smooth, circular duct, 90° elbow, bend radius/duct diameter $= 1.0$	0.22
Smooth, circular duct, 45° elbow, bend radius/duct diameter $= 1.0$	0.13
Three-piece, circular duct, 90° elbow, bend radius/duct diameter $= 1.0$	0.42
Mitred, circular duct, 90° elbow	1.2
Mitred, rectangular duct, 90° bend	1.2
Mitred, rectangular duct, 90° bend with turning vanes	0.12–0.18
Mitred, rectangular duct, 45° bend	0.34
Smooth, rectangular duct, 90°, height/width $= 1.0$, radius/width $= 1.0$	0.21
Circular duct expansion, 30° cone angle, $Re = 2 \times 10^5$, $A_d/A_u = 4$	0.36
Circular duct contraction, 30° cone angle, $A_u/A_d = 4$	0.04
Supply, rectangular, 90° swept tee, swept radius/width $= 1.0$, $A_b/A_d = 1.0$, $A_b/A_u = 1.0$, $\dot{V}_b/\dot{V}_u = 0.5$	Branch: 0.32, main: 0.06
Extract, rectangular, 90° swept tee, swept radius/width $= 1.0$, $A_b/A_d = 1.0$, $A_b/A_u = 1.0$, $\dot{V}_b/\dot{V}_u = 0.5$	Branch: 0.13, main: 0.23
Supply, rectangular, 90° abrupt tee, $A_b/A_u = 1.0$, $\dot{V}_b/\dot{V}_u = 0.5$	1.27
Supply, rectangular, 90° abrupt tee to circular branch, $A_b/A_u = 1.0$, $\dot{V}_b/\dot{V}_u = 0.5$	1.26
Rectangular duct expansion, expansion angle $= 30°$, $A_d/A_u = 4$	0.5
Rectangular duct contraction, contraction angle $= 30°$, $A_u/A_d = 4$	0.04

Source: Adapted from Pita (2002).

After selecting a suitable velocity for the system main supply duct, the resulting pressure loss/metre length of duct is calculated. This pressure drop/metre length is maintained along the entire length of the ductwork system by adjusting (i.e. reducing) the duct diameter accordingly when the duct flow rate reduces along the main run as a result of takeoff. This process can be facilitated by the use of a propriety duct sizing chart. See Figure 3.10. The construction of the chart is as follows:

- The vertical axis carries values of duct volume flow rates (l/s)
- The horizontal axis carries values of duct pressure drop/metre length (Pa/m). A line of constant pressure drop/metre length of duct will therefore be a vertical line on the chart.
- Lines of constant duct diameter (mm) run SW to NE
- Lines of duct constant flow velocity (m/s) run NW to SE

Table 3.3 Recommended limiting velocities in ventilation ducts (m/s).

Duct system element	Limiting velocity (m/s)	
	Public buildings	Industrial plant
Outside air intake	2.5–4.5	2.5–6
Filters	1.5–1.75	1.75
Heat exchangers	2.5–3.0	3–3.5
Main supply ducts	5.0–8.0	6–11
Branch supply ducts	3.0–6.5	4.0–9
Supply registers and grilles	1.0–2.0	1.5–2.5

Source: Adapted from Pita (2002).

Use of the chart is relatively simple. Having identified the flow rates (l/s) in each duct section and specified a pressure drop per metre run (Pa/m) to be maintained throughout the system, i.e. the line of equal pressure drops, it is a simple matter to read off a suitable duct diameter at the intersection of these two parameters. The associated duct velocity can also then be read off the chart to check for noise considerations.

Unlike the Moody diagram, design aids such as these are not universal being specific to a particular duct material and cross section. The most commonly encountered aid applies to the use of circular, galvanised steel duct. Correction factors applied to the pressure loss are required for other materials and cross sections.

- Static regain method

The term static regain (Pa) is a reference to the principle of energy conservation, remembering that the total pressure is the sum of static pressure and velocity pressure. See Eqs. (1.39) and (1.40). Thus, lowering the air velocity by manipulating, i.e. increasing, the duct size will result in an increase or *regain* of static pressure. Theoretically:

$$(SP_2 - SP_1)_{Ideal} = VP_1 - VP_2 \qquad (3.49)$$

In reality, the ideal static regain will not be achieved due to the frictional resistance of the duct expansion thus:

$$(SP_2 - SP_1)_{Actual} = (VP_1 - VP_2) - 0.6K\overline{V}_2^2 \qquad (3.50)$$

This recovery of static pressure can be used to overcome system frictional losses and, in theory, lower fan size and hence energy consumption. However, this saving may come at the expense of additional design, material, and installation cost.

3.6.3 Fan Sizing

The fan flow rate will be the sum of the individual system ventilation requirements.

The fan must also overcome the maximum resistance to flow offered by the system (as well as provide the air with momentum).

The air distribution route with the greatest resistance to flow is termed the **index run**. For a simple linear distribution system with a number of equal flow takeoffs, the index run

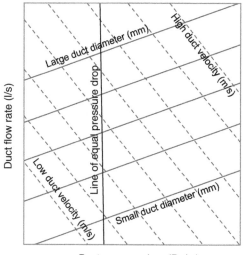

Figure 3.10 Stylised duct sizing chart.

will likely comprise the route to the outlet most distant from the fan. However, for a complex system, each route should be checked in turn to ascertain the index run. Carrying this task out manually can become tedious and is therefore best carried out using a proprietary software.

3.6.4 Fan-System Characteristics and Matching

It was shown in Eq. (3.42) that the total head of the system is proportional to the square of the velocity, and because the volume flow rate is directly proportional to the velocity, hence the pressure head of the system is related to the square of the system flow rate thus:

$$\Delta P_{sys} = c_1 \dot{V}^2 + c_2 \tag{3.51}$$

where c_1 and c_2 are constants. Graphically, this is a second-order polynomial.

Figure 3.11 displays a notional fan performance, with a range of system pressure drop characteristics plotted on the same diagram.

Reviewing Eq. (3.42), it will be seen that the shape of the system curvature may vary if the system is modified, say, by increasing or decreasing the length of the duct, changing the duct size, adding or taking away some of the fittings or even the cleanliness of the system. In graphical terms, this has the effect of rotating the system curve either clockwise (decreased resistance) or anticlockwise (increased resistance) about the curve's origin.

The point of intersection of a given fan curve with a specific system curve is known as the *operating point or condition (OP)*. Figure 3.11 also indicates the effect of varying the resistance on operating point.

The general equation for calculating the power input ($\dot{W}$, W) required *at the operating point* is given by:

$$\dot{W} = \frac{\dot{V}(FTP)}{\eta_{overall}} \tag{3.52}$$

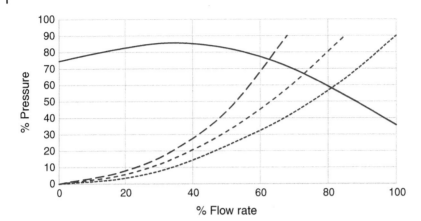

------- Decreased resistance   - - - - Baseline resistance   — — - Increased resistance

Figure 3.11 Fan/system performance curves.

where

$$\eta_{\text{overall}} = \eta_{\text{FTE}}\eta_{\text{mech drive}}\eta_{\text{motor}} \tag{3.53}$$

3.6.5 Fan Laws

Assuming the air density remains constant, the performance of a fan incorporated in a system of ventilation is governed by the following laws:

Flow rate Law

$$\frac{\dot{V}_2}{\dot{V}_1} = \frac{N_2}{N_1} \tag{3.54}$$

where $\dot{V}$ is the volumetric flow rate and N is the rotational speed of the fan.

Pressure Rise Law

$$\frac{\Delta P_2}{\Delta P_1} = \left(\frac{N_2}{N_1}\right)^2 \tag{3.55}$$

where ΔP is the pressure across the fan.

The combined effect of speed change on the fan pressure–flow rate characteristic is shown in Figure 3.12. Essentially with a change in original fan speed, the baseline characteristic experiences a parallel shift – upwards for a speed increase and downwards for a speed decrease.

Fan Power Law

$$\frac{\dot{W}_2}{\dot{W}_1} = \left(\frac{N_2}{N_1}\right)^3 \tag{3.56}$$

where $\dot{W}$ is the power absorbed by the fan.

It is a rare thing to be able to specify a fan that exactly matches the ventilation demand when turned on for the first time.

In the past, it was quite common to make extensive use of dampers, throttling valves, by-pass systems, and pressure relief valves to modulate fan output to a level that matches

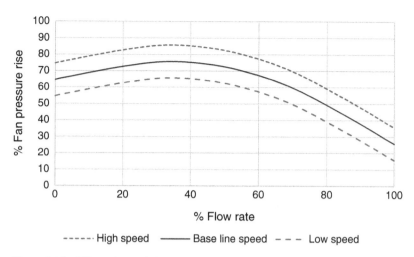

Figure 3.12 Effect of speed change on fan performance curve.

demand. However, this often moves the fan away from its maximum efficiency and thus incurs unwelcome energy wastage. Modern practice is to employ fan speed control to regulate delivery.

3.7 Worked Examples

Worked Example 3.1 Air Change Calculation

A hall, having dimensions of $30\,\text{m} \times 20\,\text{m} \times 4\,\text{m}$ shown in Figure E3.1, is occupied by a maximum of 100 people at any one time. The ventilation system operates on supply rates of 12 l/s of fresh air plus 12 l/s of recirculated air per person. A consultant suggests that a fan with capacity to provide three air changes per hour will satisfy the need of the building. Verify this recommendation.

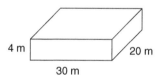

Figure E3.1 Building dimensions for Worked Example 3.1.

Solution

Given: $V = 30 \times 20 \times 4 = 2400\,\text{m}^3$, 100 occupants, $\dot{V}_{\text{fresh}} = 12\,\text{l/s person}$, $\dot{V}_{\text{recirc}} = 12\,\text{l/s person}$, $N_{\text{Proposed}} = 3\,\text{ACH}$

Find: N_{actual}

$$\dot{V}_{\text{tot}} = (\dot{V}_{\text{fresh}} + \dot{V}_{\text{recirc}}) \times \text{number of occupants}$$
$$= (0.012 + 0.012) \times 100$$
$$= 2.4\,\text{m}^3/\text{s}$$

$$\dot{V} = \frac{VN_{actual}}{3600}$$

$$2.4 = \frac{(30 \times 20 \times 4)N_{actual}}{3600}$$

$$N_{actual} = 3.6 \, ACH$$

This is higher than the proposed figure of 3 ACH.

Worked Example 3.2   Use of the Decay Equation

A room having an initial CO_2 concentration of 400 ppm is ventilated with $0.5 \, m^3/s$ of outside air having the same CO_2 concentration. The room has dimensions of $20 \, m \times 10 \, m \times 4 \, m$ and contains 20 people exhaling $1 \times 10^{-5} \, m^3/s$ person of CO_2.

Determine the resulting CO_2 concentration (ppm) in the room after two hours of occupation.

Solution

Given: $c_o = c_a = 400 \, ppm$, $V = 20 \times 10 \times 4 = 800 \, m^3$, $\dot{V} = 0.5 \, m^3/s$, $V_c = 1 \times 10^{-5} \, m^3/s$ person, $t = 2 \, hours = 2 \times 60 \times 60 = 7200 \, seconds$

Occupancy: 20 people

Find: c after two hours

$$n = \dot{V}t/V$$

$$= 0.5 \times 7200/800$$

$$= 4.5$$

$$c = ([V_c \times 10^6/\dot{V}] + c_a)(1 - e^{-n}) + c_o e^{-n}$$

$$= ([1 \times 10^{-5} \times 20 \times 10^6/0.5] + 400)(1 - e^{-4.5}) + 400 e^{-4.5}$$

$$= (400 + 400)(0.989) + 4.44$$

$$= 796 \, ppm$$

Worked Example 3.3   Buoyancy-Driven Natural Ventilation

A classroom is to operate with stack effect ventilation.

Both the air inlets and outlets have an area of $0.5 \, m^2$.

The outlets are 3 m above the inlets.

The temperature inside the classroom is 20 °C and that outside is 10 °C. See Figure E3.3. The classroom contains 25 students each requiring a ventilation rate of 15 l/s person.

How much (%) will the zero emission ventilation system contribute to this requirement?

Take C_d for inlets and outlets as 0.6.

Solution

Given: $A_1 = A_2 = 0.5 \, m^2$, $T_i = 20°C = 293 \, K$, $T_o = 10°C = 283 \, K$, $\Delta T = 10 \, K$, $T_m = (293 + 283)/2 = 288 \, K$, $H = 3 \, m$, $C_d = 0.6$

Find: $\dot{V}_{stack}$ as a%

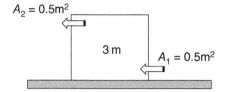

$A_2 = 0.5\text{m}^2$

3 m

$A_1 = 0.5\text{m}^2$

Figure E3.3 Data for Worked Example 3.3.

Required ventilation rate:

$$25 \times 15/1000 = 0.375\ \text{m}^3/\text{s}$$

$$\frac{1}{A_{\text{stack}}^2} = \frac{1}{A_1^2} + \frac{1}{A_2^2}$$

$$= \frac{1}{0.5^2} + \frac{1}{0.5^2}$$

$$A_{\text{stack}} = 0.354\ \text{m}^2$$

$$\dot{V}_{\text{stack}} = C_d A_{\text{stack}} \left(\frac{2gH\Delta T}{T_m} \right)^{0.5}$$

$$= 0.6 \times 0.354 \times (2 \times 9.81 \times 3 \times 10/288)^{0.5}$$

$$= 0.302\ \text{m}^3/\text{s}$$

i.e. $0.302/0.375 = 0.808 \cong 81\%$ of the student's requirements.

Worked Example 3.4   Wind-Driven Natural Ventilation

Estimate the wind-driven natural ventilation rate (m^3/s) for the mono-sloped building, shown in Figure E3.4, at a wind velocity of 2.5 m/s.

The air inlet area on the upwind side is $1\ \text{m}^2$. Outlets having an area of $0.75\ \text{m}^2$ are positioned on the roof. The opening area discharge coefficient may be taken as 0.6 in both cases.

Solution

Given: $A_1 = 1\ \text{m}^2$, $A_2 = 0.75\ \text{m}^2$, $\overline{V}_{\text{wind}} = 2.5\ \text{m/s}$, $C_d = 0.6$, $\Delta\mathbb{C}_P = +0.7 - 0 = 0.7$

Find: $\dot{V}_{\text{wind effect}}$

$$\frac{1}{A_{\text{wind}}^2} = \frac{1}{A_1^2} + \frac{1}{A_2^2}$$

$$= \frac{1}{1^2} + \frac{1}{0.75^2}$$

$$A_{\text{wind}} = 0.60\ \text{m}^2$$

$$\dot{V}_{\text{wind effect}} = C_d \left(\sum A_{\text{wind}} \right) \overline{V}_{\text{wind}} (\Delta\mathbb{C}_P)^{0.5}$$

$$= 0.6 \times 0.60 \times 2.5 \times 0.7^{0.5}$$

$$= 0.753\ \text{m}^3/\text{s}$$

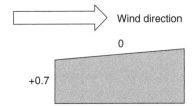

Figure E3.4 Data for Worked Example 3.4.

Worked Example 3.5 Effect of Duct Cross-Sectional Geometry

You are given the task to select a suitable duct for a main air supply to a building of dimensions $10\,\text{m} \times 6\,\text{m} \times 3\,\text{m}$ requiring 5 air changes per hour (ACH).

The two ducts available are:

(a) A circular cross section of 0.25 m diameter
(b) A square duct of 0.25 m side

Determine the velocity (m/s) for each duct and comment on the result.

Solution

Given: $N = 5\,\text{ACH}$, $V = 10 \times 6 \times 3 = 180\,\text{m}^3$

(a) $d = 0.25\,\text{m}$
(b) $w = 0.25\,\text{m}$

Find: $\overline{V}$

$$\dot{V} = \frac{VN}{3600}$$

$$\dot{V} = \frac{180 \times 5}{3600}$$

$$= 0.25\,\text{m}^3/\text{s}$$

(a) For a circular duct:

$$\dot{V} = \overline{V}A$$

$$0.25 = \overline{V} \times \left(\frac{\pi}{4}\right) \times 0.25^2$$

$$\overline{V} = 5.1\,\text{m/s}$$

(b) For a square duct:

$$\dot{V} = \overline{V}A$$

$$0.25 = \overline{V} \times 0.25 \times 0.25$$

$$\overline{V} = 4\,\text{m/s}$$

The square duct results in a lower velocity and hence lower power consumption.

Worked Example 3.6 Darcy Friction Factor

A 150 mm circular plastic duct is to carry an airflow of 25 l/s. If the duct can be considered *smooth*, use the data supplied to determine its Darcy friction factor.

Data

Air density: 1.2 kg/m³, air dynamic viscosity: 1.8×10^{-5} kg/ms.

Solution

Given: $D = 0.15$ m, $\dot{V} = 25\,\text{l/s} = 0.025\,\text{m}^3/\text{s}$, $\rho = 1.2\,\text{kg/m}^3$, 1.8×10^{-5} kg/ms

Find: f

$$A = \pi D^2/4 = \pi 0.15^2/4 = 0.017\,67\,\text{m}^2$$

$$\overline{V} = \dot{V}/A = 0.025/0.017\,67 = 1.41\,\text{m/s}$$

$$Re = \frac{\rho \overline{V} D}{\mu} = \frac{1.2 \times 1.41 \times 0.15}{1.8 \times 10^{-5}} = 14\,100,\ \text{i.e.turbulent}$$

$$f = 1/\left(1.8 \log \frac{Re}{6.9}\right)^2$$

$$= \frac{1}{\left(1.8 \log \frac{14\,100}{6.9}\right)^2}$$

$$= 0.0281$$

Worked Example 3.7   An Index Run Resistance Calculation

The supply for a ventilation system has a total flow rate of 4.2 m³/min in a 200 mm circular main duct. The flow is split *equally* into two 150 mm ducts serving two zones (1, 2) as shown in Figure E3.7.

The straight duct lengths are also shown in the diagram.

Assuming the supply to zone 2 (**1–2–3–5**) to be the *index run* and the straight length of any takeoff branches to be negligible, use the data supplied to determine the attendant system pressure drop (Pa).

Data

Minor loss coefficients *reference upstream velocity:*

$$K_{\text{junction takeoff}} = 0.48, K_{\text{junction st through}} = 0.06, K_{\text{damper}} = 0.2$$

$$K_{\text{bend}} = 0.2, K_{\text{grill}} = 3.6, K_{\text{reducer}} = 0$$

Ducts have been sized to provide a *linear length* pressure loss of 0.4 Pa/m.

Solution

Given: $\dot{V}_{\text{main duct}} = 4.2\,\text{m}^3/\text{min} = 0.07\,\text{m}^3/\text{s}$, Fitting K factors , Straight duct pressure loss $= 0.4\,\text{Pa/m}$

Find: Index run system resistance (1–2–3–5)

Calculating duct flows and areas:

$$\dot{V}_{\text{zone 1}} = \dot{V}_{\text{zone 2}} = \frac{\dot{V}_{\text{main duct}}}{2} = 2.1\,\text{m}^3/\text{min} = 0.035\,\text{m}^3/\text{s}$$

$$D_{\text{Main duct}} = 0.2\,\text{m},\ A_{\text{main duct}} = \pi 0.2^2/4 = 0.0314\,\text{m}^2,$$

$$D_{\text{zone 1}} = D_{\text{zone 2}} = 0.15\,\text{m},\ A_{\text{zone 1}} = A_{\text{zone 2}} = \pi 0.15^2/4 = 0.0177\,\text{m}^2$$

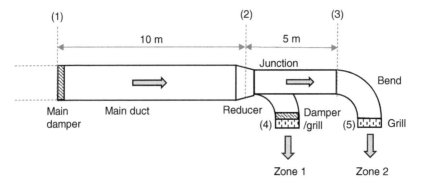

Figure E3.7 Data for Worked Example 3.7.

Calculating duct velocities:

$$\dot{V}_{\text{main duct}} = \overline{V}_{\text{main duct}} A_{\text{main duct}}, \quad 0.07 = \overline{V}_{\text{main duct}} \times 0.0314, \quad \overline{V}_{\text{main duct}} = 2.23 \, \text{m/s}$$

$$\dot{V}_{\text{zones 1,2}} = \overline{V}_{\text{zones 1,2}} A_{\text{zones 1,2}}, \quad 0.035 = \overline{V}_{\text{zones 1,2}} \times 0.0177, \quad \overline{V}_{\text{zones 1,2}} = 1.98 \, \text{m/s}$$

Calculating velocity pressures:
Commencing with the grill at zone 2 and working back towards supply point

$$\Delta P_{\text{grill(5)}} = 0.6 K_{\text{grill}} \overline{V}^2 = 0.6 \times 3.6 \times 1.98^2 = 8.47 \, \text{Pa}$$

$$\Delta P_{\text{bend(3–5)}} = 0.6 K_{\text{bend}} \overline{V}^2 = 0.6 \times 0.2 \times 1.98^2 = 0.47 \, \text{Pa}$$

$$\Delta P_{\text{st duct(2–3)}} = \text{Straight duct pressure loss} \times \text{linear length (2 – 3)}$$
$$= 0.4 \times 5 = 2 \, \text{Pa}$$

$$\Delta P_{\text{junction(2),flow zone 2}} = 0.6 K_{\text{junction(2),flow to zone 2}} \overline{V}^2$$
$$= 0.6 \times 0.06 \times 1.98^2 = 0.14 \, \text{Pa}$$

$$\Delta P_{\text{junction (2),flow zone 1}} = 0.6 K_{\text{junction (2),flow to zone 1}} \overline{V}^2$$
$$= 0.6 \times 0.48 \times 1.98^2 = 1.13 \, \text{Pa}$$

$$\Delta P_{\text{reducer (2)}} = 0$$

$$\Delta P_{\text{st duct(1–2)}} = \text{Straight duct pressure loss} \times \text{linear length (1 – 2)}$$
$$= 0.4 \times 10 = 4 \, \text{Pa}$$

$$\Delta P_{\text{main damper}} = 0.6 K_{\text{damper}} \overline{V}^2$$
$$= 0.6 \times 0.2 \times 2.23^2 = 0.6 \, \text{Pa}$$

Index run resistance: $8.47 + 0.47 + 2 + 0.14 + 1.13 + 0 + 4 + 0.6 = 16.81 \, \text{Pa}$

Worked Example 3.8 Static Regain
In an attempt to generate static pressure, a duct having a flow rate of 600 l/s increases in diameter from 0.3 to 0.5 m via an expansion piece having a minor loss coefficient of 0.5 (with reference to velocity in the larger section). See Figure E3.8.

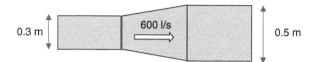

Figure E3.8 Data for Worked Example 3.8.

Determine the theoretical (i.e. ignoring the expander frictional loss) and actual static regain (Pa).

Solution

Given: $D_1 = 0.3$ m, $D_2 = 0.5$ m, $\dot{V} = 600 \, 1/s = 0.6 \, m^3/s$, $K = 0.5$

Find: $(SP_2 - SP_1)_{ideal}$, $(SP_2 - SP_1)_{actual}$

$$A_1 = \pi D_1^2/4 = \pi 0.3^2/4 = 0.070\,69 \, m^2$$

$$A_2 = \pi D_2^2/4 = \pi 0.5^2/4 = 0.1963 \, m^2$$

$$\overline{V}_1 = \dot{V}/A_1 = 0.6/0.070\,69 = 8.49 \, m/s$$

$$\overline{V}_2 = \dot{V}/A_2 = 0.6/0.1963 = 3.06 \, m/s$$

$$(SP_2 - SP_1)_{Ideal} = VP_1 - VP_2$$
$$= 0.6(8.49^2 - 3.06^2) = 37.6 \, Pa$$

$$(SP_2 - SP_1)_{Actual} = (VP_1 - VP_2) - 0.6K\overline{V}_2^2$$
$$= 0.6(8.49^2 - 3.06^2) - 0.6 \times 0.5 \times 3.06^2$$
$$= 37.6 - 2.8$$
$$= 34.8 \, Pa$$

Worked Example 3.9 Fan/System Matching

A fume extraction system has an extraction hood connected to the main duct by a rectangular duct 400 mm by 600 mm and of length 35 m. It is proposed that a fan with the characteristic shown in Table E3.9a be used to operate the system.

Assuming that the Darcy friction factor (f) for the duct remains constant at 0.015.

Take the density of extracted air as 1.2 kg/m³. Determine:

(a) The power consumption (W) by the fan at the operating point
(b) The annual operating cost (£/annum) if it is working 10 hours per day over 300 days per year, assume electricity cost 10p per kWh.

Solution

Given: $L = 35$ m, $f = 0.015$, $\rho = 1.2 \, kg/m^3$, $height_{duct} = 0.4$ m, $width_{duct} = 0.6$ m, Operating hours per day = 10 h/d, operating days per year 300 per annum, Elec cost = 0.1 £/kWh

Find: E, £/d

Calculate the hydraulic diameter:

$$D_h = \frac{4A}{P} = \frac{4 \times (0.4 \times 0.6)}{2 \times (0.4 + 0.6)} = 0.48 \, m$$

Table E3.9a Data for Worked Example 3.9.

Flow rate $\dot{V}$ (m³/s)	0	1	2	3	4
Pressure rise ΔP_{FTP} (Pa)	150	168	156	105	45
Overall efficiency $\eta_{overall}$ (%)	28	65	78	75	50

Table E3.9b Results for Worked Example 3.9.

Flow rate $\dot{V}$ (m³/s)	0	1	2	3	4
Pressure rise ΔP_{FTP} (Pa)	150	168	156	105	45
Overall efficiency $\eta_{overall}$ (%)	28	65	78	75	50
Duct velocity $\overline{V}$ (m/s)	0	4.166	8.333	12.5	16.666
System pressure ΔP_{sys} (Pa)	0	11.4	45.6	102.5	182.2

Calculate the system pressure:

$$\Delta P_{sys} = \left(f \frac{L}{D_h} \right) \left(\rho_i \frac{\overline{V}^2}{2} \right) + \sum K \left(\rho_i \frac{\overline{V}^2}{2} \right)$$

$$= \left(0.015 \times \frac{35}{0.48} \right) \left(1.2 \times \frac{\overline{V}^2}{2} \right) + 0$$

$$= 0.656\overline{V}^2$$

To complete, calculate the duct velocities at each of the fan flow rates using the continuity equation. Then using the velocities, calculate the associated system pressure drop.
Table E3.9b results:

The plot is on a pressure vs. flow rate basis both the fan and system data. Additionally,
plot the efficiency vs. flow rate curve for the fan. See Figure E3.9. At the operation point
(i.e. the intersection of the fan and system curves), read:

$$\Delta P = 105 \text{ Pa}, \quad V = 3.0 \text{ m}^3/\text{s}, \quad \text{and } \eta = 74\%$$

(a) *Calculate the power consumption:*

$$\dot{W} = \frac{\dot{V}\Delta P}{\eta}$$

$$= \frac{3.0 \times 105}{0.74}$$

$$= 426 \text{ W}$$

(b) *Calculate the annual cost:*

$$\text{annual cost} = \text{power} \times \text{duration} \times \text{unit cost}$$

$$= 0.426 \times 10 \times 300 \times 0.1$$

$$= £\,127.8 \text{ per annum}$$

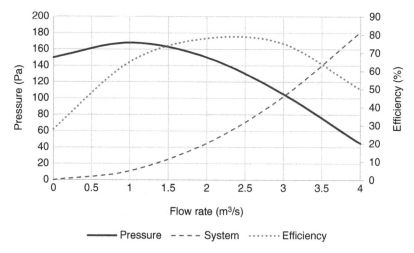

Figure E3.9 Performance curves for Worked Example 3.9.

Worked Example 3.10 Fan Laws

A fan absorbs 2.3 kW of power and discharges 2.5 m³/s when the impeller angular velocity is 1000 revolutions per minute. In order to reduce energy consumption, it was suggested to reduce the impeller angular velocity to 500 revolutions per minute, calculate the discharge in m³/s and the power absorbed for this new condition, and hence estimate the percentage saving on power.

Solution

Given: $\dot{W}_1 = 2.3\,\text{kW}$, $\dot{V}_1 = 2.5\,\text{m}^3/\text{s}$, $N_1 = 1000\,\text{rpm}$, $N_2 = 500\,\text{rpm}$

Find: $\dot{V}_2, \dot{W}_2, \%\Delta E$

Using the flow rate law:

$$\dot{V}_2 = \dot{V}_1 \times \frac{N_2}{N_1}$$
$$= 2.5 \times \left(\frac{500}{1000}\right)$$
$$= 1.25\,\text{m}^3/\text{s}$$

Using the fan power law:

$$\dot{W}_2 = \dot{W}_1 \times \left[\frac{N_2}{N_1}\right]^3$$
$$= 2.3 \times \left(\frac{500}{1000}\right)^3$$
$$= 0.287\,\text{kW}$$

Percentage saving (providing the accompanying drop in flow rates is acceptable):

$$\%\text{saving} = ((\dot{W}_1 - \dot{W}_2)/\dot{W}_1) \times 100$$
$$= ((2.3 - 0.287)/0.287) \times 100$$
$$= 87.5\%$$

3.8 Tutorial Problems

3.1 A gym having dimensions of 30 m × 20 m × 4 m high is occupied by a maximum of 60 people. The ventilation system operates on supply rates of 12 l/s fresh air and 12 l/s of recirculated air per person.
Calculate the number of air changes per hour.
[Answer: 2.16 ACH]

3.2 An arena is to be built to accommodate 5000 people. Prior to occupation, it is assumed that the arena is filled with ambient air having a CO_2 concentration of 400 ppm. The volume of the arena is 100 000 m³. A ventilation consultant says based on a CO_2 generation rate of 1×10^{-5} m³/s person that a total fresh air flow rate of 10 l/s person will be sufficient to keep the inside CO_2 concentration below 1000 ppm after three hours of occupation.
Is his assertion correct?
[Answer: No, CO_2 concentration is 1395 ppm after three hours of occupation]

3.3 A mono-slope building has the pressure coefficients shown in Figure P3.3 with wind direction normal to the higher facade. Air inlets are located on the windward face with air outlets on the roof.

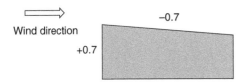

Figure P3.3 Data for tutorial Problem 3.3.

The inlet area on the windward face is 0.6 m² and the exit area on the roof is 0.5 m². Using a value of 0.6 for the discharge coefficient of all openings, estimate the wind-driven ventilation (m³/s) for the building at a wind velocity of 2 m/s.
[Answer: 0.545 m³/s]

3.4 Using the data supplied, determine the combined wind and stack effect ventilation rate (m³/s) for the building shown in Figure P3.4 having low-level inlet louvres on

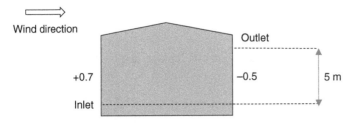

Figure P3.4 Data for tutorial Problem 3.4.

the upwind façade, high-level outlet louvres on the downwind façade and pressure coefficients indicated.

Data

Both the air inlets and outlets have an area of 0.6 m^2.

Interior building temperature: 21 °C

Outside ambient temperature: 12 °C.

C_d for inlets and outlets: 0.6

Wind velocity: 4 m/s

[Answer: 1.2 m^3/s]

3.5 A ventilation system for an occupied space is designed to provide 8 air changes per hour. The building size is 10 m × 5 m × 4 m.

   (a) If the current duct system has a diameter of 0.25 m, determine the capacity (m^3/min) of the fan required for the duty and the speed of air in the duct.

   (b) What size duct do you suggest in order to restrict the speed of air in the supply duct to 5 m/s?

   [Answers: (a) 26.6 m^3/min, 9.05 m/s, (b) 0.336 m.]

3.6 A duct has a 0.5 m × 0.3 m section. Determine its equivalent circular hydraulic diameter (m).

   [Answer: 0.375 m]

3.7 The (Swamee–Jain) equation, supplied, has been proposed as providing a non-iterative value of the Darcy friction factor.

$$f = 0.25 \left[\log \left(\frac{k/D}{3.7} + \frac{5.74}{Re^{0.9}} \right) \right]^{-2}$$

Using the data supplied, estimate the value of the friction factor for air flowing under the conditions described.

Data

Air density: 1.2 kg/m^3, air dynamic viscosity: 1.8×10^{-5} kg/ms

Duct air velocity: 1.5 m/s, duct diameter: 0.3 m, duct absolute roughness: 0.05 mm

[Answer: 0.024]

3.8 A small ventilation supply system comprises:

3 m of horizontal straight duct

A 90° rounded horizontal bend ($K = 0.5$)

2 m of horizontal straight duct

A 90° rounded vertical bend ($K = 0.5$)

A control damper, fully open ($K = 0.3$)

1 m of vertical straight duct

All at 0.3 m nominal diameter.

The system is terminated by:

An expansion piece, 0.45 m diameter ($K = 0.4$, based on the velocity in smaller area)

An outlet grill, 0.45 m diameter ($K = 6$, based on the velocity in expansion outlet).

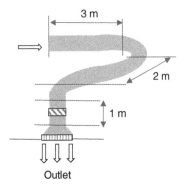

3 m

2 m

1 m

Outlet

Figure P3.8 Data for tutorial Problem 3.8.

See Figure P3.8.
If the flow velocity in the 0.3 m diameter duct is 2.5 m/s and all straight duct lengths have a nominal pressure drop of 0.4 Pa/m, determine the system frictional resistance (Pa)
[Answer: 13.2 Pa]

3.9 The pressure rise (Pa) vs. flow rate (m³/s) characteristic for a *fan* is given by:

$$\Delta P_{fan} = -79\ 396\dot{V}^2 + 2292\dot{V} + 366$$

The fan is installed in a ductwork *system* described by the following characteristic:

$$\Delta P_{system} = 30\ 000\dot{V}^2$$

Predict the approximate operating point (Pa, m³/s) for the combination.
(Hint: Use the quadratic equation)
[Answers: 144 Pa, 0.07 m³/s]

3.10 A fan running at 2900 rpm has a pressure rise of 275 Pa whilst delivering 80 l/s of air. Under these conditions, the absorbed power is 75 W. If the fan rotational speed is reduced to 1725 rpm, determine the new performance.
[Answers: 97.3 Pa, 47.6 l/s, 15.8 W]

4

Psychrometry and Air Conditioning

4.1 Overview

In addition to oxygen, nitrogen, and small amounts of trace gases, ambient air also contains water vapour. The properties of water vapour are quite different from those of dry air. For example, at atmospheric pressure (101.325 kPa), water has a liquid–solid phase change at 0 °C (273 K) and a liquid–vapour phase change at 100 °C (373 K). This is very different from the phase change temperatures and pressures of air's gaseous components which occur well outside of normal ambient conditions.

Furthermore, in a gas–water vapour mixture, at any given temperature, it is the contributory or *partial* pressure of the water vapour and not the total pressure that determines the phase of the water.

The behaviour of air–water vapour mixtures (commonly known as *moist* or *humid* air) is of great importance in air conditioning and its study is known as *psychrometry* or *hygrometry*.

This chapter deals with properties of humid air and the use of the Psychrometric chart to determine them. It describes the principles of the different air-conditioning processes and demonstrates, by examples, the various calculations needed by air-conditioning engineers.

Learning Outcomes

- To appreciate the importance of water vapour and phase change in psychrometry.
- To be aware of the definitions of the principle properties describing humid air.
- To understand air-conditioning processes and carry out associated load calculations.
- To have an appreciation of typical air-conditioning plant and equipment.

4.2 Psychrometric Properties

4.2.1 Pressure

For ambient air calculations, let P_{wv} denote the partial pressure of the water vapour. If P_a is the total pressure of the mixture, then the partial pressure of the dry air (P_{da}) is given by

Building Services Engineering: Smart and Sustainable Design for Health and Wellbeing, First Edition.
Tarik Al-Shemmeri and Neil Packer.

Dalton's law:

$$P_{da} = P_a - P_{wv} \tag{4.1}$$

In general, the amount of water vapour that can be held by an air sample increases with increasing temperature.

For any given temperature, there is, however, an upper limit to the water vapour content in an atmospheric air sample. At the limit, the air is said to be **saturated**.

The component pressure of the water vapour in an air sample at this maximum condition is termed the **saturated (vapour) pressure** (P_{sat}, Pa).

The saturation pressure for water can be obtained from standard tables or be calculated using the following formula:

$$P_{sat} = 610.78e^{(17.27T/(T+237.3))} \tag{4.2}$$

where T is the air temperature (°C).

For an unsaturated vapour, P_{wv} is lower than the saturation pressure and the vapour behaves as a perfect gas.

4.2.2 Temperature

The temperature of an air–water vapour mixture read from an ordinary liquid-in-glass thermometer is known as the **dry-bulb temperature** (T_{db}, °C).

If an unsaturated mixture is allowed to pass through an insulated vessel containing water, the mixture becomes saturated under adiabatic conditions and cooled due to evaporation of the liquid. The lowered exit temperature is known as the **wet-bulb temperature** (T_{wb}, °C).

4.2.3 Water Content

A number of parameters are in use.

- **Moisture content** (ω, kg_{wv}/kg_{da}) is the ratio of the mass of water vapour, m_{wv}, to the mass of air, m_{da}, i.e.

$$\omega = m_{wv}/m_{da} \tag{4.3}$$

Treating water vapour and air as perfect gases:

$$m_{wv} = P_{wv}V/R_{wv}T \quad \text{and} \quad m_{da} = P_{da}V/R_{da}T$$

Then substituting into Eq. (4.3):

$$\omega = \frac{\dfrac{P_{wv}V}{R_{wv}T}}{\dfrac{P_{da}V}{R_{da}T}} = \frac{R_{da}P_{wv}}{R_{wv}P_{da}}$$

Using $R_{da} = 0.287$ kJ/kg K and $R_{wv} = 0.4615$ kJ/kg K and substituting Eq. (4.1) to eliminate P_{da} gives:

$$\omega = 0.622\frac{P_{wv}}{P_a - P_{wv}} \tag{4.4}$$

Note that the terms **humidity ratio** and **mixing ratio** are also used in lieu of moisture content.

A further term, the **specific humidity** is also in usage. This term is defined as the ratio of the mass of water vapour, m_{wv}, to the mass of *humid air*, i.e. m_a, in a sample.

Outside of tropical climates, its value is within a few percentage of that of the moisture content and they are often interchangeable.

- **Relative humidity** (ϕ) is defined as the ratio of partial pressure of the vapour (P_{wv}) in an air sample to the vapour pressure at saturation (P_{sat}) conditions:

$$\phi = \frac{P_{wv}}{P_{sat}} \tag{4.5}$$

The term relative humidity is often confused with the state of wetness/dryness of air. Air at 10 °C dry bulb and 90% relative humidity is drier than air with a relative humidity of 30% at a temperature of 42 °C dry bulb. In fact, air of the former sample contains half the quantity of water vapour in the latter sample.

There exists a relationship between relative and specific humidity:

$$\omega = \frac{0.622\phi \cdot P_{wv}}{P_a - \phi P_{wv}} \tag{4.6}$$

The term **percentage saturation** (μ, %) is the ratio of actual moisture content in an air sample to that of a saturated sample. Again, within the range of states used in air conditioning, it is virtually indistinguishable with relative humidity ϕ.

[Using a volume flow rate ($\dot{V}$, m^3/s) at the outset of the above analyses results in mass flow rate ($\dot{m}$, kg/s) solutions, but is equally valid.]

4.2.4 Condensation

The **dew point temperature** (T_{dp}, °C) of an unsaturated mixture is the temperature to which the mixture must be cooled at constant pressure in order to become saturated. At this point, condensation commences. A useful correlation providing an estimate of dew point temperature in terms of dry-bulb temperature and relative humidity is:

$$T_{dp} = \phi^{1/8}(112 + 0.9T_{db}) + 0.1T_{db} - 112 \tag{4.7}$$

4.2.5 Energy Content

The **specific enthalpy** (h, kJ/kg) of the air–water vapour mixture is the energy contained per kg of air. Assuming the air to be a perfect gas:

$$h = C_{p,da}T + \omega h_{wv} \tag{4.8}$$

Taking $C_{p,air}$ as 1.006 kJ/kg K and using $h_{wv} = 2501 + 1.805T_{db}$, this can be written as:

$$h = 1.006T_{db} + \omega(2501 + 1.805T_{db}) \tag{4.9}$$

4.2.6 Mass Flow and Volume

The total mass of a humid air sample will be the sum of the mass of dry air plus that of any water vapour, i.e.:

$$m_a = m_{da} + m_{wv} \tag{4.10}$$

Divide through by m_{da}, using Eq. (4.3) and collecting terms:

$$m_a = m_{da}(1 + \omega) \tag{4.11}$$

Under comfort conditions however, say 18–24 °C db and 30–70% relative humidity, the water vapour will typically contribute no more than a few percentage towards the total mass and is sometimes ignored in terms of total air flow rate.

However, the water flow rates arising from moisture addition/extraction processes cannot be ignored.

The air mass flow rate ($\dot{m}_a$, kg/s) required for air conditioning will be a function of the room heat load ($\dot{Q}_{room}$, kW), notwithstanding fresh air ventilation requirements.

If the room supply (S) and resulting room (R) conditions (i.e. at room exit) are known then:

$$\dot{Q}_{room} = \dot{m}_a C_p(\Delta T_{R-S}) \tag{4.12}$$

And the required air mass flow rate is given by:

$$\dot{m}_a = \dot{Q}_{room}/C_p(\Delta T_{R-S})$$
$$= \dot{Q}_{room}/(\Delta h_{R-S}) \tag{4.13}$$

The **specific volume** (v, m³/kg) of the air–water vapour mixture is the volume occupied by a unit mass of humid air. In air flow terms:

$$v = \frac{\dot{V}_a}{\dot{m}_a} \tag{4.14}$$

4.3 The Psychrometric Chart

From knowledge of the dry- and wet-bulb temperatures, the relative humidity and other properties for air–water vapour mixtures may be determined from a Psychrometric chart. The chart is based on a standard pressure of 1.013 25 bar, but data tables are available for other pressures.

The Psychrometric chart provides an excellent representation of the interrelationships between the previously discussed properties in an isometric form.

Use of the chart is facilitated by familiarisation with its scales/axes/lines of constant property as follows:

- Dry-bulb temperature (°C, *read from the horizontal bottom axis*)
- Wet-bulb temperature (°C, *inclined lines at ~45° with horizontal axis*)

- Percentage saturation (%, *read at horizontal top and lower LHS vertical axis, taken to be equivalent to relative humidity φ*)
- Dew-point temperature (°C, *read at the 100% percentage saturation curve where the dry-bulb and wet-bulb temperatures are equal*)
- Specific volume (m³/kg, *inclined lines at ~60° with horizontal axis*)
- Moisture content (kg/kg$_{DRY\ AIR}$, *read from vertical axis on RHS, taken to be equivalent to the specific humidity*)
- Specific enthalpy (kJ/kg, *stepped or linear scales peripheral to chart main body*)

  For a given point on the chart, the correct *stepped scale* enthalpy is evaluated by rotating a line centred on the point until its extension reads identical values of enthalpy on the two opposing stepped scales.

  Linear enthalpy scales are often complete with lines of constant enthalpy and can be read directly.

  The actual value of enthalpy is unaffected by the scale form.

  The format of a (stepped enthalpy scale) psychrometric chart is shown in Figure 4.1. An example of a linear enthalpy scale chart is supplied in Appendix A.

Given any two properties, an air state point can be marked on the chart and other properties readily ascertained.

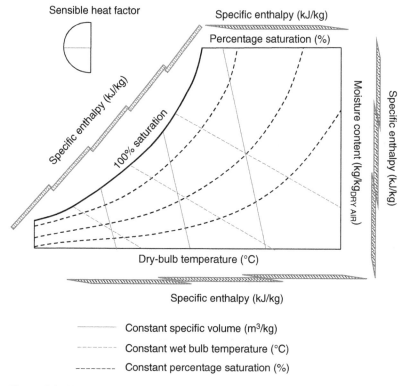

Figure 4.1 The psychrometric chart.

4.4 Air-Conditioning Processes

The prime objective of air conditioning is to provide and maintain the desired condition of the internal environment by supplying the air at the correct predetermined temperature, humidity, air purity, and air velocity for the comfort of occupants or for the correct operation of an industrial process or to maintain the quality and life of a stored product. In order to achieve the above goal, the air must undergo, depending on the application and time/season, a series of psychrometric processes which may take any combination of the following six processes:

(a) **Sensible heating or sensible cooling**

As the name implies, sensible heating and cooling are two opposite processes where only the dry-bulb temperature of the humid air is either increased or decreased, respectively. Therefore, these two processes are represented by a horizontal line of constant specific humidity directed from left to right for heating or from right to left for cooling. Figure 4.2 illustrates a sensible cooling process.

The quantity of heat added to (+) or removed from (−) the air by an indirect heat exchanger coil (sometimes called a '*battery*') is calculated by:

For Heating :

$$(+)\dot{Q}_{12} = \dot{m}_a(h_2 - h_1) \tag{4.15}$$

For Cooling :

$$(-)\dot{Q}_{12} = \dot{m}_a(h_2 - h_1)$$

i.e.

$$\dot{Q}_{12} = \dot{m}_a(h_1 - h_2) \tag{4.16}$$

(b) **Adiabatic mixing of two streams of air**

Consider the situation when an air stream 1 meets and mixes adiabatically with air stream 2 as shown in Figure 4.3.

The law of mass conservation for air flow gives:

$$\dot{m}_{a1} + \dot{m}_{a2} = \dot{m}_{a3} \tag{4.17}$$

And for moisture:

$$\omega_1 \dot{m}_{a1} + \omega_2 \dot{m}_{a2} = \omega_3 \dot{m}_{a3} \tag{4.18}$$

Hence

$$\omega_3 = \frac{\dot{m}_{a1}\omega_1 + \dot{m}_{a2}\omega_2}{\dot{m}_{a1} + \dot{m}_{a2}} \tag{4.19}$$

Similarly, a consideration of energy flow conservation gives:

$$\dot{m}_{a1}h_1 + \dot{m}_{a2}h_2 = \dot{m}_{a3}h_3$$

i.e.

$$h_3 = \frac{\dot{m}_{a1}h_1 + \dot{m}_{a2}h_2}{\dot{m}_{a1} + \dot{m}_{a2}} \tag{4.20}$$

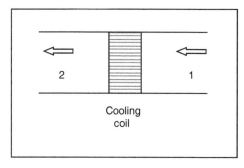 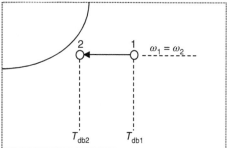

Figure 4.2 Sensible cooling process.

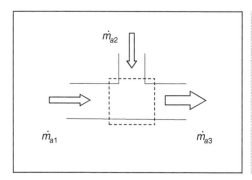 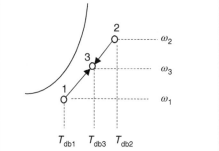

Figure 4.3 Adiabatic mixing of two air streams.

The above equation can also be written in terms of the dry-bulb temperatures and hence the resulting mixed stream temperature can be predicted as:

$$T_{db3} = \frac{\dot{m}_{a1} T_{db1} + \dot{m}_{a2} T_{db2}}{\dot{m}_{a1} + \dot{m}_{a2}} \tag{4.21}$$

When two air streams mix adiabatically, the mixture state (3) lies on the straight line which joins the constituent state points 1 and 2. Moreover, the position of the mixture state point is such that the line is divided *inversely* as the ratio of the masses of air in the constituent air streams, i.e.:

$$\frac{\text{length } 1-3}{\text{length } 2-3} = \frac{\dot{m}_{a2}}{\dot{m}_{a1}} \tag{4.22}$$

Note that for the duration of the Covid-19 pandemic (2020), mixing of fresh and recirculated air was deemed to be inadvisable and systems to be run on full external air where possible.

(c) **Moisture addition**

Three processes are available to increase air moisture content:

(i) Steam injection

The injection of steam into an airflow is used to increase the moisture content of the air while keeping the dry-bulb temperature nearly constant! Consider the addition of $\dot{m}_{\text{steam}}$ kg/s of steam to $\dot{m}_a$ kg/s of air as shown diagrammatically in Figure 4.4.

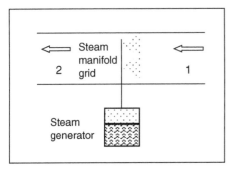 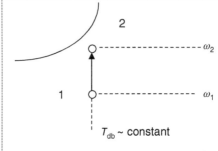

Figure 4.4 Steam spray process.

The analysis will consider the following laws:

Conservation of water vapour mass flow

Mass of initial water vapour in the air at entry plus water vapour added = mass of water vapour in the air at exit.

$$\dot{m}_{a1} + \dot{m}_{a2} = \dot{m}_{a3}$$

$$\dot{m}_a \omega_1 + \dot{m}_{steam} = \dot{m}_a \omega_2$$

Hence:

$$\omega_2 = \omega_1 + \frac{\dot{m}_{steam}}{\dot{m}_a} \qquad (4.23)$$

Conservation of energy flow

$$\dot{m}_1 h_1 + \dot{m}_{steam} h_{steam} = \dot{m}_2 h_2$$

and

$$h_2 = \frac{\dot{m}_a h_1 + \dot{m}_{steam} h_{steam}}{\dot{m}_a + \dot{m}_{steam}} \qquad (4.24)$$

(ii) Ultrasonic vibration

A ceramic plate placed in a water reservoir is vibrated at ultrasonic frequencies of up to 2.4 MHz. The vibrations set up waves in the water and the mechanical agitation results in the release of water droplets at the water/air interface. As a result of the high frequency compression/rarefaction and the cavitation process, moisture transfer is further enhanced.

(iii) Spray atomisation

This is an adiabatic process in which water is supplied from a reservoir at temperature T_2 with enthalpy h_{f2}. The water is sprayed onto some form of wetted pack or matrix in the path of the air flow. Air flowing across the surface of the pack evaporates the water and increases its moisture content.

Alternatively, atomised water may be sprayed directly into the air.

Ideally, the air becomes saturated (100% RH) at the temperature of the water T_2. The outlet temperature of the air T_2 then corresponds to the wet-bulb temperature of the air at inlet (see Figure 4.5).

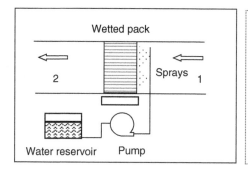 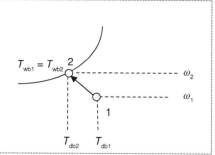

Figure 4.5 Humidification by water spray process.

In other words, this is a constant wet-bulb temperature process.
An energy balance between sections 1 and 2 yields:

$$\dot{m}_a h_1 + \dot{m}_a(\omega_2 - \omega_1)h_{f2} = \dot{m}_a h_2$$

The above equation can be rewritten to determine the final enthalpy of the air (h_2):

$$h_2 = h_1 + (\omega_2 - \omega_1)h_{f2} \tag{4.25}$$

The mass flow of water picked up by the air in this process is:

$$\dot{m}_w = \dot{m}_a(\omega_2 - \omega_1) \tag{4.26}$$

It is worth noting that the second term is so small, the process is considered to be, practically, one of constant enthalpy.

There is a sensible heat loss accompanied by a latent heat gain, but the process is almost neutral as the net heat exchange is minute.

In addition to sizing, the specification of moisture addition plant demands careful consideration be given to its health aspect and in particular the potential for the growth and release of bacterial organisms from the equipment. Of primary concern is *Legionella pneumophila*. This is a bacterium commonly found in nature in ponds, lakes, and soils where concentrations are, in general, not hazardous to human health.

However, high concentrations can build up in water tanks and pipes having a nutrient source, calm conditions, and a temperature in the range 20–45 °C. Spraying fine water droplets laden with the bacterium into an air supply must be avoided as deeply inhaling the aerosol can result in pneumonia and can have fatal consequences in some sections of the population. The elderly and those with respiratory or immunological problems are particularly susceptible. In the media, outbreaks of this disease have been given names such as *Legionnaire's disease* or *Pontiac fever*. The probability of its occurrence can be greatly reduced at the design stage by, for example, specifying materials and geometries that will not support a biofilm substrate, using UV disinfection and other purification systems and not oversizing reservoirs. Once in operation, a preventative maintenance schedule must be enforced to include measures such as flushing, draining down, and control system checks.

(d) **Dehumidification**

In order to reduce both the dry-bulb temperature and the moisture content, three methods may be employed:

(i) Cooling Coil

An indirect heat exchanger with refrigerant or chilled water inside pipes provides a sensible cooling in the first instance. If cooled to dew point (T_{dp}), further cooling will cause condensation of the water vapour in the air on the coils surface. See Figure 4.6.

This water is collected resulting in dryer, cooler air at the outlet.

Consider a mass balance between sections 1 and 2:

If $\dot{m}_w$ is the mass of water condensed by the coil then:

$$\dot{m}_w = \dot{m}_{wv1} - \dot{m}_{wv2}$$
$$= \dot{m}_a(\omega_1 - \omega_2) \tag{4.27}$$

An energy balance between sections 1 and 2 yields:

$$\dot{Q} = \dot{m}_a h_2 + \dot{m}_w h_f \tag{4.28}$$

(ii) Dehumidification with a chilled water spray.

This process is linear between the initial state and the final state which, ideally, will reach the saturation curve (or ***apparatus dew point 2'***). See Figure 4.7.

However, 100% contact between the air and the chilled water is not guaranteed. To accommodate this, two 'efficiency' parameters are commonly defined in enthalpy terms:

– Contact factor

$$CF = \frac{h_1 - h_2}{h_1 - h_{2'}} \tag{4.29}$$

– Bypass factor

$$BF = \frac{h_2 - h_{2'}}{h_1 - h_{2'}} \tag{4.30}$$

Note that:

$$CF + BF = 1 \tag{4.31}$$

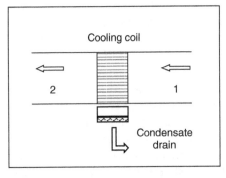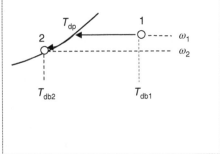

Figure 4.6 Dehumidification by cooling coil.

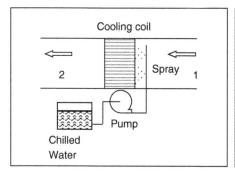
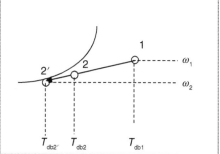

Figure 4.7 Dehumidification by chilled water spray.

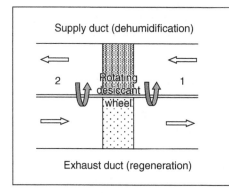
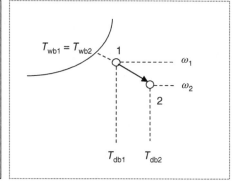

Figure 4.8 Desiccant dehumidification.

Again, earlier comments on the health-related aspects of the generation of water spray and aerosols for air-conditioning plant should be considered.

(iii) Desiccant dehumidification

Desiccant driers operate by exposing the air to a matrix coated in a moisture absorbent or hydrophilic material. To provide continuous operation, a regenerative process is required where the absorbent is heated or reactivated to release its captured moisture making the absorbent available for reuse.

Heat is usually available in the exhaust air from an occupied space and, where supply and exhaust ducts can be brought into close proximity, these systems take the form of a slowly rotating wheel. The process is one of constant wet bulb.

The effectiveness of the system is dependent on regeneration of the moisture in the exhaust duct. See Figure 4.8.

(e) **Fan and duct loads**

The state of the air leaving the final temperature/humidity modifying element of the air-conditioning plant may not be the same as that entering the controlled space. If the supply fan motor sits inside the duct, then it may contribute some sensible heating to the air.

Again, if the supply duct itself is long and uninsulated, it may gain or lose sensible heat with its surroundings further modifying the room supply temperature.

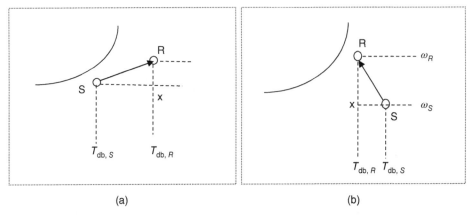

Figure 4.9 Examples of the room line. (a) Typical Summer (Northern hemisphere) – sensible and latent heat gain. (b) Typical Winter (Northern hemisphere) – sensible heat loss and latent heat gain.

(f) **The room loads and Room Ratio Line (RRL)**

The controlled space itself will be subject to a range of thermal influences.

In the Northern hemisphere *winter*, fabric heat losses and infiltration may be significant, whereas in the Northern hemisphere *summer*, solar input will be apparent.

Furthermore, heat and moisture transfers from the controlled space occupants and their activities will occur all year.

If the desired room condition (R) is to be achieved at the room exit point, then the room supply point (S) will be set either higher or lower, depending on circumstances.

A line (S–R) on the psychrometric chart joining the supply and room condition will have a slope dictated by the extent of the sensible and latent heat loads experienced by the room. See Figure 4.9 for examples.

The ratio of the sensible load to total (i.e. sensible + latent) load is termed the **Room Ratio Line** (RRL, 0–1).

Some psychrometric charts helpfully provide an on-chart *Sensible Heat Factor (SHF)* '*protractor*' to gauge the room line slope directly from the cycle. Alternatively, some charts provide an additional SHF scale positioned to the right-hand side of the chart. Of course, the RRL can always be calculated with reference to sensible and latent enthalpy changes at points R, S, and x.

The sensible and latent contributions from other psychrometric processes can also be evaluated by ratio. In this case, the term Sensible Heat Factor (SHF) or ratio is used directly.

4.5 Air-Conditioning Cycles

In real air-conditioning systems, depending on the application and time/season, the air must undergo a series of psychrometric processes forming a cycle.

In a climate with distinct seasons, there are two main cycles representing winter and summer. Nominally, in a northern hemisphere temperate zone such as the United Kingdom:

In winter – the requirement is for a supply of air which has been cleaned and heated. However, as the warming lowers the relative humidity, some form of humidifying plant, such as a spray or a steam injector, is necessary. Post-humidification heating may also be required. The desired indoor atmospheric condition typically lies in the temperature range of 18–22 °C.

In summer – the requirement is for a supply of air which has been cleaned and cooled. As the cooling increases the relative humidity, some form of dehumidifying plant may be an essential. This dehumidifying is generally accomplished by exposing the air to cold surfaces or cold spray, condensing the excess moisture and leaving the air at a lower temperature. The air then requires heating to achieve a more agreeable relative humidity. The desired indoor atmospheric condition typically lies in the temperature range of 21–24 °C.

In both cases, a relative humidity of about 40–60% and a high degree of air purity are essential.

4.5.1 Air-Conditioning Plant Variations

It is entirely possible that all the above processes could be achieved in a centralised plant room and, from there, the conditioned air ducted to its point of use. See Figure 4.10.

The above might be suitable for a space comprising a single zone, but for buildings with multiple zones having different requirements, there are variations on this approach providing degrees of localised or terminal conditioning.

The alternatives each have their own advantages/disadvantages.

- Terminal reheat
  This is the simplest form of satisfying different zonal requirements. It comprises a heat exchanger at the final outlet of the supply ductwork. Terminal heat is only able

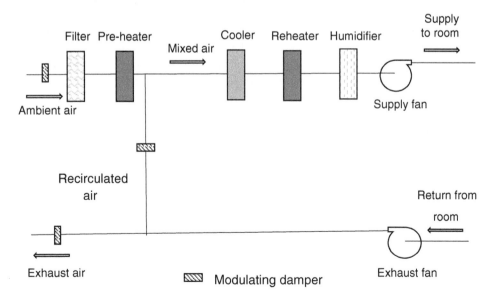

Figure 4.10   Generalised components of typical single zone centralised air-conditioning plant.

to accommodate changes in room sensible load, and there is a danger of reheating previously cooled air with unacceptable consequences in terms of energy use and cost.

- Fan Coils

  Fan coil units are essentially localised mini air conditioners comprising a fan and heat exchanger in a bespoke chamber which may be installed horizontally in a ceiling void or vertically in the form of a floor-mounted cabinet. In all cases, the unit draws in room air for conditioning prior to returning to the space. Fan coils are quite versatile. Some units have separate coils for heating and cooling. Conditioned fresh air may be supplied to the unit from a central or local source for mixing. Alternatively, the conditioned space may have a separate supply of fresh air. Humidity control is not paramount. Flow control within the unit may be effected by directional dampers. Some units utilise heat recovery. An example is provided in Figure 4.11.

- Dual duct systems

  In this arrangement, each room terminal unit will be supplied with both hot and cold air via two separate ducts. See Figure 4.12a. The temperatures of both supplies are dependent on ambient conditions. Each terminal will then facilitate a mixing process of the two streams as described earlier, the total flow remaining constant. The contribution to the mixture from each stream will depend on the required room condition. This is facilitated by a room thermostat controlling a modulating damper in the hot and cold ducts providing good supply temperature control. The resulting humidity will be that of the mixture. The main disadvantage of this arrangement is energy efficiency as previously processed air has its temperature raised or lowered before final delivery. For this reason, these systems have fallen out of favour.

- Induction unit systems

  This arrangement is a combination of centralised and local conditioning. The dehumidification requirement is carried out centrally and air ducted to local supply induction units. Each unit comprises a box containing a set of air nozzles and an air-to-water heat exchanger. See Figure 4.12b. The nozzles deliver the centralised air supply at a high enough velocity (and consequential low pressure) to entrain room air through an integrated (hot water and/or chilled water) heat exchanger. The combination air is then directed into the room. Best performance is achieved when the heat exchanger

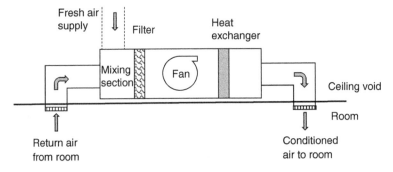

Figure 4.11 Schematic of ceiling void mounted fan-coil system.

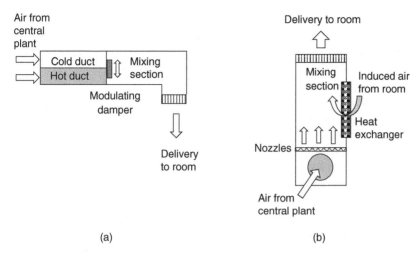

Figure 4.12 Schematic of (a) dual duct and (b) induction unit systems.

comprises separate heating coil, cooling coil, and attendant circuits. This complexity, of course, adds to capital and maintenance costs.

Again, this technology can be found in some buildings but has largely been superseded.

- Variable air volume (VAV) systems

Sensible heating/cooling loads are a function of air flow and temperature. Previously described systems attempt to meet the requirement by controlling the temperature. VAV systems take the alternative approach, i.e. throttling the air flow. Again, functions such as fresh/recirculation mixing, heating, cooling, and humidity may be controlled from a central point. Locally, however, in its simplest form, each VAV air outlet will vary its air flow by damper or other flow restrictor according to the room or zone requirement. The system supply/extract fans will respond to this variation by either fan speed control or fan inlet guide vane modulation, realising fan power energy savings. The reduced volume flow does have some disadvantages. For example, care must be taken not to reduce the air flow below the minimum ventilation requirements, and indoor pollutant dilution may also be reduced.

VAV systems are extensively employed for new designs and have a range of sophistication related to application. Modifications to the simple damper approach include:

(a) Fan-assisted VAV

Essentially, these forms mix (flow-restricted) air from the central plant with room air from the ceiling space before suppling to the room. An integral fan helps to maintain air movement in the room. The fan can be incorporated in either a series or parallel arrangement.

(b) Induction VAV (ceiling plenum)

Another device mixing air from the central plant with ceiling void air. However, the fan is dispensed with and instead the void air is drawn into the unit as a result of the high-velocity/low-pressure (Bernoulli) principle.

In both cases as the cooling load reduces the central plant, air volume flow is throttled and the recirculated volume flow increases, maintaining the total flow and never
allowing the central plant supply flow to drop below minimum fresh air ventilation
requirements. Terminal reheating is optional for the room heating load.

- Self-contained/packaged systems

Distributing conditioned air around a building in ductwork requires space. Local
self-contained/packaged systems serving individual zones remove this requirement,
and they are used extensively in smaller applications. Packaged systems may be roof
mounted, ceiling void mounted, or free standing. A packaged unit will comprise all
necessary fans, filters, and refrigeration/heating equipment. The heat exchangers form
an integral part of a heat pump/refrigeration circuit removing the need for any water
circuits.

More flexibility and reduced plant noise may be achieved by removing the heat sink coil,
fan and compressor to a remote position outside the building. This is the essence of *split
cooling systems*. With appropriate sizing and control, it is possible to link multiple indoor
units to a single outside unit. Schematics of a packaged roof-mounted unit (a) and a split
system (b) are shown in Figure 4.13.

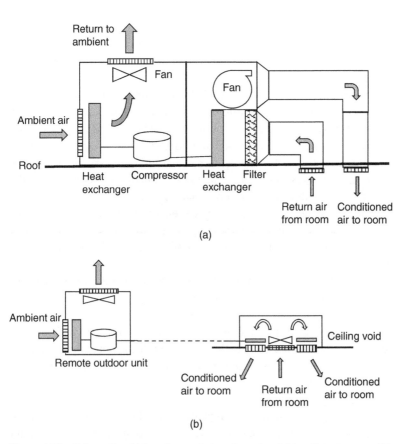

Figure 4.13 Schematic of (a) roof mounted packaged and (b) split system conditioners.

4.6 Worked Examples

A psychrometric chart (*courtesy of* http://flycarpet.net) is supplied in Appendix A.
The user's attention is drawn to the following:

- All values extracted from the chart are subject to visual interpretation
- The term humidity ratio (g/kg) is favoured over moisture content (kg/kg); hence, a small conversion will be required for the purposes of calculation.
- Wet-bulb temperature isotherm values are determined by equating with dry-bulb temperature at the 100% saturation line.

Worked Example 4.1   Calculating Air Properties
A room having dimensions of 5 m × 5 m × 3 m is kept at 25 °C dry bulb 75% RH and 100 kPa.
Determine:

(a) the partial pressure of air and water vapour (Pa)
(b) the mass of air and water vapour (kg)
(c) the moisture content of the air (kg_{wv}/kg_{da})
(d) the specific enthalpy of the air (kJ/kg)

Take R_a as 287 J/kg K and R_{wv} as 461.5 J/kg K.

Solution
Given: $V = 5 \times 5 \times 3 = 75\,m^3$, $T_{db} = 25\,°C$, $\phi = 75\% = 0.75$, $P_a = 100\,kPa$.
   Find: $P_a, P_{wv}, m_a, m_s, \omega, h$

(a) Partial pressures

$$P_{sat} = 610.78e^{(17.27T/(T+237.3))}$$

$$= 610.78e^{(17.27 \times 25/(25+237.3))}$$

$$= 3167.5\,Pa$$

$$\phi = \frac{P_{wv}}{P_{sat}}$$

$$0.75 = \frac{P_{wv}}{3167.5}$$

$$P_{wv} = 2375.6\,Pa$$

$$P_{da} = P_a - P_{wv}$$

$$= 100\,000 - 2375.6$$

$$= 97\,624\,Pa$$

(b) Mass calculation using the ideal gas equation

$$m_{da} = \frac{P_{da}V}{R_{da}T} = \frac{97.624 \times 10^3 \times 75}{287 \times 298} = 85.6\,kg$$

$$m_{wv} = \frac{P_{wv}V}{R_{wv}T} = \frac{2.37 \times 10^3 \times 75}{461.5 \times 298} = 1.296\,kg$$

(c) Moisture content

$$\omega = 0.622 \frac{P_{wv}}{P_a - P_{wv}}$$

$$= 0.622 \times \frac{2375.6}{100\,000 - 2375.6}$$

$$= 0.0151 \text{ kg}_{wv}/\text{kg}_{da}$$

(d) Specific enthalpy

$$h = 1.006 T_{db} + \omega(2501 + 1.805 T_{db})$$

$$= (1.006 \times 25) + 0.0151(2501 + (1.805 \times 25))$$

$$= 63.7 \text{ kJ/kg}$$

Worked Example 4.2   Using the Psychrometric Chart

Air at 101.325 kPa has a dry-bulb temperature of 25 °C and a relative humidity of 30%. Determine, using the psychrometric chart, the moisture content (or humidity ratio, $\text{kg}_{wv}/\text{kg}_{da}$), enthalpy (kJ/kg), specific volume (m³/kg), and the dew point temperature (°C).

Solution

Given: $T_{db} = 25°$ C, RH $= 30\%$, $P_a = 101.325$ kPa.

Find: ω, h, v, T_{dp}, see Figure E4.2

Locate the condition at the intersection of the dry-bulb and relative humidity lines. Read off the following values:

$$\omega = 0.006 \text{ kg}_{wv}/\text{kg}_{da}$$

$$h = 40 \text{ kJ/kg}$$

$$v = 0.853 \text{ m}^3/\text{kg}$$

$$T_{dp} = 6.5°\text{C}$$

Worked Example 4.3   Sensible Cooling

Calculate the heat (kW) that must be removed from air (without affecting the moisture content) in order to reduce its temperature from 25 °C dry bulb, 40% RH to 15 °C.

Assume that the room requirements are satisfied by two air changes per hour and the room dimensions are 6 m × 6 m × 4 m.

Solution

Given: $T_{db1} = 25°$ C, $\phi = 40\% = 0.4$, $N = 2$ ACH, $V = 6 \times 6 \times 4 = 144$ m³, $T_{db2} = 15°$ C, $\omega_1 = \omega_2$

Find: $\dot{Q}_{12}$, Figure E4.3

The volume flow rate is given by:

$$\dot{V}_a = \frac{\text{volume} \times N}{3600} = \frac{(6 \times 6 \times 4) \times 2}{3600} = 0.08 \text{ m}^3/\text{s}$$

Use the initial dry-bulb temperature and relative humidity to locate state point 1 on the chart and read off the specific volume $v_1 = 0.855$ m³/kg.

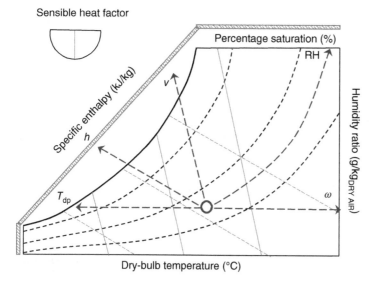

Figure E4.2 Air properties using the psychrometric chart for Worked Example 4.2.

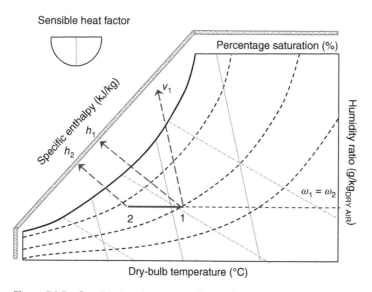

Figure E4.3 Sensible heating process for Worked Example 4.3.

Then, the mass flow rate of humid air is given by:

$$\dot{m}_a = \frac{\dot{V}_a}{v_1} = \frac{0.08}{0.855} = 0.0935 \, \text{kg/s}$$

Locate state point 2 using the final dry-bulb condition and constant moisture content
Read off the specific enthalpies $h_1 = 45 \, \text{kJ/kg}$ and $h_2 = 35 \, \text{kJ/kg}$

Then, the heat removed by the cooler is given by:

$$\dot{Q}_{12} = \dot{m}_a(h_1 - h_2)$$
$$= 0.0935(45 - 35)$$
$$= 0.935 \text{ kW}$$

Worked Example 4.4 A Mixing Process

Return air from a conditioned space at 25 °C, 50% relative humidity, and a mass flow of 2 kg/s, mixes with outside air at 10 °C dry bulb and 35% RH, flowing at 1 kg/s. What is the condition of the mixture?

Assume a constant pressure of 101 325 Pa in all sections.

Solution

Given: $T_{db1} = 25°C$, $\phi_1 = 50\%$, $\dot{m}_1 = 2$ kg/s, $T_{db2} = 10°C$, $\phi_1 = 35\%$, $\dot{m}_2 = 1$ kg/s,

Find: T_{db3}, ϕ_3, ω_3, h_3, v_3, Figure E4.4

Locate state conditions 1 and 2 at the points of the dry-bulb condition and relative humidity intersection

Read off the specific enthalpies $h_1 = 50.3$ kJ/kg and $h_2 = 16.7$ kJ/kg

Connect state points 1 and 2 with a straight line.

Locate state point 3 on the line using the mass ratio

Read off the following values:

$$T_{db3} = 20°C$$

$$\phi_3 = 45\%$$

$$\omega_3 = 0.0074 \text{ kg}_{wv}/\text{kg}_{da}$$

$$h_3 = 39.14 \text{ kJ/kg}$$

$$v_3 = 0.84 \text{ m}^3/\text{kg}$$

Worked Example 4.5 Cooling with Dehumidification

In a cooling and dehumidification process, 2 kg/s of humid air at 43 °C dry bulb, 20% RH are passed across a cooling coil, leaving at 14 °C dry bulb, 70% RH.

Determine by using the chart:

(a) the sensible heat removed (kW), the latent heat removed (kW), and the sensible heat factor
(b) the mass of water vapour condensed (kg/s)
(c) the coil bypass factor

Solution

Given: $T_{db1} = 43°C$, $RH_1 = 20\%$, $\dot{m}_a = 2$ kg/s, $RH_2 = 70\%$

Find: $(-)\dot{Q}_{sensible}$, $(-)\dot{Q}_{latent}$, SHF, $\dot{m}_w$, BF, see Figure E4.5

Locate state conditions 1 and 2 at the points of the dry- and relative humidity condition intersection.

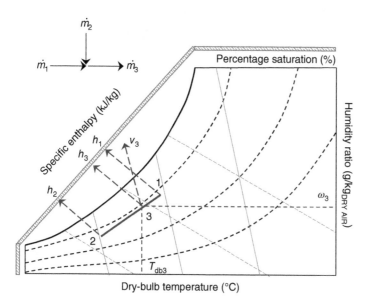

Figure E4.4 Mixing of two airstreams.

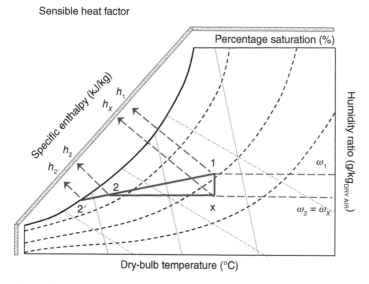

Figure E4.5 Humidification by water.

Read off the following values:

$$h_1 = 71 \text{ kJ/kg}, \; h_x = 61 \text{ kJ/kg}, \; h_2 = 31 \text{ kJ/kg}, \; h_{2'} = 20 \text{ kJ/kg}$$

$$\omega_1 = 0.011 \text{ kg}_{wv}/\text{kg}_{da}, \; \omega_2 = 0.007 \text{ kg}_{wv}/\text{kg}_{da}$$

(a) Heat transfers

$$(-)\dot{Q}_{sensible} = \dot{m}_a(h_2 - h_x)$$

$$\dot{Q}_{sensible} = \dot{m}_a(h_x - h_2)$$

$$= 2(61 - 31)$$

$$= 60 \text{ kW}$$

$$(-)\dot{Q}_{latent} = \dot{m}_a(h_x - h_1)$$

$$\dot{Q}_{latent} = \dot{m}_a(h_1 - h_x)$$

$$= 2(71 - 61)$$

$$= 20 \text{ kW}$$

$$\text{SHF} = \frac{Q_{sensible}}{Q_{sensible} + Q_{latent}} = \frac{60}{60 + 20} = 0.75$$

(b) Condensed water

$$\dot{m}_w = \dot{m}_a(\omega_1 - \omega_2)$$

$$= 2(0.011 - 0.007)$$

$$= 8 \times 10^{-3} \text{ kg/s}$$

(c) Coil bypass factor

$$\text{BF} = \frac{h_2 - h_{2'}}{h_1 - h_{2'}} = \frac{31 - 20}{71 - 20} = 0.215$$

Worked Example 4.6 Steam Humidification

Steam at 100 °C is injected into an air stream at 21 °C dry bulb, 50% relative humidity at the rate of 1 kg steam to 150 kg air.

Determine the condition of the air at the end of this process.

Assume the steam specific enthalpy is 2675.8 kJ/kg.

Solution

Given: $T_{db1} = 21 °C$, $\phi_1 = 50\%$

Moisture additions 1 kg steam per 150 kg of air, i.e. $(\Delta\omega) = 1/150$

Enthalpy of steam at 100 °C, $h_{steam} = 2675.8 \text{ kJ/kg}$

Find: ϕ_2, ω_1, and h_2, see Figure E4.6

Locate state condition 1 at the point of the dry-bulb temperature and relative humidity intersection.

Read off $\omega_1 = 0.0077 \text{ kg}_{wv}/\text{kg}_{da}$, $h_1 = 40 \text{ kJ/kg}$ Then

$$\omega_2 = \omega_1 + \Delta\omega$$

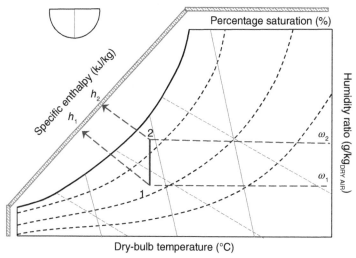

Figure E4.6 Humidification by steam.

$$= 0.0077 + (1/150)$$
$$= 0.0144 \ kg_{wv}/kg_{da}$$

The final enthalpy (2) of the humid air and steam mixture can be calculated from an energy balance:

$$h_2 = \frac{\dot{m}_a h_1 + \dot{m}_{steam} h_g}{\dot{m}_a + \dot{m}_{steam}}$$
$$= \frac{(150 \times 40) + (1 \times 2675.8)}{150 + 1}$$
$$= 57.6 \ kJ/kg$$

Assuming the steam addition is isothermal, i.e. no increase in final dry-bulb temperature, the final condition is located by reading vertically upward from condition 1 to the position of h_2.

Read off $\phi_2 = 90\%$

Worked Example 4.7   Heating and Humidification

A heating and humidification application utilises outdoor air at 4.5 °C dry bulb and 60% RH which is heated sensibly to 20 °C dry bulb and then humidified by saturated steam at 110 °C before being supplied to a space at 60% RH.

If the airflow rate is 5 m³/s, determine the system's sensible and latent heat loads (kW).

Assume the pressure is 1 atm and constant throughout the system.

Solution

Given: $T_{db1} = 4.5°C$, $\phi_1 = 60\%$, $\dot{V}_a = 5 \, m^3/s$, $\phi_3 = 60\%$, $T_{db2} = 20°C$, $T_{steam} = 110°C$.

Find: $\dot{Q}_{sensible}$, $\dot{Q}_{latent}$, see Figure E4.7

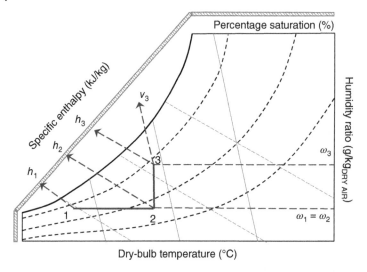

Figure E4.7 Air conditioning cycle for Worked Example 4.7.

- Locate state condition 1 using the dry-bulb temperatures and relative humidity intersection.
- State condition 2 is found by constructing a line horizontally onto the dry-bulb temperature at 2.
  Read off $h_1 = 12\,\text{kJ/kg}$, $h_2 = 28\,\text{kJ/kg}$
- Assuming the steam addition is isothermal, i.e. no increase in final dry-bulb temperature, the final condition is located by reading vertically upward from condition 1 to the position of ϕ_3.

Read off $h_3 = 42\,\text{kJ/kg}$ and $v_3 = 0.84\,\text{m}^3/\text{kg}$
Then, the mass flow rate of humid air is given by:

$$\dot{m}_a = \frac{\dot{V}_a}{v_3}$$
$$= \frac{5}{0.84}$$
$$= 5.95\,\text{kg/s}$$

Heat transfers:

$$\dot{Q}_{\text{sensible}} = \dot{m}_a(h_2 - h_1)$$
$$= 5.95(28 - 12)$$
$$= 93\,\text{kW}$$

$$\dot{Q}_{\text{latent}} = \dot{m}_a(h_3 - h_2)$$
$$= 5.95(42 - 28)$$
$$= 84\,\text{kW}$$

Worked Example 4.8 Winter Air Conditioning (Northern Hemisphere)

A winter air-conditioning cycle consists of the following processes:

Air is taken from outside at 0 °C dry bulb, 80% RH is preheated to 23.5 °C dry bulb.

It is then passed through a humidifier using spray water. At the exit, the air flow leaving the washer is measured to be 2.5 m³/s, saturated at 10 °C.

Air is then reheated before finally being delivered by a fan to the room, which is maintained at 20 °C dry bulb. Determine using the chart:

(a) the water make-up rate in the washer (kg/s)
(b) the duty of the preheater (kW)
(c) the duty of the reheater (kW)

Solution

Given: $T_{db1} = 0\,°C$, $\phi_1 = 80\,\%$, $T_{db2} = 23.5\,°C$, $\dot{V}_{a3} = 2.5\,m^3/s$, $T_{db3} = 10°C$, $\phi_3 = 100\%$, $T_{db4} = 20°C$

Find: $\dot{m}_w$, $\dot{Q}_{preheater}$, $\dot{Q}_{reheater}$, see Figure E4.8

- Locate state condition 1 using the dry-bulb temperatures and relative humidity intersection. State condition 2 is found by constructing a line horizontally to the dry-bulb temperature at 2.
- Extend a line from 2 to the condition 3 as specified by its wet bulb temperature at saturation conditions
- Extend a line from condition 3 horizontally to the dry-bulb temperature at 4.

Read off $\omega_1 = \omega_2 = 0.002\ kg_{wv}/kg_{da}$, $\omega_3 = 0.0076\ kg_{wv}/kg_{da}$, $v_3 = 0.81\ m^3/kg$, $h_1 = 5\ kJ/kg$, $h_2 = 27\ kJ/kg$, $h_3 = 29\ kJ/kg$, $h_4 = 38\ kJ/kg$

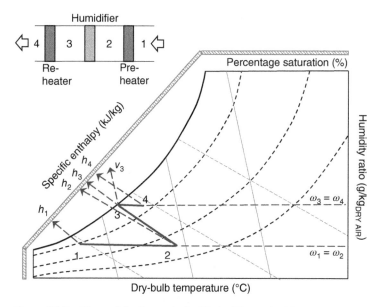

Figure E4.8 Air conditioning cycle for Worked Example 4.8.

Then, the mass flow rate of humid air is given by:

$$\dot{m}_a = \frac{\dot{V}_a}{v_a} = \frac{2.5}{0.81} = 3.086 \text{ kg/s}$$

(a) Water added

$$\dot{m}_w = \dot{m}_a(\omega_3 - \omega_2)$$
$$= 3.086(0.0076 - 0.002)$$
$$= 0.017 \text{ kg/s}$$

(b) Preheater duty

$$\dot{Q}_{\text{preheater}} = \dot{m}_a(h_2 - h_1)$$
$$= 3.086(27 - 5)$$
$$= 67.9 \text{ kW}$$

(c) Reheater duty

$$\dot{Q}_{\text{reheater}} = \dot{m}_a(h_4 - h_3)$$
$$= 3.086(38 - 29)$$
$$= 27.7 \text{ kW}$$

Worked Example 4.9 Summer Cooling (Northern Hemisphere)

In summer, a room (R) is to be maintained at 20 °C dry bulb and 50% RH when the outside air (O) is at 26 °C, 50% relative humidity. Outside air is mixed with (re-circulated) room air in the ratio of three parts recirculated air to one-part fresh air.

The air after mixing (M) is passed over a cooling coil with an apparatus *dew point* of 4 °C and a coil contact factor of 0.9.

On leaving the cooling coil, the air passes over a reheating coil and is then delivered to the room by the fan.

The mass flow rate of air supplied to the room (S) is 2 kg/s at 15 °C db.

Determine using the chart:

(a) the cooling coil load (kW)
(b) the heating coil load (kW)
(c) the rate of water vapour condensed (kg/s)

Solution

Given: $T_{\text{dbR}} = 20°$ C, $\phi_R = 50\%$, $T_{\text{dbO}} = 26°$ C, $\phi_O = 50\%$, $\dot{m}_R/\dot{m}_O = 3$, $T_{\text{dp}} = 4°$C, CF $= 0.9$, $\dot{m}_a = 2$ kg/s, $T_{\text{dbS}} = 15°$C

Find: $\dot{Q}_{\text{cooling}}$, $\dot{Q}_{\text{heating}}$, $\dot{m}_w$, see Figure E4.9

• Locate state conditions O and R at the points of the dry-bulb condition and relative humidity intersection.

Read off the specific enthalpies $h_O = 53$ kJ/kg and $h_R = 38$ kJ/kg

Use an energy balance to find the mixture enthalpy thus:

$$h_M = \frac{\dot{m}_O h_O + \dot{m}_R h_R}{\dot{m}_O + \dot{m}_R} = \frac{\dot{m}_O h_O + 3\dot{m}_O h_R}{\dot{m}_O + 3\dot{m}_O} = \frac{h_O + 3h_R}{4}$$

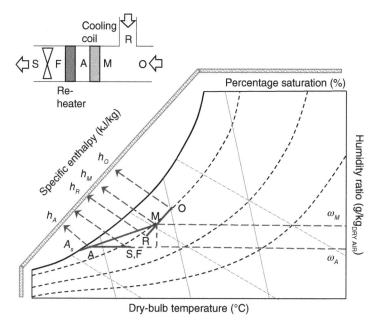

Figure E4.9 Air conditioning cycle for Worked Example 4.9.

$$= \frac{53 + (3 \times 38)}{4} = 41.75 \text{ kJ/kg}$$

Connect state points O and R with a straight line.

Locate state point M on the line using the calculated enthalpy.

Read off $\omega_M = 0.008 \text{ kg}_{wv}/\text{kg}_{da}$

- Extend a line from point M to a value of 4 °C on the 100% saturation curve (As). Since the contact factor for the cooler is 0.9, locate a point 90% of the way along this line and label this as A. This is the entry point for the reheater.

Read off $\omega_A = 0.005 \text{ kg}_{wv}/\text{kg}_{da}$, $h_A = 17 \text{ kJ/kg}$

- Extend a line horizontally from point A to S at a dry-bulb temperature of 15 °C. This is the supply condition for the room.

Read off $h_S = 28 \text{ kJ/kg}$

(a) cooling load

$$(-)\dot{Q}_{cooler} = \dot{m}_a(h_A - h_M)$$

$$\dot{Q}_{cooler} = \dot{m}_a(h_M - h_A)$$
$$= 2(41.75 - 17)$$
$$= 49.5 \text{ kW}$$

(b) reheat load

$$\dot{Q}_{reheater} = \dot{m}_a(h_S - h_A)$$
$$= 2(28 - 17)$$
$$= 22 \text{ kW}$$

(c) water condensed

$$\dot{m}_w = \dot{m}_a(\omega_M - \omega_A)$$
$$= 2(0.008 - 0.005)$$
$$= 0.006 \text{ kg/s}$$

Worked Example 4.10 Summer Cooling (Tropical Climate)

In a tropical climate, a typical air-conditioning application provides air at 20 °C and 60% relative humidity, while the ambient air is at 30 °C and 90% relative humidity. The process consists of cooling the air *until saturation*, to produce condensation followed by a reheating process.

If the conditioned space has a volume of 1000 m³ requiring four air changes per hour, determine:

(a) the temperature to which the air should be cooled (°C)
(b) the quantity of water removed (kg/s)
(c) the amount of reheating required (kW)

Solution

Given: $T_{db1} = 30\,°C$, $\phi_1 = 90\,\%$, $V = 1000\,m^3$, $T_{db2} = 20\,°C$, $\phi_2 = 60\,\%$, $N = 4\,\text{ACH}$

Find: T_{dp}, $\dot{Q}_{reheater}$, $\dot{m}_w$, see Figure E4.10

Locate state conditions 1 and 2 using the dry-bulb temperature and relative humidity intersections.

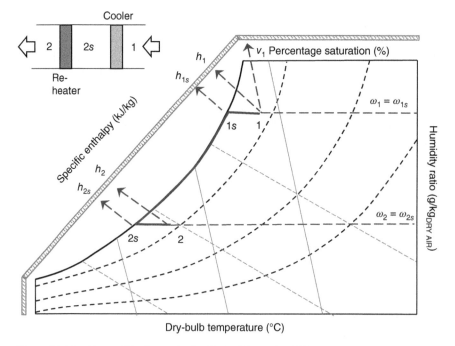

Figure E4.10 Air conditioning cycle for Worked Example 4.10.

(a) To initiate condensation, the air temperature should be reduced to dew point.
Extend a line horizontally from 1% to the 100% saturation line.
Travel down the 100% relative humidity curve until a point is reached horizontal with the final condition 2.

Read off the temperature at this point, $T_{dp} = 12°C$

(b) Water removal
The volume flow rate is given by:

$$\dot{V}_a = \frac{\text{volume} \times N}{3600} = \frac{1000 \times 4}{3600} = 1.111 \text{ m}^3/\text{s}$$

Read off $v_1 = 0.893 \text{ m}^3/\text{kg}$

Then, the mass flow rate of humid air is given by:

$$\dot{m}_a = \frac{\dot{V}_a}{v_1} = \frac{1.111}{0.893} = 1.244 \text{ kg/s}$$

Read off $\omega_1 = 0.024 \text{ kg}_{wv}/\text{kg}_{da}$, $\omega_2 = 0.009 \text{ kg}_{wv}/\text{kg}_{da}$

$$\dot{m}_w = \dot{m}_a(\omega_1 - \omega_2)$$
$$= 1.244(0.024 - 0.009)$$
$$= 0.0187 \text{ kg/s}$$

(c) Condensation will proceed down the 100% relative humidity line until the air reaches the moisture content specified by the final condition.
Extend a line is horizontally from this point on the saturation line (2') to the final condition (2).

Read off $h_2 = 42 \text{ kJ/kg}$, $h_{2'} = 24 \text{ kJ/kg}$

Then

$$\dot{Q}_{reheater} = \dot{m}_a(h_2 - h_{2s})$$
$$= 1.244(42 - 34)$$
$$= 9.952 \text{ kW}$$
$$\cong 10 \text{ kW}$$

4.7 Tutorial Problems

All values extracted from the chart are subject to visual interpretation

4.1 The ambient air inside a building is maintained at 25 °C dry bulb and 40% relative humidity. Determine by calculations:
(a) The moisture content $(\text{kg}_{wv}/\text{kg}_{da})$
(b) The enthalpy of the ambient air at this state (kJ/kg).
[Answers: 0.0079 $(\text{kg}_{wv}/\text{kg}_{da})$, 45.2 kJ/kg]

4.2 Repeat the solution of Problem 4.1 using the chart.
[Answers: 0.008 (kg_{wv}/kg_{da}), 45 kJ/kg]

4.3 Calculate the heat (kW) that must be added to air (without affecting the moisture content) in order to increase its temperature from 5 °C dry bulb 90% RH to 20 °C. Assume that the room requires four air changes per hour, the room dimensions are 6 m × 4 m × 3 m.
[Answer: 1.5 kW]

4.4 Air at 23 °C and 50% relative humidity is mixed adiabatically at a mass ratio of 1 : 1 with another stream of air at 7 °C, saturated.
What is the condition of the mixture?
Assume constant pressure of 101.325 Pa in all sections.
[Answers: 15 °C db, 75% RH]

4.5 Moist air having a relative humidity of 80% at 30 °C dry bulb temperature is cooled by a cold-water spray. This causes saturation followed by condensation, the process is assumed to take place adiabatically with an exit dew point of 17 °C. The resulting saturated air is then heated sensibly to produce the required conditions of 60% relative humidity at 25 °C.
The pressure is atmospheric, and the rate of conditioned air is $2\,m^3/s$.
Determine the heater load (kW).
[Answer: 18.6 kW]

4.6 Steam at 100 °C is injected into an air stream at 25 °C dry bulb and 15% RH. If the moisture content at the end of the process is to be limited to 0.010, determine the mass of air that can be handled per kg of steam.
[Answer: 143 kg_a/kg_s]

4.7 A living room is to be maintained at 20 °C dry bulb and 50% relative humidity. The room has two air changes per hour, and its dimensions are 6 m × 6 m × 3 m. There are four people dissipating 100 W each (70% sensible, 30% latent).
Determine using the chart, the condition of the supply air.
Assume air pressure is atmospheric throughout.
[Answers: 16°C drybulb 55% RH]

4.8 A winter air-conditioning plant provides an office (30 m × 12 m × 5 m) with fresh air conditioned to 20 °C dry bulb with 40% relative humidity at a rate of 2 ACH. The outside air on a winter day has a 7 °C dry bulb and 30% relative humidity.
The plant consists of a 1.0 kW fan, a preheater, a water spray humidifier (20 °C), and a reheater.
If the system air is atmospheric throughout, determine using the psychrometric chart:
(a) the preheater load (kW)

(b) the humidification load (kg/s)
(c) the reheater load (kW)
[Answers: $16\,kW$, $5 \times 10^{-3}\,kg/s$, $13.6\,kW$]

4.9 An air-conditioning plant is designed to maintain a room at 20 °C dry bulb, 50% RH, with an air supply to the room of 1.5 kg/s at 14 °C dry bulb, 60% RH. The design outside air conditions are 27 °C dry bulb and 70% RH. The plant consists of a mixing chamber for recirculated and fresh air, a cooling coil supplied with chilled water, heating coil, and supply fan. The mass ratio of recirculated air to fresh air is 1 : 1; the cooling coil has an apparatus dew point of 5 °C.
Use the psychrometric chart, determine:
(a) the contact factor for the chiller
(b) the rate of water condensed in the chiller (kg/s)
(c) the sensible heat ratio (SHR) for the room.
[Answers: 0.88, $9 \times 10^{-3}\,kg/s$, 0.6]

4.10 $1.5\,m^3/s$ of moist air at 28 °C dry bulb, 47% RH and 1.013 25 bar flows across a cooler coil and leaves at 12 °C dry bulb and $0.0082\,kg_{wv}/kg_{da}$ moisture content. Determine:
(a) the cooler apparatus dew point (°C)
(b) the contact factor and bypass factor
(c) the cooling load (kW)
[Answers: $T_{dp} = 11\,°C$, 0.11, 0.89, $42\,kW$]

5

The Building Envelope

5.1 Overview

Energy transfer through a building space is influenced by the materials used; by geometric factors such as size, wall thickness, and orientation; and by the existence of internal heat sources and by climatic factors. System design requires each of these factors to be examined and the impact of their interactions to be carefully evaluated.

Generally, there are four heat transfer sources/sinks applicable to any building:

Solar irradiation: Solar heat gain due to transmission of solar energy through a transparent building component or absorption by an opaque building component.

Fabric heat exchange: Heat loss or heat gain due to a temperature difference across a building element.

Ventilation heat exchange: Heat loss or heat gain due to the controlled (or otherwise) ingress of outside air into a conditioned space and the egress of conditioned air from the space.

Internal heat gains (or casual heat gains): Heat and moisture transfer due to the processes taking place within a space (lights, people, equipment, etc.).

The conditions that provide a thermally comfortable environment for building occupants have been discussed earlier in Chapter 2. The challenge for an indoor climate control system is to maintain this requirement in the face of variations in building use and the external environment occurring on the other side of the building envelope.

Learning Outcomes

- To appreciate the importance of changes in the external climatic environment
- To be familiar with fundamental heat and mass transfer phenomena
- To be able to quantify energy transfers across and within a structure
- To be able to evaluate the thermal contribution associated with building occupancy

5.2 Variation in Meteorological Conditions

The external parameters of the greatest interest are temperature, humidity, wind, and solar irradiation.

Building Services Engineering: Smart and Sustainable Design for Health and Wellbeing, First Edition.
Tarik Al-Shemmeri and Neil Packer.

5.2.1   Temperature and Humidity

Across the globe, external temperature can vary enormously with a range of factors such as longitude, latitude, altitude, oceanic proximity, and solar irradiation. In the polar regions, the mean winter temperature can drop below −40 °C whilst in an inter-tropic area, such as Saudi Arabia, mean summer temperatures can exceed +40 °C.

No one geographical factor can be taken as a guide.

For example, the typically encountered external design temperature *range* in Cardiff, Wales, UK (Latitude 51.29N, Longitude 3.1W, Altitude ~10 m) is of the order of 25 °C.

The annual relative humidity rarely drops below 70% and can approach 90%.

Calgary, Alberta, Canada (Latitude 51.03N, Longitude 114.04W, Altitude ~1000 m), i.e. at *approximately the same latitude,* has an external temperature range approaching 2.5× that of the figure for Cardiff.

Its average annual relative humidity is of the order of 50% with a range of 40–60%.

5.2.2   Wind

Wind is simply the movement of air as a result of atmospheric pressure differences. On a global scale, air movement is the combined result of differential solar heating and the rotational motion of the planet. Locally, wind velocity varies with topography and atmospheric temperature gradients. As a result of surface drag, wind velocity also increases with elevation.

Continuing with the above comparison, Cardiff has a typical annual average wind velocity range of 3–8 m/s, whilst Calgary is a little calmer with a range of 3–5 m/s.

5.2.3   Solar Irradiation

At the outer edge of our atmosphere, approximately 1.5×10^8 km distant from the Sun, on a surface perpendicular to the Sun, the solar input varies from about 1325 to 1420 W/m² depending on the actual orbital radius. An average figure of 1367 W/m², known as the **solar constant**, is commonly used in calculations.

The amount of energy reaching the surface will be reduced below the solar constant as a result of:

- Atmospheric reflection
- Atmospheric absorption by O_3, H_2O, O_2, and CO_2
- Rayleigh scattering (small particles)
- Mie scattering (large particles)

Even after atmospheric absorption and reflection, significant solar energy reaches the Earth's surface.

The extent of the effects is related to light wavelength and particulate or molecule dimensions. Typically, in terms of wavelength, solar radiation reaching the Earth's surface comprises 60% infrared, 39% light, and 1% ultraviolet.

The calculation of local solar input is complex depending on:

- Orientation and tilt of the receiving surface
- Latitude
- Time of year
- Time of day
- Shading
- Localised ratio of direct/diffuse radiation

Returning to the two cities mentioned above, Cardiff has a typical annual horizontal surface solar irradiance of 1000 kWh/m^2 whilst Calgary is a little sunnier at around 1300 kWh/m^2.

Ideally, engineering plant would be designed and specified to cope with local climate extremes.

In reality, under normal circumstances, climate control systems are not sized to accommodate the absolute worst case as this would render the plant grossly oversized, inefficient, and uneconomic. Instead, a compromise is reached on external design conditions. Ultimately, this will involve the probability of an acceptable number of days per annum when the installed plant may not be able to achieve the most desirable indoor conditions. Hence, a decision will need to be taken on the values of local 'summer highs and winter lows'.

Climatic data for design purposes can be found in national data bases, global internet sources, and engineering handbooks.

5.3 Heat Transfer

5.3.1 Conduction

Conduction is a heat transfer mechanism that takes place through a solid medium of finite thickness and surface area. Consider a solid material that is heated on one side, while the opposite side is subjected to the ambient condition. As the heated side is at a higher temperature than the ambient, the molecules adjacent to the source of heat are energised, *vibrating about their mean positions*. The molecular vibration propagates away from the source and hence there will be a flow of energy from the hot surface to the cool surface.

Nineteenth-century French physicist Jean Baptiste Joseph Fourier found that the rate of heat transfer by conduction through a material is proportional to:

(a) the cross section A
(b) the temperature difference dT
(c) the distance, dx

Hence,

$$\dot{Q}_{cond} \propto (A, dT/dx)$$

The '*constant*' of proportionality associated with this relationship is known as the thermal conductivity of the material through which the heat transfer is taking place. Thermal

Table 5.1 Approximate thermal conductivities of common building materials @ 300 K.

Material	Thermal conductivity (k, W/m K)
Air	0.025
Brick	0.3–0.84
Concrete block:	
Aerated	0.24
Lightweight density	0.73
High density	1.31
Glass	1.05
Glass fibre	0.035
Granite	3.49
Mineral wool insulation	0.034
Polyurethane foam	0.028
Polystyrene (expanded)	0.035
Sandstone	1.8
Steel	45
Water	0.6
Wood	0.1–0.3

Source: Adapted from Davies (2004).

conductivity is not actually a constant but is in fact dependant on temperature. The effect for air was discussed in Section 1.5.4.

The thermal conductivities of some common building materials are given in Table 5.1.

Table 5.1 includes values for air and water. The former can play a large part in the effectiveness of insulating materials. The latter is less welcome inside the building envelope but is still present under many practical circumstances resulting in a range of thermal conductivity values.

A range of values are also in evidence for natural materials where densities can vary with source. Non-homogeneity factors, e.g. grain direction can also be important.

Materials having values of less than 0.04 W/m² K are generally regarded as heat flow inhibitors or insulants. The rate of conductive heat transfer ($\dot{Q}_{cond}$, W) is calculated by

$$\text{i.e.} \quad \dot{Q}_{cond} = -kA\frac{dT}{dx} = -kA\frac{T_2 - T_1}{x_2 - x_1} = kA\frac{T_1 - T_2}{x_2 - x_1} \tag{5.1}$$

The minus sign appears due to the fact that the temperature is decreasing with increased distance from the hot surface.

The importance of thermal conductivity in heat transfer studies is extended to the situation where a heated (or cooled) solid surface ($T_{surface}$) interfaces with a fluid (liquid or gas) at a different temperature (T_{bulk}).

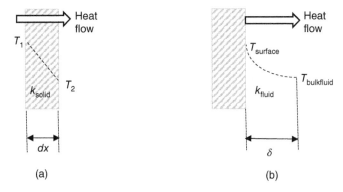

Figure 5.1 Temperature gradients (a) in a solid, (b) at a solid–fluid interface.

This situation is common throughout the building envelope.

At distances close to the surface, it is assumed that conduction continues for a short distance (δ, m) into the adjacent fluid (i.e. the *film* layer) until the bulk fluid temperature is reached.

A comparison with the temperature gradient in a solid is shown in Figure 5.1.

Applying Fourier's law:

$$\dot{Q}_{cond} = -kA\frac{T_{surface} - T_{bulk}}{\delta} \qquad (5.2)$$

The ratio k/δ is termed the film convection heat transfer coefficient (h_c, W/m²K). Hence,

$$\dot{Q}_{interface} = h_c A(T_{surface} - T_{bulk}) \qquad (5.3)$$

5.3.2 Convection

This type of heat transfer is associated with the motion of a fluid past a solid boundary. In general, the rate of convective heat transfer is governed by Eq. (5.3) which is commonly known as Newton's Law. Low values of heat transfer coefficient are associated with motion resulting from differences in temperature related, fluid density, i.e. *free convection*. In the presence of a driving force such as the wind, *forced convection* prevails and higher values are anticipated.

Determination of the value of the convective heat transfer coefficient (h_c, W/m²K) for a set of local circumstances is not a simple matter. It will vary with the prevailing fluid properties, surface geometry, and the exposure of the surface to fluid movement.

For simple geometries, the convective heat transfer coefficient can be determined from empirical correlations utilising the dimensionless quantities described in Chapter 1.

Some examples of correlations for a *flat surface geometry* are given below.

Free convection

- Vertical geometry/Laminar flow:

$$Nu = 0.59(GrPr)^{0.25} \qquad 10^4 < (GrPr) < 10^9.$$

- Vertical geometry/Turbulent flow:

$$Nu = 0.13(GrPr)^{0.33} \qquad 10^9 < (GrPr) < 10^{12}.$$

- Horizontal geometry/Laminar flow:

$$Nu = 0.54(GrPr)^{0.25} \qquad 10^5 < (GrPr) < 2 \times 10^7.$$

- Horizontal geometry/Turbulent flow:

$$Nu = 0.14(GrPr)^{0.33} \qquad 2 \times 10^7 < (GrPr) < 3 \times 10^{10}.$$

Forced convection (any geometry)

- Laminar flow:

$$Nu = 0.664Re^{0.5}Pr^{0.33} \qquad Re < 10^5, 0.6 < Pr < 10.$$

- Turbulent flow:

$$Nu = 0.037Re^{0.8}Pr^{0.33} \qquad Re > 10^5, Pr > 0.6.$$

Other simpler correlations (Hens 2007) for building envelope heat transfer coefficients based, solely, on air velocity (i.e. forced convection) include the following:

For air velocities $< 5\, m/s$

$$h_c = 3.9\overline{V} + 5.6 \tag{5.4}$$

For air velocities $> 5\, m/s$

$$h_c = 7.2\overline{V}^{-0.78} \tag{5.5}$$

5.3.3 Radiation

All bodies radiate thermal energy. If two bodies are at different temperatures and the space between them is unoccupied (or even under a vacuum), the hotter body will still emit radiation to the cooler one. The term '*Blackbody*' is in use to describe a perfect absorber/emitter. The term 'grey' is used to describe a non-black surface. Note that in neither case is the term related to the surface's colour.

The net outcome of heat exchange by radiation between two bodies (1, 2) is given by the Stefan–Boltzmann equation:

$$\dot{Q}_{rad} = \varepsilon\sigma F_{12}A_1(T_1^4 - T_2^4) \tag{5.6}$$

where

ε (0–1) is the emissivity and describes how well a radiating surface approximates a perfect emitter/absorber.

σ is the Stefan–Boltzmann constant having a value of 5.67×10^{-8} W/m^2 T^4

T is surface temperature (K)

F_{12} is known as the *view* or *shape factor* and is determined by the spatial relationship between the emitting and absorbing surface and specifies the proportion leaving one surface that directly impinges on another, as well as accounting for any reflected radiation.

Table 5.2 Typical emissivity of common building surfaces @300 K.

Material	Typical emissivity ε
Oxidised aluminium	0.2–0.3
Brick	0.93
Glass	0.94
Paint	0.82–0.9
Plaster	0.91
Natural stone	0.93 (marble)
Wood	0.9 (planed oak)

Source: Adapted from Welty et al. (2008).

Emissivity is wavelength/temperature dependant and also highly dependent on surface roughness or finish. Examples of building material emissivities are given in Table 5.2.

The product of shape factor and emissivity is often termed the *configuration factor (CF)*. Hence,

$$\dot{Q}_{rad} = \sigma CF_{12}A_1(T_1^4 - T_2^4) \tag{5.7}$$

The simplest instance of configuration factor applies for the assumption of a 'grey' surface (1) in "black" surroundings (2). In this case,

$$CF_{12} = \varepsilon_1 \tag{5.8}$$

Again, for two parallel flat surfaces (of infinite extent, emissivity ε_1, ε_2), A_1 will be equal to A_2 and any radiation emitted by one of the surfaces will be intercepted by the other; therefore, $F_{12} = 1$. Under these circumstances, it can be shown that

$$CF_{12} = \frac{1}{\dfrac{1}{\varepsilon_1} + \dfrac{1}{\varepsilon_2} - 1} \tag{5.9}$$

In general, however, evaluation of shape and configuration factors can be complex for anything other than the simplest geometries and solutions are often presented graphically.

Often it is expedient to define a radiative heat transfer coefficient (h_r, W/m^{2}K) analogous to the convective heat transfer coefficient h_c where

$$h_r = CF_{12}\sigma(T_1^2 + T_2^2)(T_1 + T_2) \tag{5.10}$$

Then radiative heat exchange can be evaluated by

$$\dot{Q}_{rad} = h_r A_1(T_1 - T_2) \tag{5.11}$$

5.4 Solar Irradiation

Predicting the location of the Sun at any given time is, of course, important. The relationship between the incoming solar energy and a receiving surface is expressed in terms of trigonometry.

The analysis becomes complex because the Earth's axis of rotation is not parallel to its solar orbital axis and the Earth's elliptical solar orbit. This is further complicated by the necessary anthropogenic imposition of global time zones and synchronisation.

Planetary time zones use Greenwich, London, UK (longitude 0°) as the prime meridian (line of longitude) or datum. Each time zone has its own *standard meridian* and covers approximately 15° of longitude. That is to say, in most cases, each time zone varies from the next by one hour.

For example, France which is one time zone to the east of the United Kingdom has its clocks set one hour in advance of those in the United Kingdom. There are, however a few variations to this rule.

For calculation purposes, time zones eastward of the Greenwich meridian have positive values and those westward, negative.

The situation is further complicated by the need of some countries to adopt *Daylight saving time* (known in the United Kingdom as British summer time [BST]) during parts of the year and artificially advance local clocks in an attempt to match everyday life to daylight patterns.

5.4.1 Solar Time

Algorithms for predicting solar radiation commence with the calculation of *solar time* as opposed to local time. The required parameters are as follows:

- **Day angle** (y) defined as $360 \times$ (day of year/number of days of year). This is the fraction of the Earth's annual solar orbit expressed in degrees.
- **Offset local time** (*OLT*) is an adjustment that resets the prevailing local clock time to account for time zone meridian and locality longitude (LONG, degrees) difference.

$$\text{OLT} = \text{local time} + 4(\text{LONG}_{\text{local}} - \text{LONG}_{\text{meridian}}) \qquad (5.12)$$

Note that the second term in Eq. (5.12) has units of minutes.

- **Equation of time** (*EOT*). A correction to adjusted local time to correct for the astronomical conditions that cause the local passage of time to vary (by up to 16 minutes 25 seconds) with that observed by the passage of the Sun.

$$\text{EOT} = 0.0066 + [7.3525 \cos(y + 85.9)] + [9.9359 \cos(2y + 108.9)]$$
$$+ [0.3387 \cos(3y + 105.2)] \qquad (5.13)$$

- **Solar time** (*ST*) is then simply the addition of OLT and the EOT.

$$\text{ST} = \text{OLT} + \text{EOT} \qquad (5.14)$$

If daylight saving is in operation, a further 60 minutes should be subtracted to provide local solar time, i.e.

$$\text{ST} = \text{OLT} + \text{EOT} - 60 \qquad (5.15)$$

5.4.2 Solar Angles

With reference to Figure 5.2, solar geometry *for a horizontal surface* is then defined by the following co-ordinate angles:

- **Hour angle** (ω). The angular displacement of the sun from its position at solar noon:

$$\omega = (12:00 - ST) \times 15 \tag{5.16}$$

- The **solar azimuth** (α_s) measured *clockwise from the north* to the compass direction of the Sun. (Readers should beware that other algorithms utilise an alternative datum.)
- The **solar altitude** (γ_s) or *solar elevation* or *sun height* indicating the angular relationship between the sun and the horizon.

The solar altitude is also used to describe the amount of air between the extra-terrestrial solar radiation and the Earth's surface by the term **air mass** (AM) which is a unit-free parameter expressing multiples of atmospheric thickness.

At the outer edge of the atmosphere, the AM is 0. When the incoming solar irradiation is passing vertically through the atmosphere to the Earth's surface, the AM has a value of unity.

The variation in AM with the solar altitude angle (γ_s) is approximated by

$$AM = 1/\sin(\gamma_s) \tag{5.17}$$

- The **solar zenith angle** (γ) measured between a vertical drawn from the point of incidence on *a horizontal surface* and the line of solar altitude.

$$\gamma = 90° - \gamma_s \tag{5.18}$$

- The **solar declination** (δ) measured between the equatorial plane of the Earth and the rotational plane of the Earth around the Sun. See Figure 5.3.

A number of correlations are available for its estimation.

Astronomical predictions of this parameter tend to be very accurate and complex. Declination values oscillate over a year by approximately $\mp 23.45°$.

On any given day (y), the solar declination $\delta(y)$ can be calculated from:

$$\delta(y) = 0.3948 - [23.2559 \cos(y + 9.1)] - [0.3915 \cos(2y + 5.4)]$$
$$- [0.1764 \cos(3y + 105.2)] \tag{5.19}$$

Knowing the latitude (LAT, degrees) at the position of irradiation then the solar altitude γ_s is given by

$$\gamma_s = \arcsin[\{(\cos \omega)\cos(LAT)(\cos \delta)\} + \{\sin(LAT)(\sin \delta)\}] \tag{5.20}$$

If solar time $\leq 12:00$ h, then the solar azimuth α_s angle is given by

$$\alpha_s = 180 - \arccos\left[\frac{(\sin \gamma_s)[\sin(LAT)] - (\sin \delta)}{(\cos \gamma_s)[\cos(LAT)]}\right] \tag{5.21}$$

If solar time $\geq 12:00$ h, then the solar azimuth α_s angle is given by

$$\alpha_s = 180 + \arccos\left[\frac{(\sin \gamma_s)[\sin(LAT)] - (\sin \delta)}{\cos \gamma_s[\cos(LAT)]}\right] \tag{5.22}$$

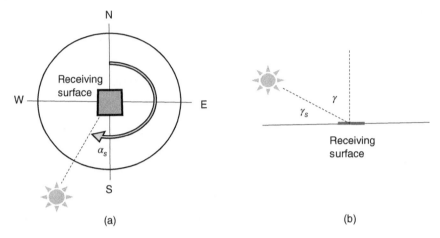

Figure 5.2 Solar angle geometries. (a) Plan and (b) elevation.

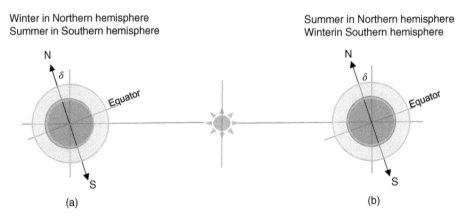

Figure 5.3 Solar declination.

In calculating the irradiation *on a tilted surface* a few more concepts are required. See Figure 5.4.

- The **surface azimuth angle** (α_t) describes the deviation of the surface from the north.
- The **inclination angle** (γ_{tilt}) describes the surface tilt or slope from the horizontal.
- The **angle of incidence** (θ_i) is formed between a line in the direction of the Sun and the normal (i.e. perpendicular) to the surface.

It can be shown that

$$\theta_i = \arccos[\{(-\cos\,\gamma_s)(\sin\,\gamma_{\text{tilt}})\cos(\alpha_s - \alpha_t)\} + \{(\sin\,\gamma_s)(\cos\,\gamma_{\text{tilt}})\}] \tag{5.23}$$

For a vertical surface $\gamma_{\text{tilt}} = 90°$ and this reduces to

$$\theta_i = \arccos[(-\cos\,\gamma_s)\cos(\alpha_s - \alpha_t)] \tag{5.24}$$

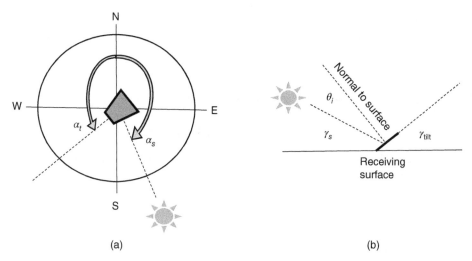

Figure 5.4 Additional solar geometries for a tilted receiving surface. (a) Plan and (b) elevation.

5.4.3 Surface Irradiation

The availability of actual, measured solar irradiation data for specific locations around the planet can still be, surprisingly, patchy. However, it is quite possible to use trigonometric modelling, empirical irradiation correlations and extrapolations to predict the solar heat arriving at a surface at any given time of year. In some cases, standard tables (or software) are available to reduce the amount of computation.

For any surface the total irradiance (I_{tot}) is the sum of the:

- Direct (or *shadow casting*) irradiance, I_{dir}
- Diffuse (*directionally undefined*) irradiance, I_{dif}
- Ground reflected irradiance, I_{ref}

i.e.

$$I_{tot} = I_{dir} + I_{dif} + I_{ref} \tag{5.25}$$

For a tilted surface, these components can be approximately evaluated as follows:

Direct radiation

$$I_{direct,tilt} = I_{direct,horizontal} \left(\frac{\cos \theta_i}{\sin \gamma_s} \right) \tag{5.26}$$

Diffuse radiation (isotropic conditions, i.e. very overcast)

$$I_{diffuse,tilt} = 0.5(I_{diffuse,horizontal})(1 + \cos \gamma_{tilt}) \tag{5.27}$$

(Ground) Reflected radiation

$$I_{reflected,tilt} = 0.5(I_{direct,horizontal} + I_{diffuse,horizontal})(1 + \cos \gamma_{tilt})A \tag{5.28}$$

In Eq. (5.28), A is the reflectivity or **Albedo** of the reflecting surface. Albedo values vary between 0 and 1 with very dark surfaces tending to zero and bright metallic surfaces tending to unity. Most ground conditions tend to the lower end of the scale with a typical urban environment having an albedo of 0.2. Painted surfaces and highly polished metal surfaces can have much higher values in the range 0.75–0.9.

5.5 Heat Losses/Gains Across the Envelope

5.5.1 Opaque Elements, i.e. Walls, Doors, Roofs, Floors, and Cavities

The general equation for heat loss/gain through building structure (wall, roof, or floor) as a result of conduction/convection is given by

$$\dot{Q}_F = UA\Delta T \tag{5.29}$$

where

A is the area through which heat is transmitted (m²)
U is the overall conductance or thermal transmittance of the structure (W/m² K)
ΔT is the temperature difference across the structure, i.e. internal to external (K)

During the heating season, heat flow will be from the building interior to the exterior and is regarded as a heat loss. During the cooling season, the reverse is generally the case and the heat flow regarded as a gain.

The overall U-value is given by

$$U = \frac{1}{\Sigma R_{th}} = \frac{1}{R_i + \Sigma\left(\frac{x}{k}\right) + R_{aircavity} + R_o} \tag{5.30}$$

The denominator in Eq. (5.30) is the sum of the structure's thermal resistances (R_{th}, K m²/W).

The term $\Sigma(x/k)$ is the sum of the conductive thermal resistances of each individual solid layer, each with its own thermal conductivity (k, W/m K) and thickness (x, m).

R_i is the thermal resistance of the inside air film and R_o is the thermal resistance of the outside air film.

The calculation of the thermal resistance of an air layer can be complex depending on thermodynamic and transport properties for the air under any given circumstances but is, generally, defined as the reciprocal of the film convective heat transfer coefficient (h_c, W/m² K) for the air layer plus the radiative heat transfer coefficient (h_r, W/m² K) for the adjacent surface, i.e.

$$R_{i,\ o} = 1/(h_c + h_r) \tag{5.31}$$

Values of external film heat transfer coefficients are usually greater than internal values due to the likelihood of greater fluid motion outside the building. Hence, external film thermal resistances tend to be lower than internal values.

The thermal resistance of any air cavity ($R_{air\ cavity}$), if present, recognises the presence of two air/surface interfaces and the intervening air layer in series. However, the flow conditions will differ from those at single internal and external interfaces. A single circulating, cross-cavity flow or a series of smaller circulating cells is possible. Again, the resulting value of the cavity thermal resistance will depend on whether the cavity is ventilated.

Table 5.3 Indicative thermal resistance values for air layers.

Building component	Heat flow direction	Thermal resistance (K m^2/W)	
Vertical surfaces (e.g. walls)	Internal	0.13	
	External	0.04	
Horizontal surfaces (e.g. floors, ceilings)	Upwards	0.1	
	Downwards	0.17	
Nonventilated cavities		**2 'grey' surfaces**	**1 reflecting surface**
Horizontal	Upwards	0.14	0.28
	Downwards	0.2	0.4
Vertical	—	0.17	0.35

Source: Data adapted from Hens (2007).

Some nominal values of air film layer related thermal resistances are supplied in Table 5.3. Further data can be found in engineering design handbooks with values that differentiate by season or degree of exposure to atmospheric elements.

For the purposes of illustration, typical design U-value ranges for *new*, low carbon emission buildings with significant heating requirements, in northern hemisphere climates are 0.1–0.3 W/m^2 K for walls and floors and 0.1–0.15 W/m^2 K for roofs.

5.5.2 Transparent Elements, i.e. Windows, Roof Lights, Light Wells, Atria

All surfaces above 0 K radiate energy and so the envelope may receive radiation from inside or outside a building. The most important radiative interaction for transparent elements will be with solar radiation.

However, glass also transmits heat via convection/conduction and so U-value calculations are again pertinent.

- Radiative considerations

  Upon interacting with a medium (solid, liquid, or gas), electromagnetic radiation (emr) may be transmitted, reflected, absorbed, or scattered. See Figure 5.5. In general, the magnitudes of these phenomena are dependent on the incoming electromagnetic wavelength and medium composition. For a range of emr incident on a medium, all four processes may occur simultaneously, as is the case for example, with the earth's atmosphere.

  (a) **Transmission** is a process describing the path propagation of the emr through the medium. Let the amount of emr (photons per second) encountering a medium be I_{in} and the amount of emr (photons per second) exiting be I_{trans}.

  The percentage transmitted is expressed as ratio of the amount of emr transmitted by a medium to that incoming (I_{in}), i.e.

  $$\%\text{transmitted} = 100(I_{trans}/I_{in}) \tag{5.32}$$

  (b) **Reflection** is the term used to describe the path reversal or redirection of emr due to contact with a medium. The percentage reflected is expressed as the ratio of the

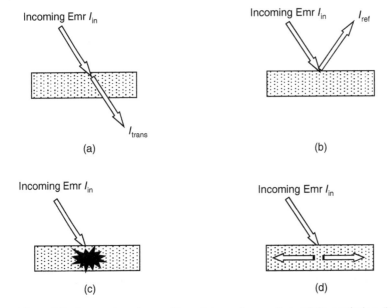

Figure 5.5 Interaction of emr with an intervening medium. (a) Transmission, (b) reflection, (c) absorption, and (d) scattering.

amount of emr reflected (I_{ref}) by a medium to that incoming (I_{in}), i.e.

$$\%\text{reflected} = 100(I_{ref}/I_{in}) \tag{5.33}$$

(c) **Absorption** entails the capture and (energy) conversion of emr. The percentage absorbed is expressed as the ratio of the amount of emr absorbed (I_{abs}) by a medium to that incoming (I_{in}), i.e.

$$\%\text{absorbed} = 100(I_{abs}/I_{in}) \tag{5.34}$$

Note that absorbed energy will heat the medium and may be reradiated at the different wavelength to that of the incoming.

When expressed as simple dimensionless ratios, (a)–(c) are referred to as the *transmittance, reflectance,* and *absorptance,* respectively.

(d) **Scattering** is a process describing the indirect propagation or path diffusion of the emr through a medium. It may be thought of as the absorption and intra-medium redirection of emr.

Consider the single glazed component in Figure 5.6 having a solar input at an angle normal to its surface.

The proportion of solar radiation transmitted *directly* through the glass is termed the direct solar energy transmittance (ET).

Here, the emr is predominantly of a short wavelength (SW).

The proportion of solar radiation *directly* reflected by the glass back into the atmosphere is termed the solar energy reflectance (ER).

The proportion of solar radiation absorbed by the glass is termed the solar energy absorptance (EA). This will result in the glazing, itself, becoming a source of convection and

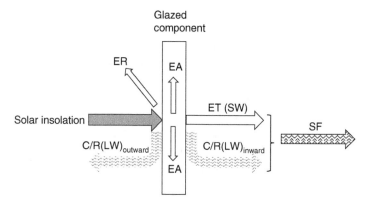

Figure 5.6 Glazing–emr interactions.

radiative heat (C/R). Reradiated wavelengths tend to be of the longer variety and may take place to both interior and exterior environments.

Therefore,

$$ET = 1 - (ER + EA) \tag{5.35}$$

The solar factor (SF) or *total transmittance* or *total solar heat gain coefficient* is the sum of the direct solar transmittance (ET) and absorptive heat reradiated or transferred via convection (C/R) transmitted into the space, i.e.

$$SF = ET + C/R \tag{5.36}$$

A typical solar factor for a single glazed window might be around 0.9, whilst for a double glazing unit this could reduce to 0.6.

Typically, low emissivity triple glazing can lower the solar factor to below 0.3.

The term **total shading coefficient** (TSC) is also in use and describes the solar factor of an arrangement compared to a standard 4 mm glass (SF = 0.87), i.e.

$$TSC = SF/0.87 \tag{5.37}$$

- Conduction/convection considerations

  Published glazing U-values must be considered carefully.

  Factors to take into account in analysis include the following:
  - Glass thickness (typically 4–6 mm)
  - Frame material, e.g. plastic, wood, or metal.
  - Percentage of window area that comprises the frame (typically 10–30%)
  - Gap size (typically 6–16 mm) between panes in multiglazed units
  - Cavity gas composition (air/inert gas, e.g. argon)
  - Glass emissivity (typically 0.9 untreated glass, 0.05–0.2 for low emissivity)
  - Presence of colours/coatings

  Some typical data for wood or PVC-U framed, *inert gas-filled* (90% argon) glazing systems are given in Table 5.4.

  Many building regulations in cold northern hemisphere climates now (2020) typically require a glazing U-value of 1.6 W/m² K or less for *new*, low carbon emission installations.

Table 5.4 Typical glazing *U*-values.

Number of panes	Cavity	Glazing pane coating	Typical *U*-value (W/m² K)
Single	None	Uncoated	4.8
Double	1 × 16 mm	Uncoated	2.6
		Low emissivity ($\varepsilon = 0.2$)	2.0
		Very low emissivity ($\varepsilon = 0.05$)	1.7
Triple	2 × 16 mm	Low emissivity ($\varepsilon = 0.2$)	1.5
		Very low emissivity ($\varepsilon = 0.05$)	1.3

Source: Data adapted from Billington et al. (2017).

5.5.3 Unsteady State Heat Transfer

A building's air conditioning control system will attempt to maintain an internal set point in the face of changing internal and, less predictably, external conditions. Consider external air temperature to be a result of insolation. The daily or *diurnal* variation is often modelled as a sinusoid. How changes like this affect a building's interior will depend on the building envelope, specifically its *thermal mass*.

Thermal mass is concerned with the absorption, storage, and release of an incident heat flow and so material density, specific heat capacity, and thermal conductivity will be of importance. It can be regarded as a measure of the extent to which a structure can *smooth out or dampen* variations in input.

These properties are often combined into two different parameters – *thermal diffusivity* (α, m²/s) and thermal effusivity (β, W s$^{1/2}$/m² K). The former is often regarded as a measure of the time to thermal equilibrium following a change in conditions, whilst the latter is indicative of heat absorption/exchange rate at an element's contact surface under changing thermal conditions.

$$\alpha = k/\rho C \tag{5.38}$$

and

$$\beta = \sqrt{k\rho C} \tag{5.39}$$

where C is the heat capacity (J/kg K) of the building element.

Both are essentially a measure of the thermal inertia of a building element.

With respect to the dampening of a cyclic heat flow, two effects are evident:

- Ratios describing a reduction or attenuation in the transmitted magnitude of an applied heat flow. Two parameters are common:
  (i) A *decrement factor* (DF, no units) that is a ratio applied to attenuation through a building element. A little care must be exercised here as, depending on source, this term has been applied to
  – the internal/external surface temperature differences

$$\mathrm{DF} = (T_{is,max} - T_{is,min})/(T_{es,max} - T_{es,min}) \tag{5.40}$$

– the internal surface/external environmental temperature differences

$$DF = (T_{is,max} - T_{is,min})/(T_{ea,max} - T_{ea,min}) \qquad (5.41)$$

– the ratio of cyclic thermal transmittance to steady-state thermal transmittance.

$$DF = (\dot{Q}_{cyc,max} - \dot{Q}_{cyc,min})/U(T_{ea,max} - T_{ea,min}) \qquad (5.42)$$

(ii) A surface factor that is another ratio describing the attenuation in the reradiated magnitude from an internal surface when subjected to an internal heat source or insolation entering the occupied space via glazing.
- Time delays (hour) in transmitted heat flow or temperature variations

In line with the above attenuation ratios, two parameters are common:

(i) *Time lag*: A delay in the transmitted variation in *peak* heat flow or temperature variation. This parameter has three variants associated with the three decrement bases described above.
(ii) *Time factor*: A time delay associated with surface factor.

The principles of attenuation and time lag referred to a sinusoidal input and the resulting wall behaviour are illustrated in Figure 5.7.

With reference to thermal mass, the thermal response of building constructions is sometimes described as 'Heavy weight' (a slow, damped response), 'Medium weight', or 'Light weight' (a fast, undamped response).

A more quantitative term used to describe the effect of thermal response is *admittance* (Y, W/m² K).

In electrical circuits experiencing an applied alternating current/voltage, the pure resistance circuit component is unaffected by the varying current–voltage relationship. Practical circuits, however, are not pure resistance. They have some electrical storage capabilities which must be taken into account and so the term *impedance* is more correctly used to describe the circuit's response. The reciprocal of impedance is admittance. Employing an electrical analogy, the thermal admittance is often described as the reciprocal of the thermal impedance.

In addition to the density, heat capacity, and thermal conductivity, the admittance therefore takes into account *cyclic* rates of heat charge and discharge at the material surface. Hence, the prevailing heat transfer coefficient and long wave emissivity will be important. The temperature parameter, in its unit, refers to the (varying) difference in temperature between a fabric surface temperature and the prevailing internal environmental temperature. As the heat load is fluctuating it is, therefore, a dynamic parameter, whose value is based on a given cycling time (typically 24 hours).

Often, the admittance is regarded as a *cyclic U*-value.

Heavyweight building materials tend towards high admittance values and lower fluctuations in internal temperature in the face of heat flow fluctuations. The admittance value for a given construction reduces with longer cycling times.

CIBSE Guide A provides, with caveats, the use of a thermal response factor (f_r) for occupied spaces:

$$f_r = \frac{\sum AY + \frac{1}{3}NV}{\sum AU + \frac{1}{3}NV} \qquad (5.43)$$

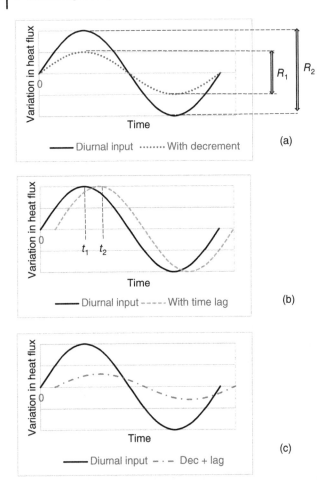

Figure 5.7 Notional effect of attenuation and time lag on heat flow through a material. (a) Attenuation $= R_1/R_2$, (b) time lag $= t_2 - t_1$, and (c) attenuation $+$ time lag.

It is suggested that, within limits, values of f_r in excess of four are indicative of a slow thermal response, i.e. heavyweight construction.

Simple cyclic models are often used to estimate space cooling loads and peak summertime temperature. However, computer modelling will be required for more complex transient variations in load, and imaginary number, matrix manipulation is required. This text, therefore, restricts itself to steady-state calculation conditions.

Thermal mass can be actively deployed at the building design stage to minimise the variation in internal conditions due to external variations. For example, a judiciously positioned thermal mass could, in the winter, absorb any available solar gain during the day to help heat a space overnight.

Alternatively, in a hot daytime/cool night climate, a thermal mass could be used to absorb unwelcome solar heat during the day and release it at night. If not required overnight, increased ventilation could be used to provide cooling of the space in preparation for the following cycle.

5.6   Moisture and Air Transfer

5.6.1   Water Vapour Generation and Control

Moisture in a building can result from a number of sources such as:

- The entry of outside air either by controlled ventilation or by leakage
- Occupants breathing and perspiration, dependent on activity level
- Washing and drying (clothes, personal, surfaces)
- Drinks and food preparation
- Fluidised processing and manufacture.

Rates of generation are available for domestic processes but less common for other applications.

Aided by both convection and diffusion, moisture will pass through the layers of a structural element down a vapour pressure gradient (high to low). Since the capacity of air to hold water vapour increases with increasing temperature, typically this will take place from the warm side to the cold side of a structure.

In colder climates, this will result in vapour diffusion from a heated interior to the colder exterior. (The direction of flow can also be completely reversed when damp external masonry walls are heated by the sun driving the water vapour inwards.)

A hot, humid exterior to air-conditioned interior vapour flow direction is more common in warmer climates.

The presence of water vapour and condensation in an internal space can be minimised by air change rate control or the use of dehumidification equipment.

The occurrence of condensation inside a structural element can be eliminated by the provision of a vapour barrier and adequate thermal insulation to maintain interstitial spaces and interfaces above the dew point temperature.

5.6.2   Vapour Pressure Gradients and Moisture Transfer

Resistance to vapour flow is analogous to heat flow resistance.

Equations (5.29, 5.30) indicate that the heat flow through a material is inversely proportional to the sum of its thermal resistances ($R_{th,total}$, m^2 K/W).

This has its vapour analogue in vapour resistance (R_{vap}, N s/kg).

(The vapour analogue for thermal conductance is termed the *permeance*.)

The vapour transfer analogue for temperature gradient (ΔT) is vapour pressure gradient (ΔP_{vap}).

Vapour resistivity (r_{vap}, N s/kg m) is a measure of the opposition to the flow of water vapour offered by a building element per unit thickness of material. The estimation of vapour resistance will vary depending on whether the layer comprises air, a solid element, or a thin film.

- *Solid layers*

    For a solid layer, the vapour resistance is the product of material vapour resistivity and material thickness (L, m), i.e.:

$$R_{vap,solid\ layer} = r_{vap}L \tag{5.44}$$

Table 5.5 Vapour resistivity (r_{vap}) of common materials.

Material	Vapour resistivity (r_{vap}, GN s/kg m)
Brick work	45–70
Light concrete	15–150
Fibre-insulated board	15–375
Plaster	30–205
Mineral wool insulant	5
Plasterboard	30–60
Polystyrene insulant	100–750
Plywood	150–2000
Timber	45–1850

Source: Adapted from Engineering ToolBox (2011).

The vapour resistivity of some common materials is shown in Table 5.5.
- *Air layer*
  A commonly used value for the vapour resistivity of still air is 5 GN s/kg m. However, an estimate for the vapour resistance of an air film can be derived from the application of a heat analogy. The following is in common usage for buildings:

$$R_{vap,\text{air layer}} = 1.62 \times 10^8 \times R_{th,\text{air layer}} \tag{5.45}$$

The vapour resistances of internal and external film are small and often ignored.
- *Thin films*, e.g. *polythene sheet, paint, and paper.*
  Use is commonly made of manufacturer's data. Values vary with perceived degree of condensation risk. Some typical values are provided in Table 5.6.
  A structure made up of layers will have a total vapour resistance equal to the sum of individual layer's resistances, i.e.

$$R_{vap,total} = \sum R_{vap,layer} \tag{5.46}$$

Water vapour flow (kg/m^2 s) can be calculated from

$$\dot{V}_{vap} = \Delta P_{vap,total} / \sum R_{vap,total} \tag{5.47}$$

Table 5.6 Typical values of thin film vapour resistance (R_{vap}).

Materials		Vapour resistance R_{vap} (MN s/g)
Gloss paints		50
Polythene sheet		150
Reinforced polypropylene sheet	Low risk of condensation	300
	Medium risk of condensation	500
	With thermal foil layer	750
Polyethylene laminated, aluminium core sheet	High risk of condensation	1000–7000

5.6.3 Prediction of Interstitial Building Fabric Condensation

Condensation of water vapour is a major source of concern for owners and occupiers of buildings. It occurs when warm moist air comes in contact with a surface which is at, or below, the local air dew point temperature. There are two distinct categories of condensation: (wall) surface condensation, and interstitial (wall internal) condensation.

The temperature drop (ΔT_{layer}) across a building element layer can be calculated with knowledge of thermal resistance:

$$\Delta T_{layer} = \Delta T_{total} \left(\frac{R_{th,layer}}{R_{th,total}} \right) \tag{5.48}$$

where $R_{th, layer}$ is the thermal resistance of a layer and $R_{th, total}$ is the total thermal resistance of the structure.

The occurrence of condensation can be predicted by a pressure analogy comprising the calculation of the actual vapour pressure profile and comparing it with the saturated vapour pressure values. Saturation vapour pressure values can be estimated by Eq. (4.2) and actual vapour pressure by Eq. (4.5).

Prediction of local changes in prevailing vapour pressure ($\Delta P_{vap, layer}$) across a building layer can be calculated from:

$$\Delta P_{vap,layer} = \Delta P_{vap,total} \left(\frac{R_{vap,layer}}{R_{vap,total}} \right) \tag{5.49}$$

Condensation will occur where vapour pressure is equal to, or exceeds its saturated value.

5.6.4 Air Transfer

The heat transfer associated with the replacement of air at the rate of $\dot{m}_a$ (kg/s) is given by

$$\dot{Q}_V = \dot{m}_a C_p \Delta T \tag{5.50}$$

where C_p is the specific heat capacity of air at constant pressure (kJ/kg K) and ΔT is the temperature difference of the inside and outside air.

The mass flow rate may be evaluated by the product of the volume flow rate $\dot{V}_a$ (m^3/s) and the density of the air ρ (kg/m^3), i.e. ($\dot{m}_a = \rho \dot{V}_a$) and hence:

$$\dot{Q}_V = \rho \dot{V}_a C_p \Delta T \tag{5.51}$$

In building design, the volume flow rate through the building is often described by the number of times that the volume of the space is changed per hour (N) as covered in Chapter 3.

Using this parameter and assuming that for air at ambient conditions: $C_p = 1005$ J/kg K and $\rho = 1.2$ kg/m^3 then:

$$\dot{Q}_V = 0.335 NV \Delta T \tag{5.52}$$

Ideally, the flow of air into a building would be a completely controlled process. However, this is not the case unless a building is airtight.

5.7 Internal Heat Gains

Building occupancy brings with it sources of heat. These are the so-called internal heat gains.

In winter, they could potentially be used to offset the building's heating requirement. In summer, however, they provide a heat load, additional to insolation, which will require removal to maintain acceptable conditions.

Examples of internal heat gains include people, operations, processes and lighting.

- People
  Table 2.1 detailed the amount of heat associated with human beings involved in a number of different activities.
- Operations and Processes
  Values of internal heat gain from building operation will vary widely, dependant on building application. For a number of common electrical devices, it is not unreasonable to use the electrical power input (W) of the technology to give *an indication* of the resulting level of heat gain to the occupied space.
  Table 5.7 provides the electrical power input for some common equipment currently found in an office environment.

Table 5.7 Typical power ratings of some common electrical appliances in the workplace.

Appliance	Input rating (W)
Working environment	
Broadband router	7–10
Laptop	20–50
Desktop computer	80–150
LCD TV	125–200
Laser printer	125–550
Plasma TV	280–450
Office copier	300–1100
Peripherals	
Water coolers	~100
Refrigerated drinks machines	125–600
Vacuum cleaner	500–1200
Microwave	600–1500
Dishwasher	1050–1500
Coffee vending machine	~2200
Kettle	2200–3000
Electric shower	7000–10 500

- Lighting

  Again it is common to assume that a high proportion if not all of the electrical input (W) to a lamp/luminaire will *ultimately* result in heat gain to the space it serves. Elevated lamp surface temperatures will transfer heat to the space by convection and radiation. Even the photons emitted as light will serve to warm up an incident surface when absorbed. Upon warming, the surface will release the received photonic energy by conduction, convection, or radiation.

5.8 Worked Examples

Worked Example 5.1   Solar Time

Determine the solar time in Milan, Italy (approximate longitude 9.19° E, Time zone +1) on the 21st June at 15:00 local time. Daylight saving is not in operation.

Solution

Given: Day = 171, Local time 15:00:00, $LONG_{local}$ = 9.19°, Time zone = +1 hence $LONG_{meridian}$ = 1 × 15 = 15° E

Find: ST

$$y = 360 \times 171/365 = 168.658$$

$$EOT = 0.0066 + [7.3525 \; \cos(y + 85.9)] + [9.9359 \; \cos(2y + 108.9)]$$
$$+ [0.3387 \; \cos(3y + 105.2)]$$
$$= 0.0066 + [7.3525 \cos(168.658 + 85.9)] + [9.9359 \; \cos((2 \times 168.658) + 108.9)]$$
$$+ [0.3387 \; \cos((3 \times 168.658) + 105.2)]$$
$$\cong -1 \text{ minute} : 24 \text{ seconds}$$

$$OLT = \text{local time} + 4(LONG_{local} - LONG_{meridian})$$
$$= 15 : 00 : 00 + 4(9.19 - 15)$$
$$= 15 : 00 : 00 - 23 \text{ minutes} : 14 \text{ seconds}$$
$$= 14 : 36 : 46$$

$$ST = OLT + EOT - 0$$
$$= 14.36 : 46 + (-1 \text{ minute}, \; 24 \text{ seconds}) - 0$$
$$= 14 : 35 : 21$$

Worked Example 5.2   Solar Angles

Estimate the solar altitude and azimuth angles at 14:00 in Florence, Italy (approximate longitude: 11.26° E, Time zone +1, latitude: 43.77° N,) on the 1st June.

Daylight saving is not in operation.

Solution

Given: Day = 151, $LONG_{local}$ = 11.26° E, LAT = 43.77° N, Local time 14:00:00, Time zone = +1 hence $LONG_{meridian}$ = 1 × 15 = 15° E

Find: γ_s, α_s

$$y = 360 \times 151/365 \cong 148.931$$

Repeating the technique of Worked Example 5.1, EOT $\cong +2$ minutes 15 seconds

$$\text{OLT} = 13 : 45 : 02$$

Then

$$\text{ST} = \text{OLT} + \text{EOT} - 0 = 13 : 42 : 47$$

$$\delta(y) = 0.3948 - [23.2559 \cos(y + 9.1)] - [0.3915 \cos(2y + 5.4)]$$
$$- [0.1764 \cos(3y + 105.2)]$$
$$= 0.3948 - [23.2559 \cos(148.931 + 9.1)] - [0.3915 \cos(297.863 + 5.4)]$$
$$- [0.1764 \cos(446.793 + 105.2)]$$
$$= 0.3948 - [-21.56] - [0.21] - [-0.17] \cong 21.889°$$
$$\omega = (12 : 00 - \text{ST}) \times 15 = (12 : 00 - 13 : 47) \times 15 = -26.8°$$

The solar altitude γ_s is given by:

$$\gamma_s = \arcsin[\{(\cos \omega) \cos(\text{LAT})(\cos \delta)\} + \{\sin(\text{LAT})(\sin \delta)\}]$$
$$= \arcsin([\cos(-26.8) \times \cos(43.77) \times \cos(21.889)] + [\sin (43.77) \times \sin(21.889)])$$
$$= 58.88°$$

Then since solar time $> 12:00$ h, then solar azimuth angle α_s is given by

$$\alpha_s = 180 + \arccos \left[\frac{(\sin \gamma_s) \sin(\text{LAT}) - \sin \delta}{(\cos \gamma_s) \cos(\text{LAT})} \right]$$
$$= 180 + \arccos \left[\frac{\sin(58.88) \times \sin(43.77) - \sin(21.889)}{\cos(58.88) \times \cos(43.77)} \right]$$
$$= 180 + 54 \cong 234°$$

Worked Example 5.3   Estimation of Heat Transfer Coefficient

The surface of a 3 m high *interior* wall in a building is at a temperature of 15 °C. Using a flat plate approximation, estimate the heat transfer coefficient for the wall when exposed to the following air conditions:

Air temperature: 19 °C, thermal conductivity: 0.025 W/m K, dynamic viscosity: 1.8×10^{-5} kg/m s, density: 1.2 kg/m³, specific heat capacity: 1005 J/kg K.

Assume that the air is still.

Solution

Given: $x = 3$ m, $T_{\text{room}} = 19°\text{C} = 292$ K, $T_{\text{wall}} = 15°\text{C} = 288$ K, $k = 0.025$ W/m K, $\mu = 1.8 \times 10^{-5}$ kg/m s, $\rho = 1.2$ kg/m³, $C_p = 1005$ J/kg K, $\beta = 1/((T_{\text{wall}} + T_{\text{room}})/2) = 1/((288 + 292)/2) = 3.44 \times 10^{-3}$ K^{-1}, $\Delta T = 292 - 289 = 3$ K

Find: h_c under free convective conditions

Evaluate whether flow is laminar or turbulent

$$Gr = \frac{\rho^2 g \beta \Delta T x^3}{\mu^2} = \frac{1.2^2 \times 9.81 \times 3.44 \times 10^{-3} \times 4 \times 3^3}{(1.8 \times 10^{-5})^2} = 1.62 \times 10^{10}$$

$$Pr = \frac{\mu C_p}{k} = \frac{1.8 \times 10^{-5} \times 1005}{0.025} = 0.724$$

$GrPr = 1.175 \times 10^{10}$ and the flow is in turbulent range.

Hence

$$Nu = 0.13(GrPr)^{0.33}$$

$$= 0.13(1.175 \times 10^{10})^{0.33}$$

$$= 274$$

i.e.

$$\frac{h_c x}{k} = 274$$

$$\frac{h_c \times 3}{0.025} = 274$$

$$h_c = 2.28 \text{ W/m}^2 \text{ K}$$

Worked Example 5.4   Radiative Heat Transfer

It is proposed that a wall contains an air cavity as part of its insulation strategy and that, for the purposes of radiative heat exchange, this arrangement can be considered as behaving like two parallel faces of infinite extent. See Figure E5.4.

Figure E5.4   Wall construction for Worked Example 5.4.

Both surfaces have an emissivity of 0.9.

If the inner surface is at a temperature of 15 °C and the outer at 10 °C, determine the rate of radiative heat transfer (W/m²) across the cavity.

Solution

Given: $\varepsilon_1 = \varepsilon_2 = 0.9$, $T_1 = 15 ° C = 288$ K, $T_2 = 10 ° C = 283$ K

  Find: $\dot{Q}_{rad}/A$

$$CF_{12} = \frac{1}{\frac{1}{\varepsilon_1} + \frac{1}{\varepsilon_2} - 1} = \frac{1}{\frac{1}{0.9} + \frac{1}{0.9} - 1} = 0.818$$

$$\dot{Q}_{rad}/A = \sigma CF_{12}(T_1^4 - T_2^4)$$

$$= 5.67 \times 10^{-8} \times 0.818 \times (288^4 - 283^4)$$

$$= 21.6 \text{ W/m}^2$$

Worked Example 5.5   Glazing Solar Factor

A single pane of coated low emissivity glass has the following specification:

Energy reflectance: 0.2
Energy absorptance: 0.1
Solar factor: 0.64

Determine the energy transmittance (%), the inward absorptive heat reradiated or transferred via convection (%), and total shading coefficient for the glass.

Solution
Given: ER = 0.2, EA = 0.19, SF = 0.64
  Find: ET, C/R, TSC

$$ET = 1 - (ER + EA)$$
$$= 1 - (0.2 + 0.19)$$
$$= 1 - 0.39 = 0.61$$
$$= 61\%$$

$$SF = ET + C/R$$
$$0.64 = 0.61 + C/R$$

$$C/R = 0.03$$
$$= 3\%$$

$$TSC = SF/0.87$$
$$= 0.64/0.87$$
$$= 0.74$$

Worked Example 5.6   Calculation of *U*-Value
A building wall consists of

103 mm brick outer shell	$k = 0.62\,\text{W/m K}$
10 mm unventilated air space	$R = 0.14\,\text{m}^2\,\text{K/W}$
90 mm of thermoset phenol insulation	$k = 0.02\,\text{W/m K}$
100 mm lightweight block inner leaf	$k = 0.2\,\text{W/m K}$
30 mm plasterboard on the inside surface	$k = 0.19\,\text{W/m K}$

The outside and inside film resistances are 0.04 and 0.13 m² K/W, respectively. See Figure E5.6.

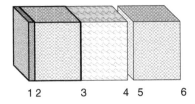

1 2   3   4 5   6

Figure E5.6   Wall construction for Worked Example 5.6.

Determine the overall U-value (W/m^2 K) for the construction.

Solution

Inside air film resistance: $R_{o1} = 0.13$ m^2 K/W

Plasterboard: $R_{12} = x_{plaster}/k_{plaster} = 0.03/0.19 = 0.158$ m^2 K/W

Inner block: $R_{23} = x_{block}/k_{block} = 0.1/0.2 = 0.5$ m^2 K/W

Insulation: $R_{34} = x_{ins}/k_{ins} = 0.09/0.02 = 4.5$ m^2 K/W

Cavity resistance: $R_{45} = 0.14$ m^2 K/W

Outer brick: $R_{56} = x_{brick}/k_{brick} = 0.103/0.62 = 0.166$ m^2 K/W

Outside air film resistance: $R_{60} = 0.04$ m^2 K/W

$$U = \frac{1}{R_i + \sum \left(\frac{x}{k}\right) + R_{air\ cavity} + R_o}$$

$$= \frac{1}{0.13 + (0.158 + 0.5 + 4.5 + 0.166) + 0.14 + 0.04}$$

$$= 0.18 \text{ W/m}^2 \text{ K}$$

Worked Example 5.7 Calculation of Building Fabric and Ventilation Heat Loss
A sports hall has the fabric specifications as shown in Table E5.7:

Table E5.7 U-values for Worked Example 5.7.

Building components	U-value (W/m^2 K)
Walls	0.5
Roof	0.45
Floor	0.45
Windows	3
Door	2

The building is 20 m × 10 m × 5 m high. The façade has 12 m^2 of glazing and an entrance door of area 6 m^2. The space is ventilated at a rate of 5 ACH.

Determine the total heat loss (kW) through the building when the inside temperature is kept at 20 °C while the average outside temperature is 0 °C.

Solution

Given: $\Delta T = 20 - 0 = 20°$ C, Building volume $= 20 \times 10 \times 5 = 1000$ m^3, $N = 5$ ACH

Find: $\dot{Q}_{total}$

It is convenient to arrange the solution in a tabular form as follows:

Building component	U-value (W/m² K)	Area (m²)	Temperature difference (°C)	Heat loss (W)
Walls	0.5	282	20	2820
Roof	0.45	200	20	1800
Floor	0.45	200	20	1800
Windows	3	12	20	720
Door	2	6	20	240
Total $\dot{Q}_F$		700		7380

$$\dot{Q}_V = 0.335 N V \Delta T$$
$$= 0.335 \times 5 \times 1000 \times 20$$
$$= 33\,500\ \text{W}$$

$$\dot{Q}_{total} = \dot{Q}_F + \dot{Q}_V = 7380 + 33\,500 = 40\,880\ \text{W}$$
$$= 40.88\ \text{kW}$$

Worked Example 5.8   Cyclic Heat Transfer

The internal (is) and external surface (es) temperatures of a building element are monitored over a 24-hour monitoring period, and the results are as shown in Figure E5.8.

Estimate the decrement factor (DF) and time lag (hours) for the element.

Solution

From the graph:

$$T_{es,max} = 28°C, \quad T_{es,min} = 14°C, \quad T_{is,max} = 24.5°C, \quad T_{is,min} = 17.5°C$$

$t_{peak,es}$ occurs six hours into monitoring period

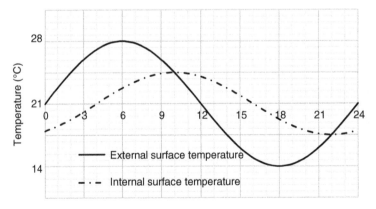

Figure E5.8   Transient temperature profile for Worked Example 5.8.

$t_{peak,is}$ occurs 10 hours into monitoring period

$$DF = (T_{is,max} - T_{is,min})/(T_{es,max} - T_{es,min})$$
$$= (24.5 - 17.5)/(28 - 14)$$
$$= 0.5$$

$$\text{Time lag} = t_{peak,int} - t_{peak,ext}$$
$$= 10 - 6$$
$$= 4 \text{ hours}$$

Worked Example 5.9   Interstitial Condensation Prediction Using Vapour Pressure
An external wall is constructed from three layers (see Figure E5.9a) for which the thicknesses and component properties are shown below in Table E5.9.

The inside air conditions are kept at 20 °C and RH at 60%.

The outside conditions are 0 °C and RH at 90%.

The convective film heat transfer resistances on the inside and outside surfaces are 0.13 and 0.06 m² K/W, respectively.

The thermal resistance of the air cavity is 0.18 m² K/W.

Table E5.9   Data for Worked Example 5.9.

Description	Thickness, x (mm)	Thermal conductivity (k, W/m K)	Vapour resistivity (r_{vap}, GN s/kg m)
Inner element	100	0.38	50
Cavity	25	—	—
Outer element	100	0.84	50

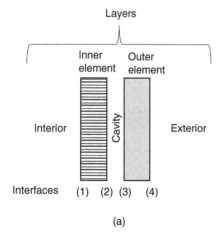

(a)

Figure E5.9a   Wall construction for Worked Example 5.9.

The vapour resistances of the internal and external air films can be ignored.

(a) Plot the saturated vapour pressure profile and the actual pressure profile for this wall and show whether condensation is likely or not.
(b) Rework the above if the cavity is filled with an insulating material of thermal conductivity $= 0.03$ W/m K.

Solution

Given: $k_{\text{inner element}} = 0.38$ W/m K, $x_{\text{inner element}} = 0.1$ m, $R_{\text{th, cavity}} = 0.18$ m^2 K/W, $x_{\text{cavity}} = 0.025$ m, $k_{\text{outer element}} = 0.84$ W/m K, $x_{\text{outer element}} = 0.1$ m, $r_{\text{vap, inner element}} = 50$ GN s/kg m, $r_{\text{vap, outer element}} = 50$ GN s/kg m, $T_i = 20°$ C, $\phi_{\text{interior}} = 60\%$, $T_o = 0°$ C, $\phi_{\text{exterior}} = 90\%$, $R_{\text{th, interior}} = 0.13$ m^2 K/W, $R_{\text{th, exterior}} = 0.06$ m^2 K/W, $R_{\text{vap, interior}} = 0$, $R_{\text{vap, outer}} = 0$

Find: Likelihood of condensation, i.e. $P_{\text{sat vap}} > P_{\text{actual}}$

(a) Calculate the individual layer and total wall thermal resistance using the denominator in Eq. (5.29)

$$R_{\text{th,total}} = R_{\text{th,interior}} + R_{\text{th,inner element}} + R_{\text{th,cavity}} + R_{\text{th,outer element}} + R_{\text{th,exterior}}$$
$$= 0.13 + (0.1/0.38) + 0.18 + (0.1/0.84) + 0.06 = 0.753 \text{ K m}^2/\text{W}$$

Calculate the temperature differences (ΔT_{layer}) across each layer using

$$\Delta T_{\text{layer}} = \Delta T_{\text{total}} \left(\frac{R_{\text{th,layer}}}{R_{\text{th,total}}}\right) = (20 - 0)\left(\frac{R_{\text{th,layer}}}{0.752}\right) = 26.596 R_{\text{th,layer}}$$

Then calculate the actual interface temperatures (T_1's) using $T_i = T_{i-1} - \Delta T_i$

Interface	(1)	(2)	(3)	(4)
T_i (°C)	16.54	9.55	4.76	1.6

Calculate the interior and exterior **saturated** vapour pressures and those at each interface using the interface tempertures and Eq. (4.2)

Interface	Internal	(1)	(2)	(3)	(4)	External
$P_{\text{sat vap}}$ (Pa)	2338	1882	1191	858	685	611

Calculate **actual** interior and exterior vapour pressures using Eq. (4.5)

$$\phi_{\text{interior}} = P_{\text{wv}}/P_{\text{sat}}, \quad 0.6 = P_{\text{wv}}/2388, \quad P_{\text{wv}} = 1403 \text{ Pa}$$
$$\phi_{\text{exterior}} = P_{\text{wv}}/P_{\text{sat}}, \quad 0.9 = P_{\text{wv}}/611, \quad P_{\text{wv}} = 550 \text{ Pa}$$

Calculate the individual layer and total wall vapour resistance using Eq. (5.45):

$$R_{v,\text{total}} = R_{v,\text{interior}} + R_{v,\text{inner element}} + R_{v,\text{cavity}} + R_{v,\text{outer element}} + R_{v,\text{exterior}}$$
$$= 0 + (50 \times 0.1) + (0.162 \times 0.18) + (50 \times 0.1) + 0 = 10.03 \text{ GN s/kg}$$

Calculate the pressure differences (ΔP_{layer}) across each layer using

$$\Delta P_{\text{layer}} = \Delta P_{\text{total}} \left(\frac{R_{\text{vap,layer}}}{R_{\text{vap,total}}} \right) = (1403 - 611) \left(\frac{R_{\text{vap,layer}}}{10.03} \right) = 78.963 R_{\text{vap,layer}}$$

Then calculate the actual interface vapour pressure (P_1's) using $P_i = P_{i-1} - \Delta P_i$

Interface	(1)	(2)	(3)	(4)
P_i (Pa)	1403	978	975	550

Plot both saturated vapour pressure and actual pressure profiles through the structure. See Figure E5.9b.

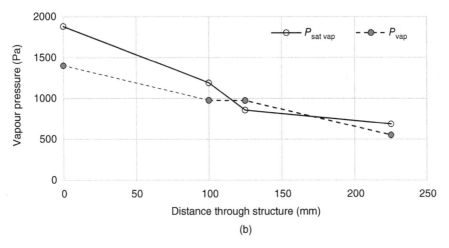

(b)

Figure E5.9b Vapour pressure profile for Worked Example 5.9a.

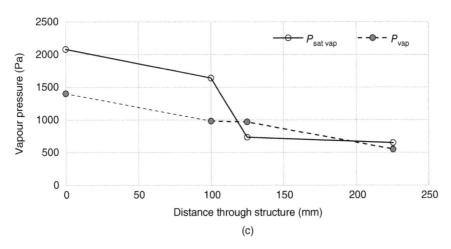

(c)

Figure E5.9c Vapour pressure profile for Worked Example 5.9b.

Examine both profiles. Anywhere that the actual vapour pressure exceeds the saturated vapour pressure there is the potential for condensation.

Examining the profile the hatched area indicates a likelihood of condensation, i.e. relative humidity of 100% in the cavity and the outer element.

(b) Recalculate the new thermal resistance based on an insulation filled cavity.

Repeat the subsequent calculations and replot. See Figure E5.9c.

Although the addition of insulation raises the temperatures at inner element interfaces (1, 2) and will reduce heat loss, condensation still occurs in the cavity as the temperature at interface 3 is lowered.

The problem could be solved by the installation of a vapour barrier on the warm side of the wall.

Worked Example 5.10   Internal Heat Gains

An office contains 20 people. Each occupant accesses their own desktop computer. The area is lit by a number of luminaires having a total electrical input of 900 W. If the heat gain associated with the occupants is 125 W/person and the computers each produce 100 W, determine the total internal heat gain (W) for the space.

Solution

Given: Number of occupants = 20, Heat generation per occupant = 125 W

Number of computers = 20, Heat generation per machine = 100 W

Lighting electrical load = 900 W

Find: $\dot{Q}_{total}$

$$\dot{Q}_{occupants} = \text{Number of occupants} \times \text{Heat generation per occupant}$$
$$= 20 \times 125$$
$$= 2500 \text{ W}$$

$$\dot{Q}_{computers} = \text{Number of machines} \times \text{Heat generation per machine}$$
$$= 20 \times 100$$
$$= 2000 \text{ W}$$

$$\dot{Q}_{lighting} = 900 \text{ W}$$

$$\dot{Q}_{total} = \dot{Q}_{occupants} + \dot{Q}_{computers} + \dot{Q}_{lighting}$$
$$= 2500 + 2000 + 900$$
$$= 5400 \text{ W}$$

5.9 Tutorial Problems

5.1 Determine the approximate solar time in Strasbourg, France (approximate longitude 7.45° E, Time zone is +1) on the 1st July at 12 : 00 local time.
Daylight saving is not in operation.
[Answer: 11 : 26 : 26]

5.2 Determine the approximate solar time, solar altitude, and solar azimuth angles at 15 : 00 in Athens, Greece (approximate longitude: 23.73° E, latitude: 37.98° N, Time zone is +2) on the 15th May. Daylight saving is not in operation.
[Answers: 14 : 38, 50.5°, 251.9°]

5.3 An engineering handbook lists the internal and external air film resistances for a plane wall as 0.13 and 0.04 m² K/W, respectively.
Determine the associated heat transfer coefficients (W/m² K).
If the plane wall has an internal air velocity of 0.25 m/s and an external value of 5.5 m/s, compare the above values with those indicated by simplified velocity-based forced convection correlations provided in the text.
[Answers: 7.7 W/m² K, (6.6 W/m² K), 25 W/m² K, (27.2 W/m² K)]

5.4 A pressed steel heating panel ($\varepsilon = 0.79$) is installed on an internal wall ($\varepsilon = 0.9$). The surface temperature of the heating panel is kept at 70 °C, while the wall surface temperature is 15 °C. See Figure P5.4
(a) Determine the radiation heat exchange per unit area (W/m²) between the two surfaces.
(b) In order to reduce the radiation between the two surfaces, a sheet of aluminium foil is placed at the back of the heating panel.
If aluminium has emissivity of 0.05, determine the percentage reduction in radiation heat transfer received by the plaster on the wall for the same temperature profile.

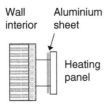

Figure P5.4 Wall construction for tutorial Problem 5.4.

[Answer: 287 W, 93%]

5.5 A triple glazed unit has the following specification:
Energy reflectance: 0.1
Energy absorptance: 0.54
Total shading coefficient: 0.49
Determine the energy transmittance (%), solar factor, and the inward absorptive heat reradiated or transferred via convection (%) for the unit.
[Answers: 36%, 0.43, 7%]

5.6 A pitched roof originally comprises a 12 mm plasterboard ceiling ($k = 0.2\,\text{W/m K}$), roof void ($R = 0.16\,\text{m}^2\,\text{K/W}$), and a tile/underlay outer skin ($R = 0.016\,\text{m}^2\,\text{K/W}$). The internal and external film heat transfer coefficients are 10 and 25 W/m^2 K, respectively.
Ignoring the effect of the supporting woodwork beams, rafters, etc., what thickness (mm) of mineral wool insulation ($k = 0.035\,\text{W/m K}$) applied to the back of the plasterboard would be required to reduce the original U-value to 0.15 W/m^2 K?
[Answer: ~220 mm]

5.7 An office space has the thermal specifications as shown in Table P5.7:
The building is 30 m × 20 m × 15 m high having a total glazed area of 520 m^2 and a door area of 8 m^2. The ventilation rate for the space is 1 ACH.
Determine the total heat loss (kW) for the building when the inside temperature is maintained at 19 °C, while the outside temperature is −3 °C.

Table P5.7 Data for tutorial Problem 5.7.

Building component	U-value (W/m² K)
Walls	0.16
Roof	0.11
Floor	0.11
Windows	2.0
Doors	1.0

[Answer: 95.711 kW]

5.8 A building has the dimensions shown in Figure P5.8 and thermal specification shown in Table P5.8.
The glazed area is 8 m^2 and the ventilation rate is 2 air changes per hour.
Use the CIBSE approach to determine the thermal response factor for the building and hence suggest its thermal weight. Ignore effect of external doors and glazing frames.
[Answers: 4.6, heavyweight]

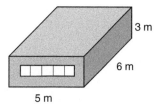

Figure P5.8 Building dimensions for tutorial Problem 5.8.

Table P5.8 Data for tutorial Problem 5.8.

Building element	Thermal transmittance (U, W/m^2 K)	Thermal admittance (Y, W/m^2 K)
Walls	0.5	4
Roof	0.25	5
Floor	0.4	5
Glazing	3.2	3

5.9 An external wall is constructed from four layers for which the thicknesses and thermal resistances of components are shown in Table P5.9.

The inside air temperature is kept at 19 °C and RH at 50%. The outside temperature is 5 °C having a relative humidity of 85%.

The convective film heat transfer resistance on the inside and outside surfaces are 0.13 and 0.06 m^2 K/W, respectively. The thermal resistance of the air cavity is 0.18 m^2 K/W.

The vapour resistance of the vapour barrier is 500 GN s/kg. The vapour resistances of the internal and external films can be ignored.

Determine the actual temperature profile and the dew point profile for this wall and estimate the *minimum* actual dew point temperature differential.

Table P5.9 Data for tutorial Problem 5.9.

Description	Thickness x (mm)	Thermal conductivity k (W/m K)	Vapour resistivity r_{vap} (GN s/kg m)
Inner element	100	0.19	50
Vapour barrier	0.3	0.5	—
Cavity	25	—	—
Outer element	100	0.84	50

[Answer: ~1.7 °C at the warm side of the cavity]

5.10 An office houses 10 personnel, each using laptop computers.

The facility has a communal office printer and water cooler.

The office is open 1800 h/annum.

Using the data supplied, estimate the average annual internal heat gain (kWh/annum) to the space.

Data

Equipment average electrical input: laptop (25 W), office printer (500 W), water cooler (100 W).

Lighting electrical input: 1000 W

Occupant heat generation: 120 W/person.

[Answer: 5490 kWh/annum]

6

Refrigeration and Heat Pumps

6.1 Overview

Refrigeration is the process of bringing a controlled environment to a temperature lower than the ambient and sits at the heart of an air conditioning system during the warmer months. Refrigeration relies on substances (*refrigerants*) which, when they evaporate, absorb heat from the surrounding air. The cooled, conditioned air is then introduced into the occupied space.

Thanks to the inventiveness of American Scientist and chemical engineer **Thomas Midgley Jr**, throughout the twentieth century, nearly all air conditioners/refrigerators used chlorofluorocarbons (CFCs) as their refrigerant. At the time, CFCs were regarded as a major advancement in cooling technology. Of course, neither Midgley nor anyone else was aware of any unintended consequence of their release into the atmosphere until the 1980s. It is now well known that CFCs in the atmosphere disrupt the production of UV absorbing Ozone. As a result, CFC production was stopped in 1995.

After the ban, nearly all air conditioning systems employed halogenated chlorofluorocarbons (HCFCs or HFCs) as a refrigerant, but these are also being gradually phased out.

Furthermore, environmental pressures have, in some cases, led to the review and reuse of early technology (*'natural'*) refrigerants such as ammonia and CO_2.

The most common types of refrigeration plant used today are *vapour compression* and *vapour absorption* systems. However, there are alternative technologies based on different thermodynamic principles which are not yet used extensively due to their complexity or limitation in providing capacity. If these limitations can be overcome, these alternatives may prove to be very attractive in terms of sustainability.

Learning Outcomes

- To be familiar with the nature and types of refrigerants.
- To be able to analyse basic refrigeration and heat pump (HP) cycles from a thermodynamic standpoint.
- To be familiar with the operation of cooling systems using a range of input energy sources.

Building Services Engineering: Smart and Sustainable Design for Health and Wellbeing, First Edition.
Tarik Al-Shemmeri and Neil Packer.
© 2021 John Wiley & Sons Ltd. Published 2021 by John Wiley & Sons Ltd.

6.2 Choice of Refrigerants

The choice of refrigerant fluid used for cooling is based on three factors:

- Thermal properties of fluid
- Costs of manufacture and operation
- Environmental impact

From a thermodynamic viewpoint, for a refrigerant to be useful, it must have the ability to evaporate at a temperature lower than the ambient, thus absorbing heat from its surroundings. Its ability to liquefy at a pressure that is not too high is also important so that it can be recirculated and reused without the expenditure of a great deal of energy.

The thermodynamic properties of refrigerants may be obtained from tables or state diagrams and charts. Use of pressure-specific enthalpy charts is most common in refrigeration although temperature-specific entropy charts are also of use in understanding processes. See Figure 6.1a,b for their generalised arrangement.

It is important to note that some tables use a unique reference point and so property values such as enthalpy may differ from a chart *for the same refrigerant*. However, as load calculations are determined by *enthalpy change*, the final calculated result is valid in both cases.

Environmental factors affecting the choice of refrigerant have become very important due to the acknowledged depletion of the ozone layer and contribution to global climate caused by, now outlawed gases, such as the CFC family (e.g. R11). Parameters have been developed to evaluate the potential damage caused by a given refrigerant:

- The *ozone depletion potential* (ODP) is related to the presence of chlorine atoms in a refrigerant. Chlorine in the upper atmosphere has a tendency to block the formation of UV scavenging ozone allowing the electromagnetic radiation to reach the surface of the planet. Its importance has led to the development of chlorine-free fluids.
- The *global warming potential* (GWP) of a refrigerant gas is measured with reference to the climate changing capacity of carbon dioxide. Generally speaking, the (per kg) GWPs of refrigerants used over the last century tend to be up to 10 000x that of carbon dioxide.
- The term *total equivalent warming impact* (TEWI) is used to evaluate the direct life cycle emissions as a result of refrigerant release and global warming gas emissions associated with system electricity consumption.

There has been an intensive effort worldwide to develop new refrigerants which are environmentally friendly yet produce the necessary refrigeration effect. It is customary nowadays to examine the GWP and ODP factors of any refrigerant before studying its thermal refrigeration properties.

Partially halogenated HCFCs, e.g. R22, were considered as promising replacement fluids in a transition period but a string of amendments to the Montreal Protocol (1987) and tighter restrictions imposed by the European Community now lead to the consideration of only chlorine-free fluids (HFCs), e.g. R134a, as acceptable alternative refrigerants. Their ODPs are zero, and they have typical GWPs of 2000 or less.

HCFCs are to be phased out in 2020 and so currently ozone-safe, chlorine-free, hydrofluorocarbons (HFCs) dominate the market.

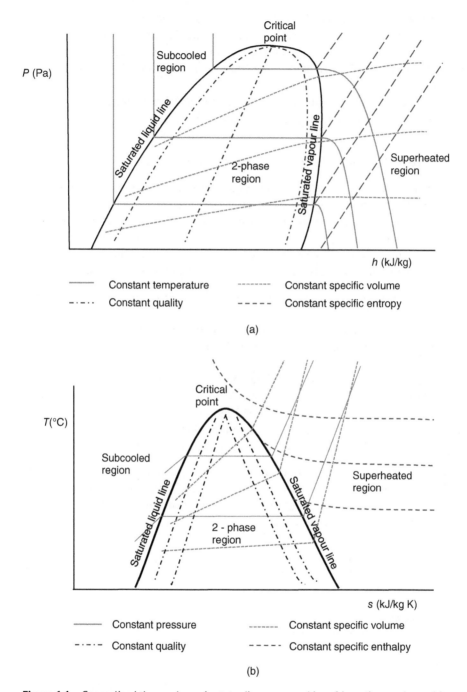

Figure 6.1 Generalised thermodynamic state diagrams used in refrigeration analyses. (a) Pressure – specific enthalpy chart and (b) temperature – specific entropy chart.

In general, it has been proposed that binary or ternary zeotropic (i.e. range of boiling point) HFC blends, e.g. R407c, be used as substitutes for the chlorine-based refrigerants. The optimum thermodynamic behaviour is obtained by varying the percentage of components. However, the negative influence of characteristics such as flammability or toxicity must also be considered.

As a result of their GWP, the 2016 (Kigali) Montreal Protocol amendments call for a HFC phase down (85% reduction by 2036) of their production and use.

6.2.1 Choice of Refrigerant for Vapour Compression Systems

At their simplest, vapour compression systems comprise two heat exchangers (evaporator and condenser), a compressor and a pressure reduction or expansion device.

The refrigerant should be a fluid that can be used within a temperature range from about $-15\,°C$ (the evaporator) up to about $30\,°C$ (the condenser). At these temperatures, the vapour pressures should not be excessively high nor below atmospheric.

The following thermophysical refrigerant characteristics are necessary or desirable:

- The critical temperature of the refrigerant must be in excess of the highest temperature in the cycle for liquefaction.
- The boiling point ($°C$) must be below the lowest temperature in the cycle for evaporation.
- The latent heat of evaporation (kJ/kg) should be high, leading to a large refrigerating effect.
- The work of compression (kJ/kg) should be small to minimise energy use
- The specific volume (m^3/kg) of the vapour should be small, leading to a compact machine. Note that specific volume is the reciprocal of density (kg/m^3)
- The specific heat capacity (kJ/kg K) of the liquid should be low.
- The pressure (bar or Pa) in the system should be low to reduce stresses.

Additionally, the fluid must be nontoxic, nonexplosive, compatible with other system materials such as lubricants and seals, noncorrosive, noninflammable, and cheap.

Table 6.1 illustrates the typical range of thermodynamic properties of HFC refrigerants suitable for vapour compression systems.

Table 6.1 Property ranges of HFC refrigerants suitable for vapour compression cycles.

Refrigerant property	Range
Boiling point @1 atm (K)	223–243
Critical temperature (K)	343–373
Critical pressure (bar absolute)	40–60
Vapour specific volume @boiling point (m^3/kg)	0.2–0.3
Latent heat of vapourisation @boiling point (kJ/kg)	220–380
Liquid specific heat capacity @20 $°C$ (kJ/kg K)	1.4–2.0

6.2.2 Choice of Refrigerant–Absorbent Pairings for Vapour Absorption Systems

The absorption cycle is, in many ways, similar to the vapour compression cycle, in that it uses a circulating refrigerant, an evaporator, a condenser, and an expansion device. However, the cycle differs in that it contains another fluid–the absorbent (or solvent) and that the compressor of the vapour compression cycle is replaced by a chemical absorption process and a thermally driven generator to desolve the two fluid components.

A pump provides the necessary circulation and pressure change at a much reduced thermodynamic work input.

Desirable working fluid pairing properties include the following:

- The refrigerant should be more volatile than the absorbent, i.e. $T_{boil, refrigerant} \ll T_{boil, absorber}$ facilitating the separation of refrigerant and absorber in the system generator and ensuring that only pure refrigerant circulates through the refrigerant circuit (evaporator–condenser–expansion valve [EV]).
- The refrigerant should exhibit high solubility with the solution in the absorber. The reverse should be the case under conditions in the generator.
- The working fluids should not undergo crystallisation or solidification at any point in the system as crystallisation will block free flow in system interconnecting pipework.

As is the case with vapour compression systems, the working fluids should be safe, chemically stable, noncorrosive, and inexpensive to manufacture.

The two most commonly used refrigerant–absorbent pairs in commercial systems are as follows:

- Water–lithium bromide (H_2O–LiBr) systems used to produce chilled water (>5 °C) for air conditioning systems. Typical commercial cooling capacities are of the order of 50 kW – 5 MW
- Ammonia–Water (NH_3–H_2O) systems are used in applications where temperatures below 5 °C are required. Cooling capacities of up to 100 kW are typically available.

6.3 Heat Pump, Refrigeration, and Vapour Compression Cycles

In '*Heat engines*', heat is received by the working fluid at a high temperature (hot reservoir) and is then rejected with the exhaust gases at a lower temperature (cold reservoir), while a net amount of work is produced by the fluid, see Figure 6.2a.

The efficiency of such a cycle is defined as the ratio of output mechanical work (energy) and the energy input as heat converted from the combustion of the fuel:

$$\eta_{Heat\ Engine} = \frac{W}{Q_1} \tag{6.1}$$

The energy balance of the system in Figure 6.2a implies that the network output is equal to the difference between the energy supplied from the fuel and the energy lost with the exhaust, hence

$$W = Q_1 - Q_2 \tag{6.2}$$

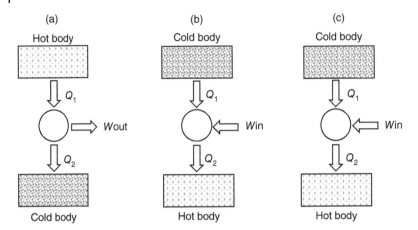

Figure 6.2 Comparison between (a) a heat engine, (b) a refrigerator, and (c) a heat pump.

hence

$$\eta_{\text{Heat Engine}} = \frac{Q_1 - Q_2}{Q_1} \qquad (6.3)$$

But the quantity of heat associated with the body temperature of the system's gases is given by

$$Q = mC_p\Delta T \qquad (6.4)$$

hence

$$\eta_{\text{Heat Engine}} = \frac{T_1 - T_2}{T_1} \qquad (6.5)$$

This is known as the Carnot efficiency, which is the maximum possible value that can be achieved in any situation.

It is possible to reverse the operation of the above system (heat engine) when external work is supplied to the machine and hence it will operate by rejecting heat to a hot reservoir. When the machines prime function under this condition is to provide a cooling effect and an environment below ambient at the cold reservoir, the machine is termed a reversed heat engine or simply a '*Refrigerator*'. See Figure 6.2b.

The term '*Heat Pump*' is given to a machine operating like a refrigerator but whose principle function is to supply heat at an elevated temperature at the hot reservoir. See Figure 6.2c.

Therefore, it is common to see a single machine with a dual function working as a heat pump and a refrigerator.

The criterion of performance of the cycle, expressed as the ratio of output/input, depends upon what is regarded as output. This implies that the efficiency definition used with heat engines is not valid for refrigerator/heat pump.

Another term is, therefore, used to express the effectiveness of a refrigerator/heat pump and is known as the coefficient of performance (COP) which is defined as the ratio of heat output to the work input. Hence, there are two different expressions for COP, one for the refrigerator (R) and the second for a heat pump (HP):

$$\text{COP}_{\text{Ref}} = \frac{Q_1}{W} \qquad (6.6)$$

and

$$COP_{HP} = \frac{Q_2}{W} \qquad (6.7)$$

But the first Law of Thermodynamics can be written as:

$$Q_1 + W = Q_2$$

Dividing by W

$$\frac{Q_1}{W} + 1 = \frac{Q_2}{W} \qquad (6.8)$$

hence

$$COP_{HP} = COP_{Ref} + 1 \qquad (6.9)$$

Since Q is proportional to temperature, the COP can be written in terms of temperatures

For a heat pump:

$$COP_{HP} = \frac{T_2}{T_2 - T_1} \qquad (6.10)$$

For a refrigerator:

$$COP_{Ref} = \frac{T_1}{T_2 - T_1} \qquad (6.11)$$

Note that

$$COP_{HP} = \frac{1}{\eta_{HP}} \qquad (6.12)$$

6.3.1 Carnot Cycle

Vapour compression refrigeration cycles are based on the well-known Carnot cycle which is a hypothetical cycle that is difficult to build in practice.

The layout of a refrigerator operating on a reversed Carnot cycle is shown in Figure 6.3

The four processes in the reversed Carnot cycle are as follows:-

1–2 Vapour is compressed isentropically from 1 to 2.
2–3 The vapour is condensed at constant pressure (and temperature) from 2 to 3.
3–4 The fluid is then expanded isentropically from 3 to 4.

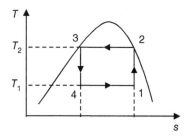

Figure 6.3 Carnot refrigeration cycle.

4–1 The fluid is evaporated at constant pressure from 4 to 1.

Based on a unit mass flow of refrigerant in the system:

Refrigeration effect:

$$q_R = h_1 - h_4 \tag{6.13}$$

Compressor work input:

$$w_{comp} = h_2 - h_1 \tag{6.14}$$

The COP:

$$COP_{Ref} = \frac{h_1 - h_4}{h_2 - h_1} \tag{6.15}$$

But enthalpy is proportional to temperature, hence

$$COP_{Ref} = \frac{T_1}{T_2 - T_1} \tag{6.16}$$

6.3.2 Ideal Vapour Compression Refrigeration Cycle

The work obtained in the expander can be used to reduce the external work required to drive the compressor. However, the greater first cost and mechanical complication of the expander are not justified in practice and a major simplification is obtained by replacing the expander by a simple throttle valve. The evaporator is also modified (its size is slightly increased) to ensure that the refrigerant leaving to the compressor is definitely in a dry vapour state. This leads to a cycle that has a lower COP but with other advantageous features.

The new cycle shown in Figure 6.4 consists of the following processes:

1–2 Isentropic compression (i.e. $s_1 = s_2$)
2–3 Condensation at constant pressure
3–4 Throttling at constant enthalpy (i.e. $h_3 = h_4$)
4–1 Evaporation at constant pressure

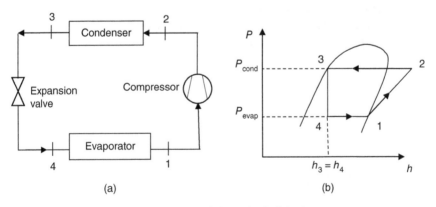

(a)   (b)

Figure 6.4 Simple refrigeration cycle. (a) Schematic, (b) P-h chart.

For this cycle, the performance is calculated for a unit mass of refrigerant, as follows:

Refrigeration effect:

$$q_R = h_1 - h_4 \tag{6.17}$$

Compressor work input:

$$w_{comp} = h_2 - h_1 \tag{6.18}$$

Hence,

$$COP_{Ref} = \frac{h_1 - h_4}{h_2 - h_1} \tag{6.19}$$

6.3.3 Practical Vapour Compression Refrigeration Cycle

The simple cycle can be further modified to improve performance by some or all of: (i) superheating (ii) under or subcooling, and (iii) use of a heat exchanger (see Figure 6.5).

(i) **Superheating**

In the simple Carnot cycle, the refrigerant enters the compressor as a vapour (condition 1 in Figure 6.4). With this improvement, the refrigerant leaving the evaporator is superheated to eliminate the possibility of wet refrigerant entering the compressor.

(ii) **Subcooling**

A further modification is the undercooling of the liquid refrigerant in the condenser (i.e. the temperature is reduced below the saturated liquid temperature at the upper pressure). The final temperature is usually a few degrees below the temperature of the coolant in the condenser. Both the superheating and undercooling increase the refrigeration effect, resulting in an improved COP.

(iii) **The use of the heat exchanger**

The use of a liquid-to-suction heat exchanger subcools the liquid from the condenser with suction vapour coming from the evaporator. The effect of using a heat exchanger results in producing a double effect of superheating and subcooling simultaneously.

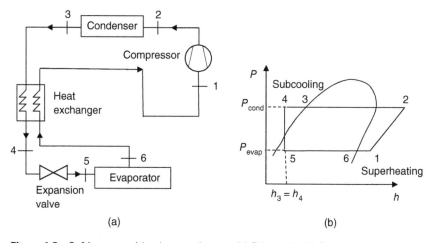

Figure 6.5 Refrigerator with a heat exchanger. (a) Schematic, (b) P-h chart.

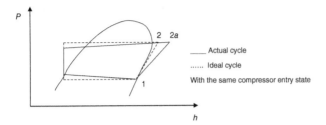

Figure 6.6 Irreversibilities in the refrigeration cycle.

6.3.4 Irreversibilities in Vapour Compression Refrigeration Cycle

In reality, the behaviour of the refrigerant during any process is not perfect. Some property changes do not occur in ways that are entirely reversible.

(a) **Pressure drop in the refrigeration cycle**

   The actual vapour-compression cycle suffers from pressure drops in the condenser and evaporator units due to friction. As a result of this pressure drop, the compression process requires more work than in the standard cycle. See Figure 6.6. It must be stressed here that these losses are usually neglected.

(b) **Actual compression**

   Due to irreversibilities in the compressor, there is an increase in the entropy of the working fluid as it goes through the compressor. A compressor works with an isentropic efficiency given by

$$\eta_{\text{isen}} = \frac{h_2 - h_1}{h_{2a} - h_1} \tag{6.20}$$

   h_1 – enthalpy at compressor inlet condition
   h_2 – ideal (100% isentropic) enthalpy at compressor outlet condition
   h_{2a} – actual (not 100% isentropic) enthalpy at compressor outlet condition

6.3.5 Multistage-Vapour Compression Refrigeration

Multistaging in refrigeration systems is employed for a number of reasons:

- To split the compression into a series of stages, this helps to reduce compression work and hence improve the performance.
- To split the expansion process, so that the cooling/refrigeration can be split into two or more zones, cold and colder, such as the domestic application, where the cold store is separated from the freezer compartment.
- Splitting the different processes means smaller units are needed, usually associated with lower manufacturing costs and more flexible maintenance.

The following examples demonstrate how the above are implemented.

(a) **Two-stage compression with intercooler**

   The vapour leaving the low pressure compressor (2), is cooled to the saturated condition (3), then further compressed in the high-pressure cylinder (4) and from there to (1) the sequence is the same as in the simple plant. See Figure 6.7.

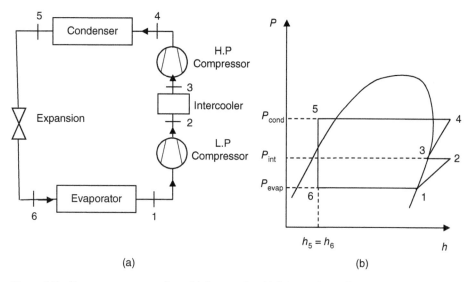

Figure 6.7 Two-stage compression with intercooler. (a) Schematic, (b) P-h chart.

Refrigeration effect:

$$q_R = h_1 - h_6 \tag{6.21}$$

Compressor's work:

$$w_{comp} = (h_2 - h_1) + (h_4 - h_3) \tag{6.22}$$

Coefficient of performance:

$$COP_{Ref} = \frac{h_1 - h_6}{(h_2 - h_1) + (h_4 - h_3)} \tag{6.23}$$

(b) Two-stage compression with flash intercooling

For every 1 kg/s of fluid passing through the HP compressor, let $\dot{m}$ kg/s pass through the evaporator. The vapour leaves the high-pressure cylinder at 4 generally in the super-heated state and passes to the condenser. From the condenser at 5 it passes to the first expansion valve V_1 and is discharged into the receiver or flash chamber. At 7, $\dot{m}$ kg/s of liquid leaves the chamber and passes through the second expansion valve V_2 into the evaporator. The vapour, which may be wet, dry saturated, or superheated, leaving the evaporator is shown dry saturated at 1. It is taken in by the low-pressure cylinder and after compression it is discharged at 2 in a superheated condition into the receiver. Here it mixes with the liquid–vapour mixture which entered the receiver after the first throttling and is cooled to the dry saturated condition. Thus, cooling is effected by a heat transfer to the liquid and causes a portion of the liquid to flash evaporate. At 3, 1 kg/s of dry saturated vapour is drawn into the high-pressure cylinder and compressed to the condenser pressure. The cycle is thereby completed. See Figure 6.8.

Consider the receiver energy balance:

$$1 \times h_6 + \dot{m} \times h_2 = 1 \times h_3 + \dot{m} \times h_7 \tag{6.24}$$

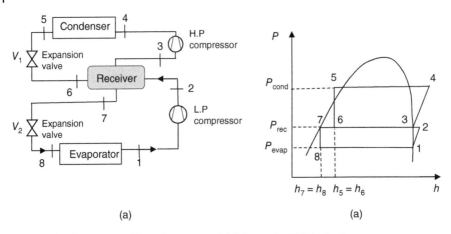

Figure 6.8 Two stage refrigeration system. (a) Schematic, (b) P-h chart.

Hence,

$$\dot{m} = \frac{h_3 - h_6}{h_2 - h_7} \tag{6.25}$$

$$\dot{Q}_R = \dot{m}(h_1 - h_8) \tag{6.26}$$

$$\dot{W}_{comp} = \dot{m}(h_2 - h_1) + 1(h_4 - h_3) \tag{6.27}$$

And

$$COP_{Ref} = \frac{\dot{m}(h_1 - h_8)}{\dot{m}(h_2 - h_1) + 1(h_4 - h_3)} \tag{6.28}$$

6.3.6 Multipurpose Refrigeration Systems with a Single Compressor

Some applications require refrigeration at more than one temperature. This could be accomplished by using a separate throttling valve and a separate compressor for each evaporator operating at different temperatures. However, such a system will be bulky and probably uneconomical. A more practical and economical approach would be to route all the exit streams from the evaporators to a single compressor and to let it handle the compression process for the entire system.

Consider, for example, an ordinary refrigerator-freezer unit. A simplified schematic of the unit and the T–s diagram of the cycle are shown in Figure 6.9. Most refrigerated goods have a high water content and the refrigerated space must be maintained above the ice-point (at about 5 °C) to prevent freezing. The freezer compartment, however, is maintained at about −15 °C. Therefore, the refrigerant should enter the freezer at about −25 °C to have heat transfer at a reasonable rate in the freezer. If a single expansion valve and evaporator were used, the refrigerant would have to circulate in both compartments at about −25 °C, which would cause ice formation in the neighbourhood of the evaporator coils and dehydration of the produce. This would not be acceptable to a user. This problem can be eliminated by throttling the refrigerant to a higher pressure (hence temperature) for use in the refrigerated

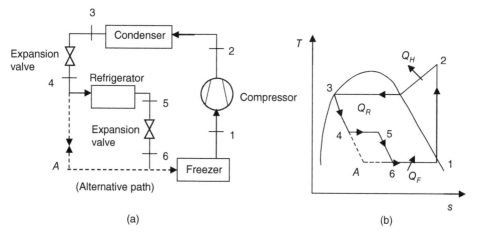

Figure 6.9 Fridge-Freezer. (a) Schematic, (b) P-h chart.

space and then throttling it to the minimum pressure for use in the freezer. The entire refrigerant leaving the freezer compartment is subsequently compressed by a single compressor to the condenser pressure.

6.4 Absorption Refrigeration

As described earlier, absorption machines work with two fluids: the **refrigerant** and a transport medium – the **absorbent** into which the refrigerant can be dissolved.

These systems are popular in applications where waste heat or renewable solar thermal energy is readily available. An absorption cycle utilizes a binary mixture of refrigerants such as ammonia–water or water–LiBr. The single effect cycle consists of an absorber, a generator or desorber, a condenser, an evaporator, and an electric solution pump, with the possibility of additional components to improve the performance, such as heat exchangers. Heat absorbed in the generator allows the refrigerant to desorb from the absorbent, creating a high-pressure vapour. In cases where a volatile absorbent is used (e.g. ammonia–water), a rectifier is needed to reduce the concentration of the volatile absorbent (e.g. water) in the vapour to the condenser.

In the absorption refrigerator, the compressor is replaced by an absorber (with the facility for heat rejection), a liquid pump, a heat exchanger, a generator (with a source of *renewable* or waste heat), and an expansion valve.

A schematic for a water/LiBr system is shown in Figure 6.10 and an enthalpy – LiBr concentration chart is shown in Figure 6.11. Notice in Figure 6.11 that LiBr mass concentrations are limited to 60–70% to avoid crystallisation.

The enthalpy – concentration relationships illustrated in Figure 6.11 can be approximated quadratic polynomials of the form:

$$h = K_1 c^2 - K_2 c + K_3$$

values of constants k's are shown in Table 6.2.

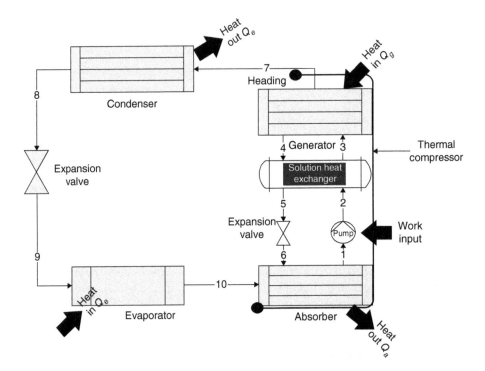

Figure 6.10 Absorption refrigeration.

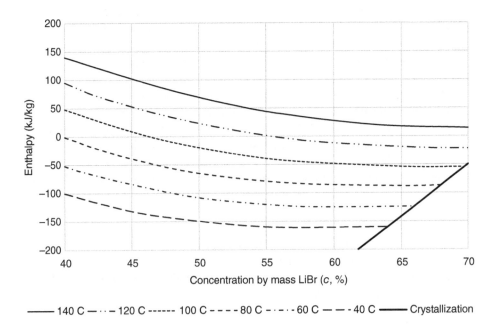

Figure 6.11 Enthalpy–% concentration property chart for LiBr.

Table 6.2 Enthalpy correlations for LiBr.

Temperature (°C)	Polynomial coefficients		
	K_1	K_2	K_3
140	0.1457	20.241	716.95
120	0.1558	20.909	679.64
100	0.1375	18.495	566.46
80	0.1605	20.375	557.67
60	0.1657	20.260	492.46
40	0.1755	20.702	446.31

6.4.1 Thermodynamic Analysis

Two laws are used to analyse each component: Laws of conservation of *mass* and *energy*:

Mass flow:

$$\sum_{in} \dot{m} = \sum_{out} \dot{m}$$ (6.29)

Energy flow (assuming kinetic and potential changes are negligible):

$$\dot{Q} - \dot{W} = \dot{m}(h_{out} - h_{in})$$ (6.30)

Evaporator 9–10 Liquid refrigerant from the expansion valve (EV) enters the evaporator, absorbs heat under isothermal and isobaric conditions leaving as a saturated vapour.

Property balances:

$$\dot{Q}_E = \dot{m}_{10}(h_{10} - h_9)$$ (6.31)

$$\dot{m}_{10} = \dot{m}_9$$ (6.32)

Condenser 7–8 Superheated refrigerant leaving the absorber is cooled under isothermal and isobaric conditions to saturated or subcooled liquid conditions at condenser exit.

Property balances:

$$\dot{Q}_C = \dot{m}_8(h_7 - h_8)$$ (6.33)

$$\dot{m}_7 = \dot{m}_8$$ (6.34)

Expansion valve (EV) 8–9 High-pressure liquid refrigerant is throttled to a low-pressure mixed phase condition at constant enthalpy.

Property balances:

$$h_9 = h_8$$ (6.35)

$$\dot{m}_8 = \dot{m}_9$$ (6.36)

Absorber 10–1 The absorber has three mass flows to consider: a vapour input from the evaporator and a weak solution from the expansion valve (EV) combining to form a rich solution output to the pump.

Property balances:

$$\text{Total flow}: \dot{m}_1 = \dot{m}_{10} + \dot{m}_6 \tag{6.37}$$

$$\text{LiBr flow}: \dot{m}_1 c_1 = \dot{m}_6 c_6 \tag{6.38}$$

$$\text{Water flow}: \dot{m}_{10} + \dot{m}_6(1 - c_6) = \dot{m}_1(1 - c_1) \tag{6.39}$$

$$\dot{Q}_A = \dot{m}_1 h_1 - \dot{m}_6 h_6 - \dot{m}_{10} h_{10} \tag{6.40}$$

Pump 1–2 Work supplied to the pump increases the rich solution enthalpy and pressure.

Property balances:

$$-\dot{W}_p = \dot{m}_1(h_2 - h_3) \tag{6.41}$$

$$\dot{m}_2 = \dot{m}_1 \tag{6.42}$$

Expansion valve (EV) 5–6 High pressure weak solution is throttled to a low pressure for return to the absorber at constant enthalpy.

Property balances:

$$h_6 = h_5 \tag{6.43}$$

$$\dot{m}_6 = \dot{m}_5 \tag{6.44}$$

Generator 3–7 The generator has three mass flows to consider: a rich solution input from the solution heat exchanger, a weak solution return to the solution heat exchanger and a saturated refrigerant output to the condenser.

Property balances:

$$\text{Total flow}: \dot{m}_3 = \dot{m}_4 + \dot{m}_7 \tag{6.45}$$

$$\text{LiBr flow}: \dot{m}_4 c_4 = \dot{m}_3 c_3 \tag{6.46}$$

$$\text{Water flow}: \dot{m}_7 + \dot{m}_3(1 - c_3) = \dot{m}_4(1 - c_4) \tag{6.47}$$

$$\dot{Q}_G = \dot{m}_7 h_7 + \dot{m}_4 h_4 - \dot{m}_3 h_3 \tag{6.48}$$

Overall the COP of the machine operating in cooling mode can be expressed as the cooling load divided by the heat input to the generator and work input to the pump:

$$\text{COP}_{\text{cooling}} = \frac{\dot{Q}_E}{\dot{Q}_G + \dot{W}_{\text{pump}}} \tag{6.49}$$

Typical values of COP cooling for LiBr machines are in the range 0.6–1.0.

In ammonia–water machines, both fluids are highly stable for a wide range of operating temperatures and pressures. However, as both ammonia and water are volatile and have contributory vapour pressures, the cycle needs an efficient water separator or *rectifier* to strip away water that normally evaporates with the ammonia. Without a rectifier, the water would accumulate in the evaporator, reducing system performance. Other disadvantages of the working fluid combination are its high pressure requirement, toxicity potential requiring an alarm/detection system, and the corrosive action on copper and its alloys.

On the positive side, crystallisation is not an issue.

Typical values of COP cooling for NH_3–H_2O machines are in the range 0.5–0.7.

6.5 Adsorption Refrigeration

An adsorption system uses multiple beds of adsorbents, such as silica gel in a silica gel water system, to provide continuous capacity and does not use any mechanical energy but only thermal energy. An adsorption refrigeration system usually consists of four main components: a solid adsorbent bed, a condenser, an expansion valve, and an evaporator.

The bed desorbs its refrigerant when externally heated and adsorbs refrigerant vapour when cooled such that it operates as a 'thermal compressor' to drive the refrigerant around the system providing space heating at the condenser or cooling at the evaporator. When the bed becomes saturated with refrigerant, it is isolated from the evaporator and connected to the condenser. The refrigerant vapour is condensed to a liquid, followed by expansion to a lower pressure in the evaporator where the low-pressure refrigerant is vaporized producing the refrigeration effect. When further heating, no longer produces desorbed refrigerant from the bed, the refrigerant vapour from the evaporator is reintroduced to the bed to complete the cycle. As described above, the system acts intermittently with the use of isolating valves.

To obtain a continuous and stable cooling effect, generally two (or multiple) adsorbent beds are used, where one bed is heated during desorption while the other bed is cooled during adsorption. To achieve high efficiency, heat of adsorption needs to be recovered to provide part of the heat needed to regenerate the adsorbent.

The thermodynamic cycle is commonly shown on an $\ln P$ vs. $-1/T$ basis. See Figure 6.12.

6.6 Stirling Cycle Refrigeration

An ideal Stirling cooler is a reversed Stirling engine. It consists of a closed-cycle regenerative heat engine with a gaseous working fluid (generally He or H_2).

The basic Stirling system cooler (Beta configuration) is depicted in Figure 6.13.

It consists of a piston, a compression space, (and heat exchanger) all at ambient temperature T_a, expansion space (and heat exchanger), and a displacer (typically 90° out of phase with the piston) all at the low temperature T_L) and a cyclic heat recovery unit or regenerator.

The cooling cycle is split in four processes:

1–2: With the displacer (D) at the cold side (C), the piston (P) moves up. The gas between the displacer and piston is compressed and heat is rejected from the hot side (H) in an isothermal process.

2–3: Displacer moves down and gas is forced from hot side (H) to cold side via regenerator (R) in an isochoric process.

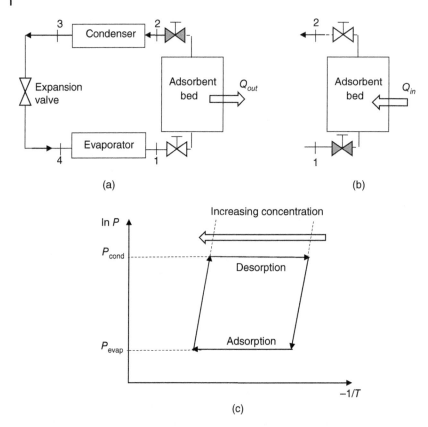

Figure 6.12 Adsorption refrigeration. (a) Adsorption stage (bed at low temperature), (b) desorption stage (bed at high temperature), and (c) thermodynamic cycle.

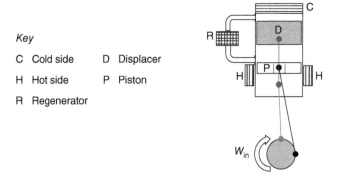

Key

C Cold side   D Displacer
H Hot side   P Piston
R Regenerator

Figure 6.13 Stirling cycle (Beta) arrangement.

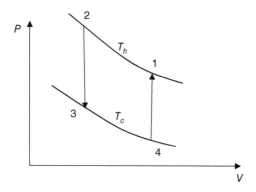

Figure 6.14 Stirling refrigeration cycle.

3–4: Piston moves down, gas is expanded, and heat is absorbed at cold side in an isothermal process.

4–1: Displacer moves up pushing gas from cold side to hot side via regenerator in an isochoric process.

The cycle is illustrated on a pressure–volume basis in Figure 6.14.

The volume V is the volume between the two pistons. In practice, the cycle is not divided in discrete steps as described above. In the ideal case, the cycle is reversible so the COP_{Ref} is equal to the Carnot COP given by $T_c/(T_h-T_c)$.

Knowledge of the energy transfers during the four processes is required in order to evaluate the COP of the machine.

Using the First law of Thermodynamics for a non-flow condition:

Compression process 1–2:

$$Q_{comp} - W_{comp} = m\Delta u \tag{6.50}$$

But for an isothermal process:

$$\Delta u = Cv\Delta T = 0 \tag{6.51}$$

And for an isothermal process, it can be shown that:

$$W_{comp} = mRT_c \ln\left(\frac{V_2}{V_1}\right) \tag{6.52}$$

Thus

$$Q_{comp} = W_{comp} = mRT_c \ln\left(\frac{V_2}{V_1}\right) \tag{6.53}$$

Expansion process 3–4:

$$Q_{exp} - W_{exp} = m\Delta u \tag{6.54}$$

But for an isothermal process

$$\Delta u = Cv \Delta T = 0 \tag{6.55}$$

Thus,

$$Q_{exp} = W_{exp} = mRT_e \ln\left(\frac{V_4}{V_3}\right) \tag{6.56}$$

Thus the net – work transfer (W_{net}) is given by:

$$W_{net} = W_{comp} - W_{exp} = mR(T_c - T_e)\ln\left(\frac{V_2}{V_1}\right) \tag{6.57}$$

And the COP is

$$COP_{Ref} = \frac{Q_{exp}}{W_{net}} = T_e/(T_c - T_e) \tag{6.58}$$

Equivalent flow rate $(\dot{m}, \text{kg/s})$, heat transfer $(\dot{Q}, \text{kW})$, and work transfer expressions can be derived with knowledge of machine rotational speed (N, rps).

6.7 Reverse Brayton–Air Refrigeration Cycle

As its name implies, a reverse Brayton cycle (or Bell-Coleman cycle) achieves a cooling effect by reversing the *gas turbine* Brayton cycle: a gas is compressed, cooled in a heat exchanger, and then expanded. An *ideal* cycle is shown in Figure 6.15. The air temperature at end of expansion process is low and can be used to cool an enclosure, either by direct contact or through a heat exchanger.

The working fluid is air which undergoes the following processes:

- 1–2 ambient air is drawn into the compressor. For a process of adiabatic reversible compression, it can be shown that:

$$T_2 = T_1\left(\frac{P_2}{P_1}\right)(\gamma - 1)/\gamma \tag{6.59}$$

where γ is a *compression* index which essentially describes the path taken by the property change.

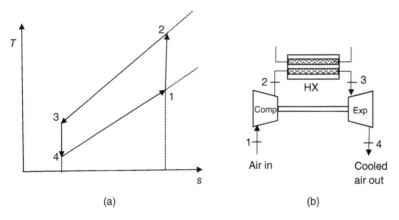

(a) (b)

Figure 6.15 Reverse Brayton (Open) cycle. (a) T-s chart, (b) schematic.

- 2–3 isobaric cooling of the compressed air with outside air in a heat exchanger;
- 3–4 the air is further cooled during an adiabatic reversible expansion in an expansion device (e.g. a turbine) described by:

$$T_4 = T_3 \left(\frac{P_4}{P_3}\right)(\gamma - 1)/\gamma \tag{6.60}$$

where γ is an *expansion* index.

Then deploying a reduced form of the flow version of the First law of Thermodynamics to each system component in turn:

$$\dot{W}_{comp} = \dot{m}Cp(T_2 - T_1)$$

$$\dot{W}_{exp} = \dot{m}Cp(T_3 - T_4) \tag{6.61}$$

$$\dot{Q}_R = \dot{m}Cp(T_1 - T_4) \tag{6.62}$$

Hence,

$$COP_{Ref} = \frac{\dot{Q}_{Ref}}{\dot{W}_{net}} \tag{6.63}$$

and

$$COP_{carnot} = \frac{T_c}{T_h - T_c} \tag{6.64}$$

6.8 Steam Jet Refrigeration Cycle

The steam jet refrigeration system relies on the pressure/velocity/evaporation temperature relationship in a fluid. The refrigerant and coolant are represented by a single fluid (water) existing under saturated, two-phase, or subcooled conditions. Like other previously described cycles, this system comprises an evaporator and condenser as the main heat transfer surfaces and an expansion valve acting as the decompressor between them. However, steam jet systems dispense with the compressor and instead have two fluid pumps, a steam generator (or boiler), a convergent-divergent nozzle and a venturi. (These last two components are sometimes termed the *ejector*.)

The purpose of the nozzle is to accelerate the steam to supersonic velocities hence reducing the local pressure. The evaporator is exposed to this low pressure. As a result, two processes now occur in the evaporator:

(a) Any evaporating or '*flashed*' fluid will be entrained into the ejector venturi. The mixed stream (steam + flashed vapour) is compressed by the velocity reduction in the venturi diffuser.

(b) Any liquid remaining in the evaporator will cool and can be pumped around the cooling circuit to satisfy the cooling load.

The evaporator therefore has four fluid streams, i.e. the input from the condenser, output to the ejector, and flow and return with the cooling load.

Assuming that no heat exchanger pressure loss the system therefore has three distinct pressures:

- A high pressure as generated by the steam boiler (8–1)
- An intermediate pressure created at the ejector outlet and experienced by the condenser (4–5)
- A low pressure downstream of the expansion valve and experienced by the evaporator (6–7)

It is worth noting that none of the thermodynamic processes described above take place isentropically. The nozzle, diffuser, and entrainment processes come with isentropic efficiency factors.

If the condenser has a condensate loss arrangement, then the expansion valve circuit is redundant and fresh make-up fluid will be necessary at the evaporator.

A schematic view of the steam jet refrigeration system and a *P–h* diagram is shown in Figure 6.16.

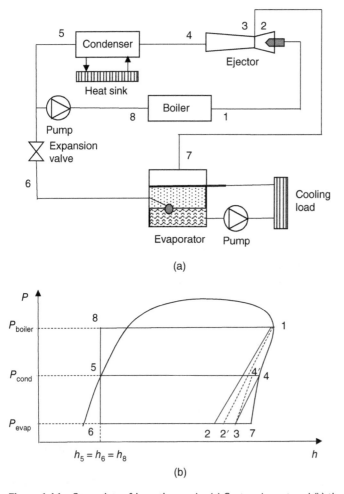

Figure 6.16 Steam jet refrigeration cycle. (a) System layout and (b) thermodynamic cycle.

An energy balance for the ejector yields:

$$\dot{m}_1 h_1 + \dot{m}_7 h_7 = (\dot{m}_1 + \dot{m}_7)h_4 \tag{6.65}$$

The cycle cooling capacity is given as:

$$\dot{Q}_R = \dot{m}_7(h_7 - h_6) \tag{6.66}$$

It can be shown that the work done in the cycle is represented by the enthalpy change between steam generator outlet and condenser outlet is given by:

$$\dot{W} = \dot{m}_1(h_1 - h_5) \tag{6.67}$$

The COP for system is given by:

$$COP_{Ref} = \frac{\dot{Q}_R}{\dot{W}} \tag{6.68}$$

6.9 Thermoelectric Refrigeration

This type of refrigeration is associated with nineteenth-century French Scientist Jean Charles Peltier who discovered that a direct electric current passing through a circuit formed by two dissimilar conductors or semiconductors, *A* and *B*, will cause a temperature difference to develop at the junctions of the two conductors. A refrigeration effect develops at the cold junction, and heat is rejected at the hot junction. See Figure 6.17. The heat produced or absorbed at each junction is proportional to the difference in the Seebeck coefficient of the two materials, the value of current and the temperature at the junction.

$$\dot{Q} = (\alpha_A - \alpha_B)IT \tag{6.69}$$

where
α is the Seebeck coefficient
I the electrical current supplied to the thermoelectric device
T is the absolute temperature of the junction.

A typical thermoelectric cooler (TEC) consists of an array made up of a multitude of p- and n-type semiconductor elements that act as the two dissimilar conductors. The array of elements is soldered between two ceramic plates, electrically in series and thermally in parallel.

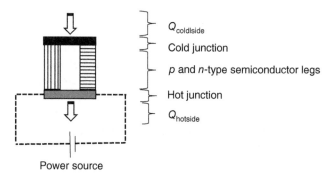

$Q_{coldlside}$

Cold junction

p and *n*-type semiconductor legs

Hot junction

$Q_{hotside}$

Power source

Figure 6.17 Thermoelectric refrigeration (single couple).

Applications for thermoelectric modules cover a wide spectrum of product areas such as simple food and beverage coolers, electronic cooling, microprocessor cooling, and similar low-capacity applications.

The temperature difference (ΔT) across the module (K or °C) is:

$$\Delta T = T_h - T_c \tag{6.70}$$

Heat pumped ($\dot{Q}_c$, Watts) by the module is:

$$\dot{Q}_c = NC[(Z(T_c^2/2)) - \Delta T] \tag{6.71}$$

where

N is the number of p–n couples

$$\text{Conduction coefficient } C = (k_p + k_n)A/L \tag{6.72}$$

$$\text{Electrical resistance } R_{\text{elec}} = (\rho_p + \rho_n)L/A \tag{6.73}$$

and

$$\text{Figure of merit } Z = (a_p - a_n)^2/R_{\text{elec}}C \tag{6.74}$$

The heat rejected by the module ($\dot{Q}_h$, W) is:

$$\dot{Q}_h = N[(\alpha_p - \alpha_n)T_h I - C\Delta T + I^2 R_{\text{elec}}/2 \tag{6.75}$$

where $I = (\alpha_p - \alpha_n)T_c/R_{\text{elec}}$

$$\dot{W}_{\text{in}} = \dot{Q}_h - \dot{Q}_c \tag{6.76}$$

The COP as a refrigerator is

$$\text{COP}_{\text{Ref}} = \frac{\dot{Q}_c}{\dot{W}_{\text{in}}} \tag{6.77}$$

6.10 Thermoacoustic Refrigeration

This device works on the principle of the changing pressure inherent to a sound wave. A sound wave is a succession of compressions and expansions. A compressed gas is heated and an expanded or rarefied gas is cooled. This effect is very small and not noticeable at normally encountered ambient conditions. At high sound pressures (~200 dB), the temperature gradient produced can be used for useful cooling.

The thermoacoustic refrigerator comprises a sound generator, a resonator tube, and a series of parallel passages or 'stack' bracketed by a hot side and a cold side heat exchangers. The unit is filled with an inert gas or mixture of gases, e.g. Argon, Helium, etc. See Figure 6.18.

The overall length of the unit is such that the sound generator sets up a standing (typically half or quarter) wave in the tube and repeated compression/expansion gas oscillations result in a temperature difference across the stack.

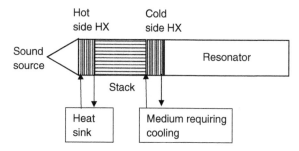

Figure 6.18 Thermoacoustic refrigeration components.

The gas essentially undergoes four processes as:

1–2 Compression and heating
2–3 Isobaric heat rejection to surroundings
3–4 Expansion and cooling
4–1 Isobaric heat absorption from surroundings

Ideally, the acoustic power necessary to set up and maintain the standing wave and temperature difference provides the work input to the system.

Any fluid to be cooled can be pumped around the cold side heat exchanger and the heat from the hot side heat exchanger transferred to a suitable heat sink.

Theoretical prediction of performance is complex requiring consideration of gas thermodynamic and transport properties (e.g. mean temperature, ratio of specific heat), wave characteristics (e.g. angular frequency, wave number, sound speed), and tube geometry (e.g. stack length, plates spacing, resonator length).

COP's of the order of unity are attainable.

6.11 Worked Examples

Note that calculations in Worked Examples 6.2–6.4 have been made with the data tables found in Appendix B. A large number of websites are available for readers wishing to practice their use of refrigerant charts.

Worked Example 6.1 Coefficient of Performance
A heat pump is used for heating purposes in winter and cooling in summer.
The COP of this heat pump *for refrigeration* is 3 and the cooling capacity is 9 kW.
Determine:

(a) The electrical power input (kW) of the heat pump if the electro-mechanical efficiency of the compressor motor is 75%.
(b) What winter heating load (kW) can the system handle?

Solution

Given: $COP_{Ref} = 3$, $\dot{Q}_c = 9\,kW$, $\eta_{motor} = 0.75$

Find: $\dot{W}_{actual}$

(a)
$$COP_{Ref} = \dot{Q}_c/\dot{W}_{ideal}$$
$$3 = 9/\dot{W}_{ideal}$$
$$W_{ideal} = 3\ kW$$

$$\dot{W}_{actual} = \dot{W}_{ideal}/\eta_{motor}$$
$$= 3/0.75$$
$$= 4\ kW$$

(b)
$$COP_{HP} = COP_{Ref} + 1$$
$$= 3 + 1 = 4$$

Then

$$\dot{Q}_h = \dot{W}_{actual} \times COP_{HP}$$
$$= 4 \times 4$$
$$= 16\ kW$$

Worked Example 6.2   Basic Vapour Compression Cycle

A refrigerator uses R134a as a working fluid. The refrigerant enters the compressor as saturated vapour at 2 bar and on leaving the condenser it is saturated liquid at 8 bar. Both pressures are absolute (Figure E6.2).

Assume that the compression process is 100% isentropic, there is no superheating or sub-cooling effects and that pressure losses are negligible.

Calculate, *using the tables in Appendix B*:

(a) the work of compression (kJ/kg)
(b) the refrigerating effect (kJ/kg)
(c) the actual COP
(d) the Carnot COP

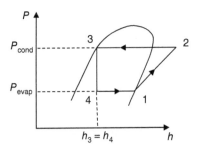

Figure E6.2   P-h chart for Worked Example 6.2.

Solution

Given: $s_1 = s_2$ (100% isentropic compression)

$$P_1 = P_4 = 2 \text{ bar}, \quad P_2 = P_3 = 8 \text{ bar (no pressure losses)}$$

Find: w_{comp}, q_R, COP_{Ref}, COP_{Carnot}
From the Table B.1 in Appendix B:

At 2 bar, saturation temperature, $T_c = -10.09\,°C = 262.91$ K
Saturation vapour conditions: $h_1 = 241.3$ kJ/kg, $s_1 = 0.9253$ kJ/kg K $= s_2$
At 8 bar, saturation temperature, $T_h = 31.33\,°C = 304.33$ K
Saturation liquid conditions: $h_3 = 93.42$ kJ/kg $= h_4$

Taking s_2 to be 0.9253 kJ/kg it can be seen, by inspection, that T_2, must have a value between the saturated vapour value and 40 °C
Using interpolation on the entropy table, T_2 is found to be 36.594 °C
Using interpolation the enthalpy at T_2, h_2 is found to be 269.92 kJ/kg

(a) *Working of compression*: $w_{comp} = h_2 - h_1 = 269.92 - 241.3 = 28.62$ kJ/kg
(b) *Refrigeration effect*: $q_R = h_1 - h_4 = 241.3 - 93.42 = 147.88$ kJ/kg
(c) *Coefficient of performance*: $COP_{Ref} = q_R/w_{comp} = 5.17$

(d) $COP_{Carnot} = \dfrac{T_c}{T_h - T_c} = \dfrac{262.91}{304.33 - 262.91} = 6.347$

The COP_{Cannot} is higher than COP_{Ref} as expected.

Worked Example 6.3   Vapour Compression with Isentropic Compression Efficiency
If the refrigeration plant in Worked Example 6.2 has an isentropic compression efficiency of 80%, determine its effect on the value of compression work (kJ/kg) and COP (Figure E6.3).

Solution

Given: $\eta_{ic} = 0.8$, $P_1 = P_4$, $P_2 = P_3$ (no pressure losses)
   Find: w_{comp}, COP_{Ref}

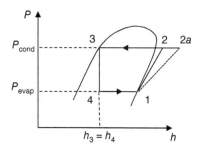

Figure E6.3   P-h chart for Worked Example 6.3.

As Worked Example 6.2, the approximate enthalpies are:

$$h_1 = 241.3 \text{ kJ/kg}, \quad h_2 = 269.92 \text{ kJ/kg(isentropic)}, \quad h_3 = h_4 = 93.42 \text{ kJ/kg}$$

The actual enthalpy the compressor outlet (h_{2a}) is given by:

$$\eta_{\text{isen}} = \frac{h_2 - h_1}{h_{2a} - h_1}$$

$$0.8 = \frac{269.92 - 241.3}{h_{2a} - 241.3}$$

$$h_{2a} = 277.08 \text{ kJ/kg}$$

(a) *Work of compression*: $w_{\text{comp}} = h_{2a} - h_1 = 277.08 - 241.3 = 35.78 \text{ kJ/kg}$
(b) *Refrigeration effect*: $q_R = h_1 - h_4 = 241.3 - 93.42 = 147.88 \text{ kJ/kg}$
(c) *Coefficient of performance*: $\text{COP}_{\text{Ref}} = q_R/w_{\text{comp}} = 4.13$

The reduction in COP_{Ref} with respect to Worked Example 6.2 is due to the compressor irreversibility.

Worked Example 6.4   Vapour Compression with Heat Exchanger
A refrigerator using R134a as a working fluid is operating between pressures of 2 bar and 8 bar (Abs) and utilises a heat exchanger. As a result of 10 °C of superheating, the refrigerant enters the compressor as superheated vapour (Figure E6.4).

At the condenser exit, the refrigerant is subcooled by 10 °C. Assume that the compression process is 100% isentropic and pressure losses are negligible.

Take the specific heat capacity for the liquid refrigerant to be 1.4 kJ/kg K.

Using the Table B.1 in Appendix B, calculate:

(a) the work of compression (kJ/kg)
(b) the refrigerating effect (kJ/kg)
(c) the COP for heating and cooling
(d) the Carnot COP

Solution
Given: $s_1 = s_2$ (100 % isentropic compression)

$$P_1 = P_4 = 2 \text{ bar}, \quad P_2 = P_3 = 8 \text{ bar (no pressure losses)}$$

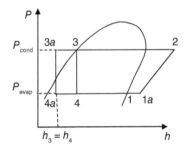

Figure E6.4   P-h chart for Worked Example 6.4.

Find: w_{comp}, q_R, COP_{Ref}, COP_{HP}, COP_{Carnot}

At 2 bar, saturation temperature $T_c = -10.09\,°C = 262.9\,K$

With 10 °C of super heat, $T_1 = -10.09 + 10 = -0.09\,°C \cong 0\,°C$

$$h_1 = 241.3\ kJ/kg,\quad h_{1a} = 250.1\ kJ/kg,\quad s_{1a} = 0.9582\ kJ/kg\ K = s_2$$

At 8 bar, **saturated liquid** conditions, $T_h = 31.334\,°C = 304.334\,K$

Taking s_2 to be 0.9582 kJ/kg it can be seen, by inspection, that T_2 must have a value between 40 and 50 °C

Using interpolation on the entropy table, T_2 is found to be 46.172 °C

Using interpolation on the enthalpy table at T_2, h_2 is found to be 280.08 kJ/kg

$$h_3 = 93.42\ kJ/kg = h_4$$

With 10 °C of subcooling, the actual condenser outlet conditions are less than the saturated liquid value:

$$h_{3a} = h_3 - C_p\Delta T$$
$$= 93.42 - (1.4 \times 10) = 79.42\ kJ/kg = h_{4a}$$

(a) *Work of compression*: $w_{comp} = h_2 - h_{1a} = 280.08 - 250.1 = 30\ kJ/kg$
(b) *Refrigeration effect*: $q_R = h_{1a} - h_{4a} = 250.1 - 79.42 = 170.68\ KJ/kg$
(c) *Coefficient of performance*: $COP_{Ref} = q_R/w_{comp} = 5.69$
   Heating effect: $q_H = h_2 - h_{3a} = 280.08 - 79.42 = 200.66\ kJ/kg$
   Coefficient of performance: $COP_{Ref} = q_H/w_{comp} = 6.69$

(d) $COP_{Carnot\text{-}ref} = \dfrac{T_c}{T_h - T_c} = \dfrac{262.9}{304.33 - 262.9} = 6.346$

$$COP_{Carnot\text{-}heating} = \dfrac{T_h}{T_h - T_c} = \dfrac{304.33}{304.33 - 262.9} = 7.346$$

Note that the COP_{Cannot} is unaffected by superheating or subcooling.

Worked Example 6.5 Ammonia Vapour Compression

A vapour compression refrigerator circulates 0.1 kg/s of ammonia between two set points; condensation takes place at 12 bar and evaporation at 2 bar (Figure E6.5).

The compression is 100% isentropic and neither superheating nor subcooling is present. *Using the Table B.2 in Appendix B, determine:*

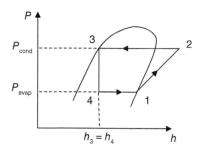

Figure E6.5 P-h chart for Worked Example 6.5.

(a) the COP_{Ref}
(b) the ice (kg/h @ 0 °C) produced from water at 18 °C

Data:
Specific heat capacity of liquid water: 4.187 kJ/kg K
Specific enthalpy of fusion of ice: 336 kJ/kg

Solution
Given: $s_1 = s_2$ (100 % isoentropic compression)

$$P_1 = P_4 = 2 \text{ bar}, \quad P_2 = P_3 = 12 \text{ bar}$$

$$\dot{m}_{NH3} = 0.1 \text{ kg/s, no superheating/subcooling}$$

$$T_{water} = 18°C, \quad T_{ice} = 0°C, \quad C_{p,water} = 4.187 \text{ kJ/kg K}, \quad h_{pc,ice} = 336 \text{ kJ/kg}$$

Find: $COP_{Ref}, \dot{m}_{ice}$
From the Table B.2 in Appendix B:

At 2 bar, saturation temperature, $T_c = -18.86°C$
Saturation vapour conditions: $h_1 = 1419.31 \text{ kJ/kg}, s_1 = 5.5969 \text{ kJ/kg K} = s_2$
At 12 bar, saturation temperature, $T_h = 30.94°C$
Saturation liquid conditions: $h_3 = 327.01 \text{ kJ/kg} = h_4$

Taking s_2 to be 5.5969 kJ/kg it can be seen, by inspection, that T_2 must have a value between 100 and 120 °C
Using interpolation on the entropy table, T_2 is found to be 114.047 °C
Using interpolation on the enthalpy table at T_2, h_2 is found to be 1692.83 kJ/kg

(a) *Work of compression*: $w_{comp} = h_2 - h_1 = 1692.83 - 1419.31 = 273.52 \text{ kJ/kg}$
   Refrigeration effect: $q_R = h_1 - h_4 = 1419.31 - 327.01 = 1092.3 \text{ kJ/kg}$
   Coefficient of performance: $COP_{Ref} = q_R/w_{comp} = 3.99$
(b) $\dot{Q}_{NH3} = \dot{Q}_{ice}$

$$\dot{m}_{NH3}q_R = \dot{m}_{ice}(C_{p,water}(T_{water} - T_{ice})) + \dot{m}_{ice}h_{pc,ice}$$

$$0.1 \times 1092.3 = \dot{m}_{ice}(4.187(18 - 0)) + \dot{m}_{ice} \times 336$$

$$\dot{m}_{ice} = 0.266 \text{ kg/s} = 958 \text{ kg/h}$$

Worked Example 6.6 Absorption Cycle
A lithium bromide water absorption refrigeration system is fitted with a preheating heat exchanger (Figure E6.6a,b, Table E6.6). The following conditions are known:

The absorber temperature is 40 °C
The condenser temperature is 50 °C
The generator temperature is 100 °C
The evaporator temperature is 5 °C.
The refrigeration capacity is 3 kW.
The pump input power is 0.5 kW.

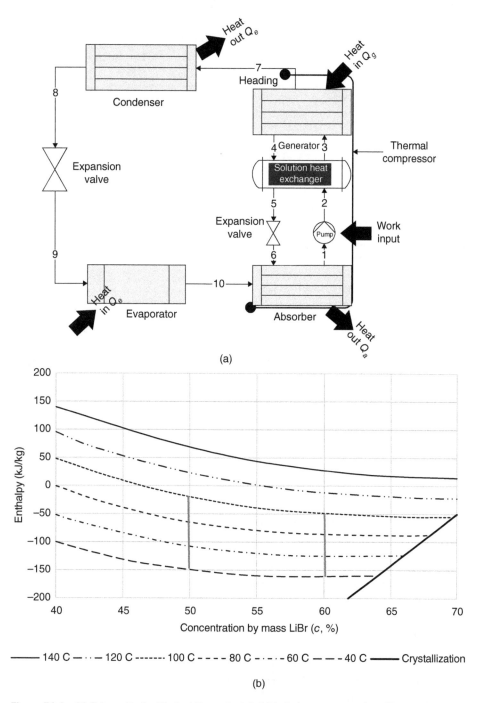

Figure E6.6 (a) Schematic for Worked Example 6.6, (b)Enthalpy concentration diagram for Worked Example 6.6.

Table E6.6 Enthalpy of water at saturation conditions for Worked Example 6.6.

	Enthalpy (h, kJ/kg)	
Temperature (°C)	Saturated liquid	Saturated vapour
5	21.0	2510.1
40	167.5	2573.5
50	209.3	2382.0
100	417.4	2675.5

The solution concentrations at the exit of absorber and generator are 50% and 60%, respectively. Assume 100% effectiveness for the solution heat exchanger, exit condition of refrigerant at evaporator and condenser to be saturated.
Using the data supplied determine:

(a) The mass flow rate in each path (kg/s)
(b) The heat transfer rates in the generator, absorber, and condenser
(c) The COP_{Ref}

Data:
Approximate enthalpy (h, kJ/kg) – LiBr concentration (c, % by mass) absorption relationships:
At 40 °C: $h = 0.1755c^2 - 20.702c + 446.31$
At 100 °C: $h = 0.1375c^2 - 18.495c + 566.46$

Solution
Given: $T_A = 40°C,\ T_C = 50°C,\ T_G = 100°C,\ T_E = 5°C,\ \dot{Q}_E = 3.0\ \text{kW},\ \dot{W}_p = 0.5\ \text{kW},$ $c_1 = 50\%,\ c_4 = 60\%$

Find: $\dot{Q}_G,\ \dot{Q}_A,\ \dot{Q}_C,\ COP_{Ref}$
Enthalpy values for the cycle:
Note that $c_1 = c_2 = c_3$ and $c_4 = c_5 = c_6$
At $T = 40°C, c_1 = 50\%$

$$h_1 = 0.1755c^2 - 20.702c + 446.31$$
$$= (0.1755 \times 50^2) - (20.702 \times 50) + 446.31$$
$$= -150.04\ \text{kJ/kg}$$

At $T = 40°C, c_5 = 60\%$

$$h_5 = 0.1755c^2 - 20.702c + 446.31$$
$$= (0.1755 \times 60^2) - (20.702 \times 60) + 446.31$$
$$= -164.01\ \text{kJ/kg}$$
$$= h_6$$

At $T = 100\,^\circ$ C, $c_3 = 50\%$

$$h_3 = 0.1375c^2 - 18.495c + 566.46$$
$$= (0.1375 \times 50^2) - (18.495 \times 50) + 566.46$$
$$= -14.54 \text{ kJ/kg}$$

At $T = 100\,^\circ$ C, $c_4 = 60\%$

$$h_4 = 0.1375c^2 - 18.495c + 566.46$$
$$= (0.1375 \times 60^2) - (18.495 \times 60) + 566.46 = -48.24 \text{ kJ/kg}$$

$h_7 = 2675.7$ kJ/kg (saturated vapour at 100°C from steam tables)

$h_8 = h_9 = 209.3$ kJ/kg (saturated liquid at 50°C from steam tables)

$h_{10} = 2510.1$ kJ/kg (saturated vapour at 5°C from steam tables)

(a) *Mass flow rates:*

$$\dot{Q}_E = \dot{m}_{10}(h_{10} - h_9)$$
$$3.0 = \dot{m}_{10}(2510.1 - 209.3)$$
$$\dot{m}_{10} = 0.0013 \text{ kg/s}$$

From mass conservation considerations at the absorber:

$$\dot{m}_{10} + \dot{m}_6 = \dot{m}_1$$
$$\dot{m}_{10}c_{10} + \dot{m}_6 c_6 = \dot{m}_1 c_1$$

Nothing that $c_{10} = 0$, solving the two previous equations simultaneously and rearranging, it can be shown that:

$$\dot{m}_1 = \dot{m}_{10}c_6/(c_6 - c_1)$$
$$= 0.0078 \text{ kg/s} = \dot{m}_2 = \dot{m}_3$$
$$\dot{m}_6 = \dot{m}_{10}c_1/(c_6 - c_1)$$
$$= 0.0065 \text{ kg/s} = \dot{m}_4 = \dot{m}_5$$

and

$$\dot{m}_7 = \dot{m}_8 = \dot{m}_9 = \dot{m}_{10}$$

(b) *Energy balances:*

At the generator

$$\dot{Q}_G = \dot{m}_7 h_7 + \dot{m}_4 h_4 - \dot{m}_3 h_3$$
$$= (0.0013 \times 2675.7) + (0.0065 \times (-48.24)) - (0.0078 \times (-14.54)) = 3.288 \text{ kW}$$

At the absorber

$$\dot{Q}_A = \dot{m}_{10}h_{10} + \dot{m}_6 h_6 - \dot{m}_1 h_1$$
$$= (0.0013 \times 2510.1) + (0.0065 \times (-164.01)) - (0.0078 \times (-150.04))$$
$$= 3.168 \text{ kW}$$

At the condenser

$$\dot{Q}_C = \dot{m}_{10}(h_7 - h_8)$$
$$= 0.0013(2675.7 - 209.3) = 3.216 \text{ kW}$$

(c) *Coefficient of performance*:

$$\text{COP}_{\text{Ref}} = \dot{Q}_E/(\dot{Q}_G + \dot{W}_{\text{pump}})$$
$$= 3.0/(3.288 + 0.5) = 0.79$$

Worked Example 6.7 Stirling Cycle Cooler
Consider the idealised model of the Stirling cooler operating between $T_C = 310$ K and $T_E = 260$ K. The maximum volume is 40 cm^3, and the minimum volume is 30 cm^3, and the maximum pressure is 1.5 MPa. The working fluid is helium which is an ideal gas, for which $R = 2.077$ kJ/kg K.
  If the speed of the machine is 200 rpm, determine:

(a) the heat absorbed (W) in the expansion space during the expansion process
(b) the rate of net work done (W).
(c) the coefficient of performance of the refrigerator.

Solution
Given: $V_1 = V_4 = 40 \text{ cm}^3$, $V_2 = V_3 = 30 \text{ cm}^3$, $P = 1.5 \times 10^6$ Pa

$$T_c = 310 \text{ K}, \quad T_E = 260 \text{ K}, \quad R = 2077 \text{ J/kg K}, \quad N = 200/60 \text{ rps}$$

Find: $\dot{Q}_E$, $\dot{W}_{\text{net}}$, COP_{Ref}

$$\dot{m} = N \times \frac{PV}{RT} = \left(\frac{200}{60}\right) \times \frac{1.5 \times 10^6 \times 30 \times 10^{-6}}{2077 \times 310} = 0.000\ 23 \text{ kg/s}$$

$$\dot{Q}_E = \dot{m}RT_e \ln\left(\frac{V_4}{V_3}\right) = 0.000\ 23 \times 2077 \times 260 \times \ln\left(\frac{40}{30}\right) = 36.2 \text{ W}$$

$$\dot{W}_{\text{net}} = \dot{m}R(T_c - T_e)\ln\left(\frac{V_1}{V_2}\right) = 0.000\ 23 \times 2077 \times (310 - 260) \times \ln\left(\frac{40}{30}\right) = 7 \text{ W}$$

$$\text{COP}_{\text{Ref}} = \frac{\dot{Q}_E}{\dot{W}_{\text{net}}} = \frac{36.2}{7} = 5.2$$

In reality, not all the heat available is utilised so the COP would be lower.

Worked Example 6.8 Brayton Refrigeration
Air enters the compressor of an ideal Brayton refrigeration cycle at a pressure of 100 kPa and a temperature of 270 K. The air volumetric flow rate is 1 m^3/s and is compressed to 400 kPa. The air temperature at the turbine inlet is 320 K.
  Determine:

(a) The net power input (kW)
(b) The refrigerating capacity (kW).
(c) The system COP
(d) The COP for its corresponding Carnot system.

  Data: for air assume $C_p = 1005$ J/kg K and $R = 287$ J/kg K, $\gamma = 1.4$

Solution

Given: $P_1 = P_4 = 100 \times 10^3$ Pa, $P_2 = P_3 = 400 \times 10^3$ Pa, $\dot{V} = 1 \, m^3/s$

$T_1 = 270$ K, $T_3 = 320$ K, $C_p = 1005 \, kJ/kg$, $R = 287 \, J/kg \, K$, $\gamma = 1.4$

Find: $\dot{W}_{net}$, $\dot{Q}_R$, COP, COP_{Carnot}

$$T_2 = T_1\left(\frac{P_2}{P_1}\right)^{(\gamma-1)\gamma} = 270\left(\frac{400}{100}\right)^{(1.4-1)/1.4} = 401.2 \text{ K}$$

$$T_4 = T_3\left(\frac{P_4}{P_3}\right)^{(\gamma-1)/\gamma} = 320\left(\frac{100}{400}\right)^{(1.4-1)/1.4} = 215.3 \text{ K}$$

$$\dot{m} = \frac{P\dot{V}}{RT} = \frac{100 \times 10^3 \times 1}{287 \times 270} = 1.290 \text{ kg/s}$$

$$\dot{W}_{comp} = \dot{m}Cp(T_2 - T_1) = 1.290 \times 1.005 \times (401.2 - 270) = 170.18 \text{ kW}$$

$$\dot{W}_{exp} = \dot{m}Cp(T_3 - T_4) = 1.290 \times 1.005 \times (320 - 215.3) = 135.73 \text{ kW}$$

(a) $\dot{W}_{net} = \dot{W}_{comp} - \dot{W}_{exp} = 169.3 - 135 = 34.45 \text{ kW}$

(b) $\dot{Q}_R = \dot{m}Cp(T_1 - T_4) = 1.290 \times 1.005 \times (270 - 215.3) = 70.89 \text{ kW}$

(c) $COP_{actual} = \dfrac{\dot{Q}_R}{\dot{W}_{net}} = \dfrac{70.89}{34.45} = 2.1$

(d) $COP_{Carnot} = \dfrac{T_c}{T_h - T_c} = \dfrac{270}{320 - 270} = 5.4$

Worked Example 6.9 Jet Refrigeration

A steam jet refrigeration plant produces chilled water. A supply of saturated steam at 10 bar is available and the condenser operates at 1 bar. The evaporator pressure is 0.1 bar, with inlet and outlet states as 0.1 and 0.9 dryness fraction, respectively. The mass flow rate of steam leaving the evaporator is 0.1 kg/s and that going through the boiler is 0.2 kg/s. Use the steam data given in Table E6.9 to calculate:

(a) The refrigeration capacity (kW)
(b) The COP for refrigeration

Solution

Given: $P_1 = P_8 = 10$ bar, $P_4 = P_5 = 1$ bar, $P_6 = P_7 = 0.1$ bar

$x_6 = 0.1$, $x_7 = 0.9$, $\dot{m}_7 = 0.1$ kg/s, $\dot{m}_8 = 0.2$ kg/s

Table E6.9 Data for jet refrigeration Worked Example 6.9.

Steam pressure (bar)	Saturation temperature (°C)	Saturated liquid enthalpy (h_f, kJ/kg)	Saturated vapour enthalpy (h_g, kJ/kg)
0.1	45.83	191.8	2583.9
1	99.63	417.4	2675.5
10	179.88	762.8	2777.1

The enthalpy (kJ/kg) of a two-phase mixture can be found from: h = sat liq enthalpy + dryness fraction (sat vapour enthalpy − sat liq enthalpy)

Find: Q_R, COP_{ref}

At 0.1 bar, $h_f = 191.8$ kJ/kg, and $h_g = 2583.9$ kJ/kg

$$h_6 = 191.8 + 0.1(2583.9 - 191.8) = 431.1 \ \text{kJ/kg}$$

$$h_7 = 191.8 + 0.9(2583.9 - 191.8) = 2344.7 \ \text{kJ/kg}$$

At 1 bar, $h_f = 417.5$ kJ/kg and $h_5 = 417.5$ kJ/kg
At 10 bar (saturated vapour state), $h_1 = 2777.1$ kJ/kg

(a) *Refrigeration capacity*:

$$\dot{Q}_R = \dot{m}_7(h_7 - h_6)$$
$$= 0.1(2344.7 - 431.1) = 191.4 \ \text{kW}$$

(b) COP_{ref}

$$\dot{W} = \dot{m}_1(h_1 - h_5)$$
$$= 0.2(2777.1 - 417.5) = 471.9 \ \text{kW}$$

$$COP_{Ref} = \frac{\dot{Q}_R}{\dot{W}} = \frac{191.4}{471.9} = 0.405$$

Worked Example 6.10 Peltier Cooler

A thermoelectric cooling system has the specification given in Table E6.10:

Couple dimensions: cylindrical with a length of 0.125 cm and a diameter of 0.1 cm.
Number of couples: 60
Junction temperatures: cold 19 °C, hot 24 °C

  Determine:

(a) The cooling capacity (*W*) of the module
(b) The COP for cooling and heating

Solution

Given: $T_h = 24\,°C$, $T_c = 19\,°C$, $L = 0.001\,25$ m, $D = 0.001$ m, $N = 60$
  Properties as shown in Table E6.10.

Table E6.10 Data for thermoelectric cooling example 6.10.

Material	Material properties		
	Seebeck coefficient (V/K)	Resistivity (Ω m)	Thermal conductivity (W/m K)
p − type	160×10^{-6}	1×10^{-5}	2
n − type	-200×10^{-6}	7×10^{-6}	2

Find: $\dot{Q}_c$, COP_{Ref}, COP_{HP}

$$A = (\pi/4)D^2 = (\pi/4)0.001^2 = 7.854 \times 10^{-7} \text{ m}^2$$

$$\Delta T = T_h - T_c = 24 - 19 = 5°\text{C}$$

Total electrical resistance:

$$R_{\text{elec}} = (\rho_p + \rho_n)L/A$$
$$= (1 \times 10^{-5} + 7 \times 10^{-6})0.00125/7.854 \times 10^{-7} = 0.0271 \ \Omega$$

Conduction coefficient:

$$C = (k_p + k_n)A/L$$
$$= (2 + 2)7.854 \times 10^{-7}/0.00125 = 2.513 \times 10^{-3} \text{ W/K}$$

Figure of merit:

$$Z = (a_p - a_n)^2/R_{\text{elec}}C$$
$$= (360 \times 10^{-6})^2/0.0271 \times 2.513 \times 10^{-3} = 1.91 \times 10^{-3} \text{ K}^{-1}$$

(a) *Cooling capacity*:

$$\dot{Q}_c = NC[(Z(T_c^2/2)) - \Delta T]$$
$$= 60 \times 2.513 \times 10^{-3}[(1.91 \times 10^{-3} \times (292^2/2)) - 5] = 11.88 \text{ W}$$

$$I = (\alpha_p - \alpha_n)T_c/R_{\text{elec}}$$
$$= (360 \times 10^{-6})292/0.0271 = 3.885 \text{ A}$$

$$\dot{Q}_h = N[(\alpha_p - \alpha_n)T_hI - C\Delta T + I^2R_{\text{elec}}/2]$$
$$= 60[(360 \times 10^{-6}) \times 297(3.91) - 2.513 \times 10^{-3}(5) + 3.885^2 \times 0.0271/2]$$
$$= 37.64 \text{ W}$$

Hence, $\dot{W}_{in} = \dot{Q}_h - \dot{Q}_c = 37.64 - 11.88 = 25.76 \text{ W}$

(b) *Coefficient of performance*:

$$\text{COP}_{\text{Ref}} = \dot{Q}_c/\dot{W}_{in} = 11.88/25.76 = 0.461$$
$$\text{COP}_{\text{HP}} = \dot{Q}_h/\dot{W}_{in} = 37.64/25.76 = 1.461$$

6.12 Tutorial Problems

6.1 A heat pump is used to provide heating for a house which has its indoor temperature controlled at 20 °C during the winter season. If the outside temperature in winter is 5 °C, and heat load through the building envelope is estimated to be 0.5 kW/K:
   (a) Determine the minimum power required to drive the heat pump in the heating season making a Carnot efficiency assumption.

(b) If the power output is the same as in part (a), determine the maximum out-door temperature for which the indoor air can be maintained at 20 °C during the summer season?

[Answers: 0.384 kW, 34.4 °C]

6.2 A refrigerator using R134a has an evaporator pressure of 2 bar and a condenser pressure of 10 bar. The refrigerant enters the compressor as a saturated vapour and undergoes an isentropic compression. Assume that no undercooling of the refrigerant takes place upon leaving the condenser and that pressure losses are negligible.

Use tables of R134a to determine:

(a) the work of the compression (kJ/kg)
(b) the refrigerating effect (kJ/kg)
(c) the COP

[Answers: 33.3 kJ/kg, 136.0 kJ/kg, 4.1]

6.3 A salesman has proposed to supply you with a system similar to that in Question 6.2 but with an added heat exchanger to be fitted between the condenser outlet and the evaporator outlet, resulting in the refrigerant being superheated by 10° before entering the compressor and accompanied by a subcooling of 10° at the condenser outlet. He claims that this will increase the COP by 20%. Do you agree or disagree? Prove it by calculating the new COP. Take $C_{p,ref}$ as 1.05 kJ/kgK.

[Answers: New COPr = 4.4 only added 8% so claims unaccepted]

6.4 A refrigerator using R134a has an evaporator pressure of 2 bar and a condenser pressure of 10 bar. The refrigerant at entry to the compressor is just dry saturated vapour, and upon leaving it has a temperature of 60 °C. Assume that no undercooling of the refrigerant takes place upon leaving the condenser and that pressure losses are negligible.

Use the tabulated data of R134a to determine:

(a) the isentropic efficiency of the compression (%)
(b) the actual work of compression (kJ/kg)
(c) the refrigerating effect (kJ/kg)
(d) the COP

[Answers: 66.6%, 50 kJ/kg, 136 kJ/kg, 2.7]

6.5 A vapour-compression refrigeration device uses ammonia as the working fluid. The saturated liquid entering the throttling valve has a temperature of 30.94 °C, and the temperature of the saturated vapour entering the compressor is −1.9 °C. The compressor has an isentropic efficiency of 90%, and the refrigeration capacity is 10 kW.

Calculate:

(a) the COP of this cycle
(b) the mass flow of ammonia (kg/h) through the device

[Answers: 6.7, 32.1 kg/h]

6.6 A Lithium bromide–water absorption refrigeration system is fitted with a preheating heat exchanger. The following conditions are known:

The absorber temperature is 40 °C

The condenser temperature is 50 °C

The generator temperature is 80 °C

The evaporator temperature is 5 °C.

The refrigeration capacity is 3 kW.

The pump input power is 0.5 kW.

The solution concentrations at the exit of absorber and generator are 0.5 and 0.6, respectively. Assume that 100% effectiveness for the solution heat exchanger, and the exit condition of refrigerant at evaporator and condenser to be saturated. Using the data supplied in Table P6.6, determine:

(a) The mass flow rate in each path (kg/s)

(b) The heat transfer rates in the generator, absorber, and condenser

(c) The COP_{ref}

Data:

Approximate enthalpy (h, kJ/kg) – LiBr concentration (c, % by mass) absorption relationships:

At 40 °C: $h = 0.1755c^2 - 20.702c + 446.31$

At 80 °C: $h = 0.1605c^2 - 20.375c + 557.67$

Table P6.6 Data for tutorial Problem 6.6.

Temperature (°C)	Enthalpy (h, kJ/kg)	
	Saturated liquid	Saturated vapour
5	21.0	2510.1
40	167.5	2573.5
80	335	2643
100	417.4	2675.5

[Answers: 0.0013 kg/s, 0.0065 kg/s, 0.0078 kg/s, $Q_G = 3.346$ kW, $Q_A = 3.168$ kW, $Q_C = 3.173$ kW, $COP_R = 0.78$]

6.7 Consider the idealised model of the Stirling cooler operating between $T_C = 330$ K and $T_E = 273$ K. The maximum volume is 40 cm³, the minimum volume is 30 cm³, and the maximum pressure is 1.5 MPa. The working fluid is air which is an ideal gas for which $R = 287$ J/kg K and $C_p = 1005$ J/kg K.

If the speed of the machine is 200 rpm, determine:

(a) The heat absorbed in the expansion space during the expansion process (W)

(b) The net work done (W).

(c) The COP of the refrigerator.

[Answers: 35.7 W, 7.45 W, 4.790]

6.8 Air enters the compressor of an ideal Brayton refrigeration cycle at 100 kPa, $T_1 = 300$ K, with a volumetric flow rate of 1 m³/s.

It is then compressed to 500 kPa. The temperature at the turbine inlet is 320 K.

Assume for air: $R = 287$ J/kg K and $C_p = 1005$ J/kgK.

Determine:

(a) The net power input (kW)

(b) The refrigerating capacity (kW).

(c) The COP of this system

[Answers: 66.75 kW, 114.34 kW, 1.71]

6.9 A steam jet refrigeration plant produces chilled water. A supply of saturated steam at 10 bar is available with the condenser operating at 1 bar. The evaporator pressure is 0.1 bar, with inlet and outlet states having 0.1 and 0.9 dryness fractions, respectively. The mass flow rate of steam entering the evaporator is 0.15 kg/s and that going through the boiler as 0.15 kg/s.

Table P6.9 Data for tutorial Problem 6.9.

Steam pressure (bar)	Saturation temperature (°C)	Saturated liquid enthalpy (h_f, kJ/kg)	Saturated vapour enthalpy (h_g, kJ/kg)
0.1	45.83	191.8	2584.7
1	99.63	417.4	2675.5
10	179.88	762.8	2778.1

Use the steam data in Table P6.9 to calculate:

(a) The refrigeration capacity (kW)

(b) The COP

[Answers: 287 kW, 0.811]

6.10 A thermoelectric cooling module consists of 50 couples or elements. Each thermoelectric element is cylindrical with a length of 0.125 cm and a diameter of 0.1 cm. The thermoelectric properties of the elements are given in Table P6.10:

Table P6.10 Data for tutorial Problem 6.10.

	Material properties		
Material	Seebeck coefficient (V/K)	Resistivity (Ω m)	Thermal conductivity (W/m K)
p – type	160×10^{-6}	1×10^{-5}	2
n – type	-200×10^{-6}	7×10^{-6}	2

The cold and hot junctions are at 20 and 25 °C, respectively. The electrical resistance of the peripheral connections contributes a further 10% to the total resistance of each element.

Determine:

(a) The cooling capacity of the module (W)

(b) The COP_{Ref}

[Answers: 8.7 W, 0.46]

7

Acoustic Factors

7.1   Overview

Sound is an elastic, molecular motion on a relatively large scale. In the presence of an intervening medium like air or water, the vibrational energy in sound is capable of being transmitted from one point to another. When the resulting sensation is transmitted to an auditory sensing mechanism, like an ear, the energy is converted to electrical impulses that are interpreted by the brain as, for example, speech, music, or just noise.

Consideration of prevailing sound level in the internal environment must be given as much regard as any operational parameter if the space is to be habitable. Without careful design, the provision of building services for other comfort parameters can be counter to acceptable sound fields.

Learning Outcomes

- To understand the nature of sound waves and attendant properties
- To quantify sound levels using a range of parameters
- To be able to evaluate sound combination, propagation, and transmission
- To suggest noise reduction measures

7.2   The Human Ear

Descriptions of the human ear usually divide it into three parts as shown in Figure 7.1.

The outer ear comprises a flap of cartilage called the *pinna*. In evolutionary terms, this structure would have been a sound directional locator and concentrator. Although still evident, these functions are however much reduced in humans. The pinna connects to the *auditory meatus* or ear canal which is a sound conduit leading to a flexible membrane called the *tympanum* or eardrum which vibrates in sympathy with incoming sound. In acoustic terms, the ear canal behaves like a pipe or tube open at one end enhancing the pressure of the sound at the tympanum.

Building Services Engineering: Smart and Sustainable Design for Health and Wellbeing, First Edition.
Tarik Al-Shemmeri and Neil Packer.
© 2021 John Wiley & Sons Ltd. Published 2021 by John Wiley & Sons Ltd.

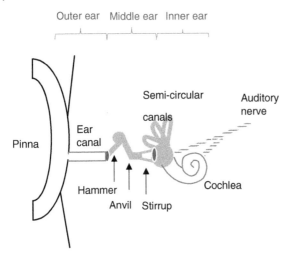

Outer ear   Middle ear   Inner ear

Figure 7.1   Schematic of the human ear.

The middle ear comprises three small connected bones or *ossicles* named in order of connection the malleus (or hammer), incus (or anvil), and stapes (or stirrup).

The malleus is connected to the inside of the tympanum and transmits any vibration sensed there onto its companion ossicles.

The middle ear interfaces with the inner ear via the stapes bone at a point known as the *fenestra ovalis* (or oval window). Due to the difference in surface contact area between tympanum ($\sim$40 mm^2 area) and fenestra ovalis ($\sim$3 mm^2 area) and lever action, the net result of this bony bridge is to reduce the amplitude of any vibration but to amplify the applied force.

The malleus and stapes bones have localised muscle and ligament connections which can, depending upon their state of contraction, limit ossicle movement and enhance or reduce sound transmission through the middle ear. When triggered by excessively loud sounds, this provides a protection mechanism against physical damage.

The inner ear comprises a structure containing an array of liquid-filled semi-circular canals associated with balance and the *cochlea* (or snail shell). The cochlea is a rigid-walled, liquid-filled helix divided along its length by membranes. One of these dividers, called the *basilar membrane*, has a hair-like fibre lining, extending from its walls towards the centre of the helix.

Vibrations from the fenestra ovalis are felt in the cochlea liquid by the membrane and sensed by a nerve cell structure known as the Organ of Corti, at their base. The nerve cells convert the fluid vibrations to nerve impulses that are sent to the brain via the *auditory nerve*.

The pressure set up in the cochlea fluid is dissipated back into the middle ear by the bulging of a circular membrane just below the fenestra ovalis called the *fenestra rotunda*.

The human ear is a very sensitive organ able to operate over a large range of sound level and frequency. The limits of its range are known as the **threshold of audibility** or hearing (at the lower end) and the **threshold of feeling** and above that the **threshold of pain** (at the upper end).

With reference to a pure tone at 1000 Hz, the human ear sound pressure range is typically from 2.2×10^{-5} to 20 Pa.

The minimum change in sound level that can be detected by the human ear (called the 'just noticeable difference') is about 2.25×10^{-5} Pa.

The sensation of hearing is also dependent on frequency.

In terms of frequency, the range of the human ear is typically from 15 to 20 000 Hz.

At low frequency (20–500 Hz), sensitivity is low requiring relatively higher sound levels for a given perception with response improving towards the top of the range.

At frequencies between approximately 500 and 2000 Hz, the ears response is relatively indifferent to sound level.

In general terms, best response of the human ear is recorded in the frequency range 2000–4000 Hz.

Above 4000 Hz sensitivity starts to become uneven and deteriorate a little up to the limit of 20 000 Hz.

In terms of discrimination, a 'trained' human ear might be able to detect a change in frequency of 1 Hz, although 3–5 Hz is more common.

It is not, however, a static receiver of sound faithfully passing the acoustic condition of the external environment to the brain. Hearing is adaptive, its sensitivity being variable depending on ambient sound level.

For example, at low ambient sound levels, the sensitivity is increased. This is the common experience of someone lying in bed at night being able to hear every structural sound at what appears to be a surprisingly loud volume.

At the other end of the scale, speaking and listening to a companion at a large and noisy sports event will require significantly more effort than holding the same conservation in, say, a restaurant.

This is because from the ear's perspective, the perception of a change in stimulus is proportional to the pre-change level. Put another way, the ear 'hears in ratios' and not simple differences.

The net result of this signal processing by the ear is perceived by the listener as the **loudness** of the sound. Loudness is essentially a subjective term based on combinations of sound level and frequency. For pure tones, sound level/frequency combinations producing equal perceptions of loudness level are expressed by an empirical unit called the **phon**.

A further unit called the **sone** is an attempt to simplify the phon scale by providing a linearization and a baseline.

Vibrations above the top of the human hearing range are termed **ultrasound**, whilst those below the lower limit are termed **infrasound**. Although 'silent' to the human ear, infrasound energy may in some cases still be felt as a pulse.

7.3 Sound Waves

7.3.1 Wave Motion

Consider a body or collection of molecules of a substance experiencing a force which results in deformation.

If upon removal of the force the body or collection of molecules returns to its original shape and position, then it may be said to be **elastic** or have **elasticity**.

Sound (in air) is the elastic oscillation of air molecules about some equilibrium or starting position due to the application of changing pressure from a source of vibration.

In ambient air, an applied force will be associated with change in pressure either above or below ambient atmospheric pressure ($\sim 1 \times 10^{-5}$ Pa).

Sound waves are **longitudinal** meaning that the particles of the medium carrying the sound oscillate or vibrate in the same direction as the line of advance of the wave.

Compare this with water waves which are described as **traverse**, meaning that displacement is in a direction perpendicular to the line of wave advancement (see Figure 7.2).

Looking at the typical range of audible sound pressures (2×10^{-5} Pa to 20 Pa), it is clear that sound is a relatively small-scale pressure phenomenon. Indeed, the displacement of molecules due to the passage of a sound wave is several orders of magnitude smaller than 1 mm.

Consider an increase in air pressure above initial condition to be a **compression** and a reduction in pressure relative to initial condition to be a **rarefaction**.

Sound being transmitted through a medium will then take the form of a continuous succession of compression and rarefactions resulting in a pressure wave. Figure 7.3 illustrates a regular pressure oscillation in air for a simple single tone wave varying in a regular sinusoidal fashion.

It is important to realise that the air is not displaced permanently by the passage of sound; however, sound waves are **progressive**, i.e. their energy content is carried forward in the direction of the wave's travel.

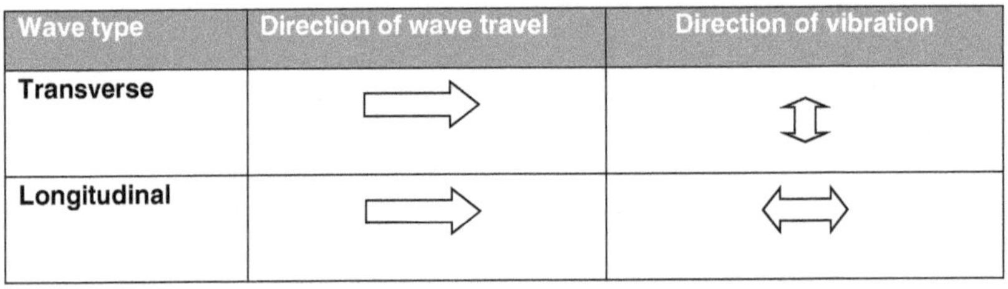

Wave type	Direction of wave travel	Direction of vibration
Transverse	$\Rightarrow$	$\updownarrow$
Longitudinal	$\Rightarrow$	$\Leftrightarrow$

Figure 7.2   Wave motion comparison.

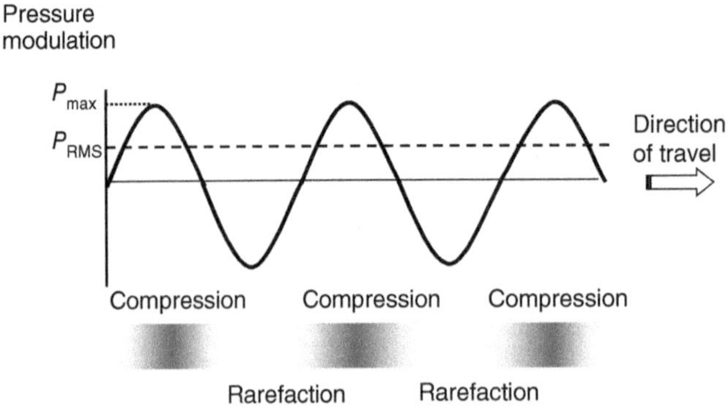

Pressure modulation

Figure 7.3   Characteristics of a simple single tone sine wave.

7.3.2 Wave Characteristics

Like electromagnetic radiation, sound waves are defined in terms of their **frequency** (f, Hz), i.e. the number of cycles passing a given point per second, a **wavelength** (λ, m) or distance between repeating points making up one complete cycle, and their **velocity** ($\overline{V}$, m/s) of transmission through a medium. The relationship for connecting these parameters is

$$\overline{V} = f\lambda \tag{7.1}$$

The time taken for one complete cycle of oscillation is termed the **period** (T, s). The period is the reciprocal of frequency:

$$T = 1/f \tag{7.2}$$

The term wavenumber (v, m^{-1}) is sometimes used to express the number of wavelengths per unit length of travel:

$$v = 2\pi/\lambda \tag{7.3}$$

'Pure' sounds, i.e. comprising a single frequency, are rare.

Some common terms used to accommodate sound complexity are **pitch**, **quality**, and **timbre**.

Pitch is the term used to describe the fundamental or dominant frequency in a complex sound.

Quality and timbre are a little more specific being related to the presence of whole number multiples of the fundamental frequency called **harmonics** or **overtones**.

These are most evident when notes of equal pitch are played on two different types of musical instrument. Again, in subjective terms, these frequencies are often said to give the sound its 'richness' or 'fullness'.

For practical purposes, sound analyses are usually made over distinct intervals of frequency called **bandwidths** having specified upper and lower frequency (cutoff) boundaries, i.e.

$$\text{Bandwidth} = f_{\text{upper}} - f_{\text{lower}} \tag{7.4}$$

An **octave** is a bandwidth having an upper frequency limit twice that of the lower frequency limit, i.e.

$$f_{\text{upper}} = 2f_{\text{lower}} \tag{7.5}$$

For large bandwidths requiring subdivision:

$f_{\text{upper}} = 2^n f_{\text{lower}}$, where n is the number of octaves.

The centre frequency (f_{centre}) of an octave is given by

$$f_{\text{centre}} = \sqrt{f_{\text{upper}} \times f_{\text{lower}}} \tag{7.6}$$

and

$$f_{\text{upper}} = 2^{n/2} f_{\text{centre}} \tag{7.7}$$

For the purposes of sound measurement, bandwidths having the following centre frequencies (Hz) have been internationally agreed.

16	31.5	63	125	250	500	1000	2000	4000	8000	16 000

For sound waves in a gas like air, velocity is strongly dependent on temperature. Recall that temperature is indicative of molecular motion, so hotter molecules are better able to transmit energy than cold molecules. In short, sound velocity increases with increasing temperature.

Assuming no heat transfer or friction associated with the wave motion, it can be shown that sound velocity (m/s) is given by

$$\overline{V} = \sqrt{\gamma RT} \tag{7.8}$$

where γ = ratio of specific heats for the gas (c_p/c_v),
R = specific gas constant for the gas (J/kg K),
T = gas temperature (Kelvin).

Alternatively, using the perfect gas law and remembering that $\rho = P/RT$, then

$$\overline{V} = \sqrt{P\gamma/\rho} \tag{7.9}$$

where P = atmospheric pressure (Pa),
ρ = atmospheric air density (kg/m^3).

7.4 Power, Intensity, and Pressure

7.4.1 The Bel

Because of the enormous range of pressure that the human ear is able to detect, expressing the magnitude of sound can become unwieldy. Coupled with the nonlinear sensitivity of the ear and its propensity to respond to a change relative to existing sound levels, making a definite statement about a sound level is difficult.

This is overcome by adopting the following. Firstly, the sound level is always expressed as a ratio relative to a baseline. For example, the sound pressure level chosen baseline is the pressure threshold of hearing, i.e. 2×10^{-5} Pa.

Secondly, it is assumed that the ear's response is logarithmic and so the logarithm (base 10) of the ratio is evaluated.

The resulting number is said to express the sound level in **bels**.

Traditionally, sound levels are expressed in decibels and so a simple allowance is finally made for the sub-multiple.

So far in the text, the generic term 'sound level' has been used. However, there are three parameters bearing the title 'sound level', and after this discovery, it should be clear that it is important to be more specific.

7.4.2 Sound Levels

7.4.2.1 Sound Power Level (L_W)
Sound power (Watts) is the rate of acoustic energy generation by a sound source and is independent of the source's surrounding environment.

It is not as easy as you might think to make a sound. The energy conversion efficiency for sound is usually very low (<1%). This is desirable for, say, a car engine but perhaps not so for musical instruments.

In terms of sound power, the audibility-to-feeling range is typically from 1×10^{-12} W to 1 W.

The threshold of audibility is used as the baseline in calculating a sound power level (L_W, dB), thus

$$L_{\dot{W}} = 10 \log \left(\frac{\dot{W}}{1 \times 10^{-12}} \right) \tag{7.10}$$

where $\dot{W}$ = sound power (Watts).

7.4.2.2 Sound Intensity Level (L_I)

Sound intensity (I) is a parameter describing the acoustic energy/sec or acoustic power passing through 1 m² of environment normal to, and at some distance from, a sound source.

Again, the threshold of audibility (1×10^{-12} W/m²) is used as the baseline in calculating a sound intensity level (L_I, dB), thus

$$L_I = 10 \log \left(\frac{I}{1 \times 10^{-12}} \right) \tag{7.11}$$

where I = sound intensity (W/m²).

Typically, the human ear is able to detect a change in sound intensity level of about 1 dB at low intensities. This decreases to 0.33–0.5 dB at higher intensities.

7.4.2.3 Sound Pressure Level (L_p)

Unlike sound power, sound pressure is dependent upon the medium of transmission and the acoustic environment. The sound pressure experienced at a point will depend not only on the characteristics of the sound source but also on factors such as the distance from the sound source, reflecting surfaces, etc.

Evaluating the magnitude of a sound pressure wave, such as that illustrated in Figure 7.3, requires some consideration since a simple average over one cycle results in a zero value. To produce a usable result, all variations of pressure amplitude over a cycle are squared producing positive values (P^2). These squared values are summed and the result is divided by the cycle period to produce a mean (P_{MS}).

A square root is then taken of this mean. The result is termed the root mean squared value (P_{RMS}). The process is illustrated in Figure 7.4.

In general, values of pressure level quoted in acoustic studies are RMS, although peak pressure levels (P_{max}) are sometimes used for short duration, 'impact' sounds.

(For a single-tone, sinusoidal wave form, the RMS value is related to the maximum pressure modulation (P_{max}) by the following: $P_{RMS} = P_{max}/\sqrt{2}$.)

Sound pressure and sound intensity are related by the following expression:

$$I = \frac{P^2}{\rho \overline{V}} \tag{7.12}$$

where ρ = density of medium through sound wave travels (kg/m³),
$\overline{V}$ = sound wave velocity (m/s).

Using the above intensity–pressure relationship and again employing the pressure threshold of audibility as the baseline, a sound pressure level (L_p, dB) can be defined thus:

$$L_P = 20 \log \left(\frac{P}{2 \times 10^{-5}} \right) \tag{7.13}$$

where P = sound pressure (Pascal).

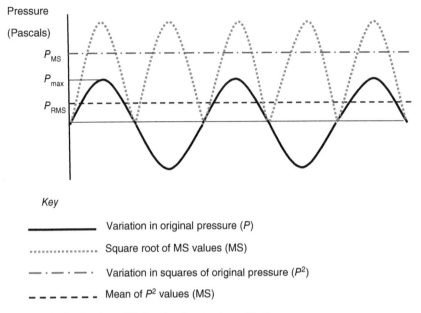

Key

——————— Variation in original pressure (*P*)

·················· Square root of MS values (MS)

— · — · — · Variation in squares of original pressure (*P²*)

— — — — — Mean of *P²* values (MS)

Figure 7.4 Production of RMS value for simple oscillating pressure wave.

(Notice that the constant has increased from 10 to 20. This is because intensity is related to the square of pressure. [Remember log $(x^n) = n$ log(x)].)

7.4.2.4 Sound-Level Interrelationships

The following interrelations between sound power, sound intensity, and sound pressure levels can be arrived at by simple substitution and a little logarithmic manipulation:

- *Sound intensity level/sound pressure level*
  With a small correction for nonstandard atmospheric conditions, sound intensity, and sound power are often regarded as numerically equal:

$$L_I = L_P - 10 \log \left(\frac{\rho \overline{V}}{400} \right) \qquad (7.14)$$

  Unless extreme atmospheric conditions prevail, the magnitude of this correction is of the order of 1 dB or less and its effect is often ignored.
- *Sound power level/sound intensity level*
  Sound power is the product of sound intensity and area.
  Remember that log$(x \times y) = $ log$(x) + $ log(y), on a decibel scale:

$$L_W = L_I + 10 \log A \qquad (7.15)$$

  where *A* is the area (m²) through which the sound wave passes.
- *Sound power level/sound pressure level*
  Substituting from above:

$$L_W = L_P - 10 \log \left(\frac{\rho \overline{V}}{400} \right) + 10 \log A \qquad (7.16)$$

Some typical sound pressures and powers are provided in Table 7.1.

Table 7.1 Typical sound pressures, powers, and their attendant levels.

Sound source	Sound pressure (P, Pa)	Sound pressure level (L_p, dB)	Sound power ($\dot{W}$, W)	Sound power level ($L_{\dot{W}}$, dB)
Jackhammer	20	120	4.1	126
Chainsaw	10	114	1.0	120
Power mower	1	94	1×10^{-2}	100
Vacuum cleaner	0.1	74	1×10^{-4}	80
Human speech	0.02	60	4.1×10^{-6}	66
Whispering	0.002	40	4.1×10^{-8}	46
Limit of audibility	0.000 02	0	1×10^{-12}	0

Source: Data adapted from Salvendy, G. (2012) *Handbook of Human Factors and Ergonomics*, John Wiley & Sons.

7.5   Laws of Sound Combination

It is commonly a requirement to determine the net effect of multiple sound sources in combination or the effect on existing total sound levels of a change in circumstances like, for example, the addition or the elimination of a sound source.

Here, a little care is required remembering that, because of their logarithmic nature, sound pressure levels in decibels cannot be operated on arithmetically in direct terms.

Sound intensities do not suffer from this drawback however, and so these are used in practice. Here is a routine for handling simple decibel operations:

- Assume sound pressure levels and sound intensity levels are numerically equal.
- Divide the sound levels (dB) of interest by 10 and generate their anti-logs.
- Perform the arithmetic appropriate to the problem (addition or subtraction).
- Take the log of the result and multiply by 10 to produce the answer in dB.

For example, for a number of sound sources (1 to n) having sound levels (dB) of L_1 to L_n operating together, the total sound pressure level (dB) would be given by

$$L_{tot} = 10 \log \left(10^{L_1/10} + 10^{L_2/10} + \ldots + 10^{L_n/10} \right) \tag{7.17}$$

One useful application of this is in sound measurement itself where a typical sound meter will measure sound pressure levels over a range of distinct bandwidths as described earlier. These measurements are then combined to produce an overall single sound level in dB(A).

Alternatively, if the contribution of a particular sound source (L_1) to total sound pressure level (L_{tot}) is required, as, for example, when determining the impact of a particular sound on existing ambient levels, then

$$L_1 = 10 \log \left(10^{L_{tot}/10} - \left[10^{L_2/10} + \ldots + 10^{L_n/10} \right] \right) \tag{7.18}$$

7.6   Sound Propagation

Sound will not travel through a vacuum. It needs a transmission medium.

The transmission medium is not an ideal, passive conduit. Passage through it and/or interaction with its boundaries affects sound pressure level.

7.6.1 Sound Attenuation

The **sound attenuation** at a point some distance from a sound source will be the difference between the source sound power and the sound pressure level at that point.

Let us look at some phenomena affecting sound pressure level in a little more detail.

(a) **Geometric dispersion or spreading**

This describes the reduction of sound pressure level with distance (r, m) from the source and is the result of simple energy density decrease due to area increase.

Underlying this effect is the relationship between sound intensity (I) and sound power $(\dot{W})$, thus

$$I = \frac{\dot{W}}{A} \tag{7.19}$$

where A is the surface area of the sound field through which the sound wave is travelling and is a function of distance from the source. Using this, the effect of increasing the source–receiver distance on sound pressure levels for some common geometries is illustrated in Figure 7.5.

(a) For a sound source radiating a spherical field, i.e. $A = 4\pi r^2$

$$L_P = L_{\dot{W}} - 20 \log r - 11 \tag{7.20}$$

(b) For a sound source radiating a hemispherical field, i.e. $A = 2\pi r^2$

$$L_P = L_{\dot{W}} - 20 \log r - 8 \tag{7.21}$$

i.e. a difference of 3 dB from the spherical case.

(c) For a sound source radiating a cylindrical field, i.e. $A = 2\pi r$

$$L_P = L_{\dot{W}} - 10 \log r - 8 \tag{7.22}$$

where $L_{\dot{W}}$ = sound power per unit length of source.

(d) For a sound source radiating a semi-cylindrical field, i.e. $A = \pi r$

$$L_P = L_{\dot{W}} - 10 \log r - 5 \tag{7.23}$$

where again $L_{\dot{W}}$ = sound power per unit length of source.

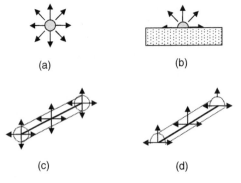

(a) (b)

(c) (d)

Figure 7.5 Sound source–receiver geometries. (a) Spherical field, (b) hemi-spherical field, (c) cylindrical field, (d) semi-cylindrical field.

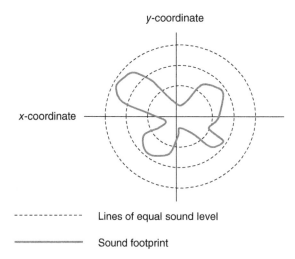

Figure 7.6 Radiation plot for complex sound source.

For a point source radiating sound spherically in a uniform way, sound pressure level has an inverse square relationship with distance. However, many sound sources are inherently directional, having geometrically complex surfaces. Furthermore, a sound source may comprise surfaces in motion relative to their surroundings, and so the sound pressure level sensed by a receiver will vary not only with distance but also with angular direction relative to the source.

Figure 7.6 illustrates an example of a 2-D sound radiation plot for a complex sound source. The sound source is at the intersection of the axes. The solid line indicates the source's directional sound output distribution or 'footprint', and the broken lines are sound contours, i.e. lines of equal sound level (dB).

Plots may be of total sound pressure level or frequency based.

Complexity like this can be accommodated by describing the degree of directionality of the source. Two parameters based on equivalent average conditions from an omni-directional, spherical field source are in use.

Directionality can be described by the ***directivity index*** (DI_θ, dB).

For a given distance from a sound source, this parameter describes the difference between the sound pressure level in a particular angle (θ) or direction of interest and the average sound pressure level of the source. It may therefore have a positive, zero, or negative value, i.e.

$$DI_\theta = L_{P,\theta} - L_{P,\text{ave}} \tag{7.24}$$

where, for n readings,

$$L_{p,\text{ave}} = 10 \log \left[\left(10^{I_1/10} + 10^{I_2/10} + \ldots + 10^{I_n/10} \right) / n \right] \tag{7.25}$$

i.e. the result can be used to adjust the geometric spreading sound attenuation relationships.

Directivity can also be described by the ***directivity index*** (DF_θ).

For a given distance from a sound source, this is the ratio of the sound intensity level in a particular angle (θ) or direction of interest to the average sound intensity level of the source, i.e.

$$DF_\theta = I_\theta/I_{ave} = P_\theta^2/P_{ave}^2$$
$$= L_{P,\theta} - L_{P,ave} \tag{7.26}$$

where, for n readings of sound intensity,

$$I_{ave} = \left(I_1 + I_2 + \ldots + I_n\right)/n \tag{7.27}$$

The conversion from directivity factor to directivity index is

$$DF_\theta = 10\log DI_\theta \tag{7.28}$$

Furthermore, an omni-directional sound source may display directionality when closely bounded by solid surfaces. In this context, directivity terms are simply accounting for the reduced sound field area relative to the spherical case (see Figure 7.7).

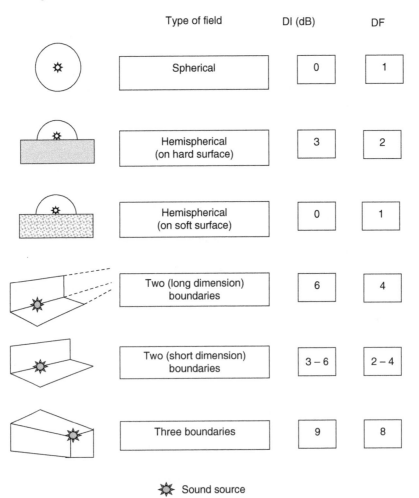

	Type of field	DI (dB)	DF
	Spherical	0	1
	Hemispherical (on hard surface)	3	2
	Hemispherical (on soft surface)	0	1
	Two (long dimension) boundaries	6	4
	Two (short dimension) boundaries	3 – 6	2 – 4
	Three boundaries	9	8

☼ Sound source

Figure 7.7 Boundary-related directionality.

(b) **Absorption/transmission**

Remember that sound is energy in transit and that passage through a transmission medium will result in energy transfer from the wave to the medium. In other words, the transmission medium will absorb some of the wave's energy through contact.

See Figure 7.8. Note that here the potential for reflection has been omitted for simplicity. Sound levels are attenuated to a relatively small extent by absorption through air itself. The relationship is not simple. The effect is strongly dependent on frequency (attenuation increasing with increasing frequency) and water vapour content (attenuation decreasing with increasing relative humidity), as well as being more weakly influenced by air temperature and pressure.

Air attenuation is a long-range factor more noticeable over kilometres rather than metres but is significant in aircraft noise control.

The frequency-dependent, wave power fraction (or percentage) absorbed by a surface is termed the **sound absorption coefficient** (%) and thus

$$\text{Absorpion coefficient} = \text{Absorbed sound power/Incident sound power}$$

The total absorption (A,m^2) for a room comprising n surfaces is calculated from

$$A = \sum_{1}^{n} (\text{Area} \times \text{Absorption coefficient}) \tag{7.29}$$

Typical values of absorption coefficient for constructional materials such as brick, concrete, and plaster are in the range 0.02–0.05. At the lower end of the frequency range, glass can realise an absorption coefficient of the order of 0.3 but this falls off at higher frequencies. Furnishings, decorative materials, and, indeed, people tend to have useful values of absorption coefficient (typically 0.3–0.6) at higher frequencies.

Three types of absorption mechanisms prevail:

- **Dissipative absorption** depending on the porosity or internal surface area of the absorbing material. Here, the motion of the sound wave will result in energy loss due to small-scale heating of the material. Examples of good dissipative absorbers include fibrous materials such as mineral wool or materials having open cellular structures.
- **Flexural vibration** in membranes or panels where the passing sound wave induces a bending deformation in the absorbing material itself.

  The magnitude of this effect will depend on the stiffness and mass of the absorbing material as well as the depth of the airspace behind it.

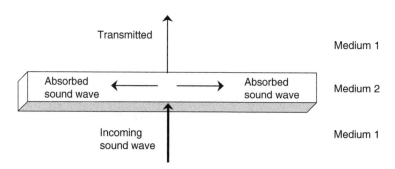

Figure 7.8 Absorption/transmission of a sound wave by a change in medium.

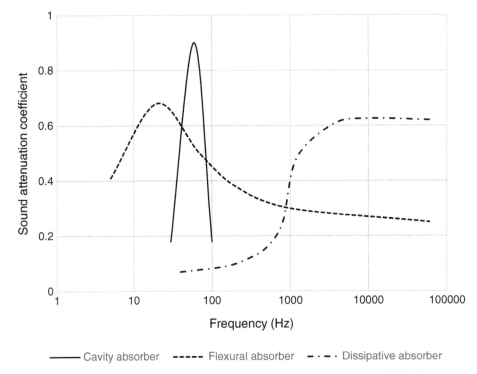

Figure 7.9 Frequency-dependent comparison of sound absorbing phenomena.

- *Cavity or Helmholtz resonation* where an incident sound wave can be absorbed by forcing the energy through a small entrance to a large closed-end volume or reservoir. The magnitude of this effect depends on the dimensions of inlet and reservoir.

  All three phenomena are frequency dependent. The general trends are shown in Figure 7.9 indicating that porous or dissipative absorbers perform better at high frequencies, whilst membrane or flexural absorbers generally perform better at lower frequencies displaying a peak sound attenuation at their natural frequency of vibration. Cavity absorbers are capable of producing high sound attenuation values over a more limited range of frequencies.

  Surfaces designed to maximise sound absorption over a range of frequencies often employ aspects of all three of these phenomena.

- **Transmission**

  The sound power transmitted by a surface will be of concern in sound insulation and may also be expressed as a fraction (or percentage) relative to the incoming wave, that is, as a *transmission coefficient*.

  Again, this is frequency dependent.

  It is more common however to express the sound insulation properties of an intervening surface in decibel terms using a *sound reduction index* as follows:

$$\text{Sound reduction index (dB)} = 10 \log \left(\dot{W}_{\text{incoming}} / \dot{W}_{\text{transmitted}} \right) \tag{7.30}$$

(c) **Reflection**

Sound reflection like light refection may be defined as the redirecting of a sound wave due to its impact with a surface where both incoming and redirected sound waves travel in the same medium.

The resulting effect to a receiver is dependent on the nature of the reflecting surface. If the surface is hard and flat, then the incoming and reflected waves are in the same plane and the angle (i) made between an incoming sound wave and a line drawn normal to the surface is equal to the angle (r) between the reflected wave and the normal. This set of circumstances is termed ***specular*** reflection (see Figure 7.10).

A sound may reach a listener by both direct and indirect routes due to reflection (see Figure 7.11). The resulting attenuation is dependent on sound frequency, sound path length difference, angle of incidence, and the nature of the reflecting surface (hard or soft).

Sound reflections lead to the concept of *reverberation time*. This is the time (s) taken for a sound from a source to, as a result of reflection, continue to live on after the source has stopped generating. It is perceived as a sound extension rather than an echo, the latter having a more distorted quality. Typically, the reverberation time is that passage of time for the sound to be reduced to one millionth of its original sound level. A reverberation time of no greater than one second is regarded as acceptable for speech and up to two seconds for musical performance. There are several application-dependent methods for estimating reverberation time. The most common is *Sabine's formula* which expresses reverberation time (t_r, s) in terms of room volume (V, m^3) and total absorption $(\mathcal{A})$.

$$t_{rev} = 0.16V/\mathcal{A} \tag{7.31}$$

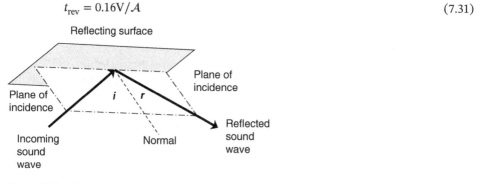

Figure 7.10   Specular reflection.

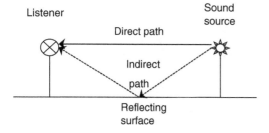

Figure 7.11   Direct and indirect sound paths.

7.7 Sound Fields

Real sound sources often have complex physical geometries and will vibrate and emit sound in a complex pattern.

Furthermore, sound waves emitted by a sound source may be modified by the passage through their medium of transmission producing variations in characteristics such as sound level and direction. The map of the resulting variations around a source is called a **sound field** (see Figure 7.12).

There are a number of sound field conditions.

7.7.1 Free Field

This is indicative of an open-air environment. In the absence of reflective or absorptive surfaces, sound produced by a source will be directly transmitted to a receiver.

7.7.2 Diffuse Field

These circumstances are more indicative of an indoor or a built-up outdoor environment. When a sound field is bounded by surfaces, sound may reach a receiver both directly and indirectly via mechanisms such as reflection from surfaces and absorption/transmission (or 'flanking') through surfaces. Under these circumstances, the sound field is described as *diffuse*. Diffuse conditions display little sound intensity directionality throughout the sound field.

7.7.3 Far-Field

The far-field is a relatively settled region in sound terms.

In the far-field, complex sound wave interactions will have reconciled their differences and a relatively coherent pattern will result. Consequently, in a far-field environment, the sound waves can be regarded as plane and one-dimensional.

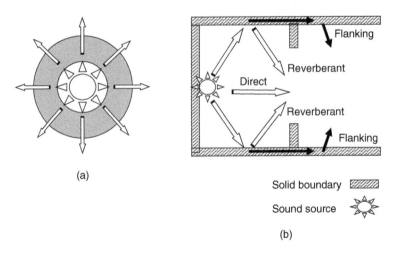

Figure 7.12 (a) Free and (b) diffuse sound fields.

Far-field conditions are said to exist if:

- the sound pressure varies in inverse proportion to the square of the distance from the source, i.e. $P \sim 1/r^2$,
- oscillations in sound pressure and particle velocities are in phase,
- the sound source radiates or appears to radiate uniformly in all directions. This is a characteristic of a *point* source, i.e. a theoretical source with nominally zero-length dimension.

In terms of system dimensions, if the radius of the sound source is r_s, the distance to the point of sound measurement is r, and the wavelength of the sound is λ, point source behaviour may be approximated if $r/r_s \gg 1$ and $r_s/\lambda \ll r/r_s$.

7.7.4 Near Field

Near-field conditions are experienced rather closer to sound sources and tend to display deviations from the simple pressure–distance inverse square. They are more disturbed areas due to the proximity of the sound-generating surfaces and the resulting sound wave interaction.

7.8 Acoustic Pollution or Noise

Acoustic pollution is difficult to define and not just a matter of sound pressure level. It is relatively straightforward to define levels and time of exposure that will result in physical damage and a reduction in the hearing capability of a receiver.

More thought is necessary when it comes to defining the nuisance value of *noise* which is just unwanted sound. It could be dependent on circumstances. For example, while relaxing in your garden on a sunny afternoon you may not be adverse to the sound of classical music coming from a neighbour's property. You may not be so tolerant of this export were it to be, say, metal or rap music. If it were it to go on until 3 a.m. in the morning, it would probably be regarded as unwanted, regardless of genre.

The prevailing acoustic indoor environment will be the resultant of sound generated by the occupants, by building activities, processes, and service provision, and as a result of the penetration of external sound.

Almost all internal environments will be subject to occupant sounds as a result of conversation. The range and level of sounds produced in a manufacturing environment is likely to be extensive and dependent on specific equipment.

Perhaps, the most common external noise source penetrating to the internal environment is transport, i.e. traffic and aviation.

Building services, e.g. air conditioning/ventilation/refrigeration systems themselves are capable of producing quite disruptive noise levels if badly specified, designed, and/or installed.

To minimise the problem, it is best to start with the sound source (e.g. fans, pumps, and compressors) and ask manufacturers to supply an acoustic/vibration footprint for machines. Isolation and insulation measures to reduce the sound transmission include acoustic enclosures, anti-vibration mounting, and flexible connections.

After source treatment, the next phase is to ensure that secondary sources do not arise as a result of bad design. For example, undersized ductwork will result in high air velocities and turbulence especially at ductwork fittings such as bends, tees, and grilles. Low velocity design, smooth transitions, and flow straighteners should be considered at the design stage.

Ancillary fluid pipework services are also important. Water flow in pipes can suffer from turbulence and pressure variation phenomena such as '*water hammer*'.

7.8.1 Effect on Humans

Sound is energy, and an overload exposure to the human ear will result in physical damage to the nerve hairs in the cochlea of the inner ear. This damage manifests itself as a loss of sensitivity over the range of audible frequencies termed a threshold shift.

A short exposure to an excess of sound energy will result in a ***temporary threshold shift***. This is the sort of hearing sensitivity loss experienced by, for example, night clubbers emerging at the end of the night from a loud environment into a more quiet area. The sensitivity loss is most evident over the 3–6 Hz frequency range. With rest in a quiet area, sensitivity is recovered over a short period of time.

With frequent exposure, the effect will become more persistent requiring extended recovery times. When exposure to high levels of noise is repeated daily over many years, damage can become irreversible and a ***permanent threshold shift*** will result.

Aside from hearing damage, elevated sound levels can produce other *physiological effects* such as changes in skin electrical conductivity, brain electrical activity, and heart and breathing rates. There is some evidence to suggest that blood pressure and blood vessel conditions are also affected. It is reasonable to assume that higher level animals are also affected in this way to some extent.

Then there are *psychological effects*. Noise can make people irritable and feel fatigued.

Finally, in a work environment, elevated noise levels can be dangerous as they interfere with speech communication. More mistakes and accidents may result. Efficiency and productivity are also adversely affected by noise.

7.8.2 Noise Standards

Standards for noise limits are based on the statistical relationship between sound level and time. Noise measurements must therefore be carried out over the monitoring time period associated with the standard.

There are many noise standard parameters in common use. Examples include the following:

(a) **Percentile ($L_{f,T}$)**

The L_f is that sound level (dBA) which is exceeded for $f\%$ of the monitoring period (T). Choosing a low value for f will indicate the presence of sound-level peak values, whilst high value f measurements will indicate more persistent, general background sound levels. Both L_{10} and L_{90}, i.e. sound levels exceeded 10% and 90% of a specified monitoring period are in use.

(b) **Equivalent continuous noise levels ($L_{eq,T}$)**

This is a time-weighted average. It is the unvarying, continuous sound level (dB) that contains the same energy content as the varying sound level over the monitoring period (T).

If a varying sound has a sound pressure level of L_1 for time t_1, L_2 for time t_2, and so on for L_n for time t_n, then the equivalent continuous noise level (L_{eq}) over the total time

(t) is given by

$$L_{eq,T} = 10\log\left[\frac{1}{T}\left(t_1\times 10^{L_1/10}+t_2\times 10^{L_2/10}+\ldots t_n\times 10^{L_n/10}\right)\right]$$ (7.32)

Note that the time units of the total monitoring period and that of the subdivisions must be the same.

Noise levels corrected for 'A' weighting are termed LA_{eq}.

A typical, acceptable maximum background equivalent continuous noise level for dwelling bedrooms or hospital wards at night would be 30 dB LA_{eq}. At the other end of the range, an acceptable maximum noise exposure at an entertainment venue might be 100 dB LA_{eq} for four hours, set for a limited number of times/annum.

(c) **SEL single events**

Sound pressure level in dB (A) that transmits the same amount of energy in one second as the varying sound.

(d) **Perceived noise level L_{pn}**

This is an index used in aircraft noise level studies. The standard is weighted to account for the high frequencies and high annoyance inherent in aircraft operation.

It is approximated by adding 13 to prevailing sound pressure levels.

The L_{pn} is sometimes used in combination with the number of flyovers in a given time to produce a so-called noise and number index (NNI).

(e) **L_{max}**

The maximum sound level (in dBA) during a sound event.

(f) **L_{peak}**

The highest pressure recorded during a short duration, potentially damaging sound.

(g) **Noise criteria (NC) or rating (NR) curves**

Sensitivity-related sound pressure level (L_p, dB) vs. octave band frequency (Hz) curves that produce standard, overall sound levels. Used for comparison with the frequency-based noise spectrum of any specified piece of equipment to determine its suitability or contribution to resulting noise levels. Some examples are given in Figure 7.13.

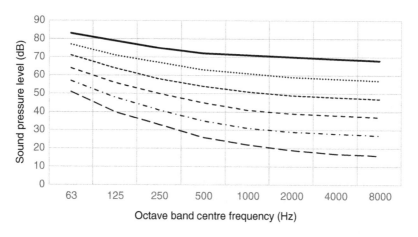

Figure 7.13 Noise criteria (NC) curves.

7.9 Worked Examples

Worked Example 7.1   Sound Wave Characteristics
A sound wave has a wavelength of 0.11 cm and a period of 3.333×10^{-4} s.
  Determine its frequency (Hz) and wavenumber.

Solution
Given: $\lambda = 11 \text{ cm} = 0.11 \text{ m}, \quad T = 3.333 \times 10^{-4} \text{ s}$
  Find: f, v

$$T = 1/f$$

$$3.333 \times 10^{-4} = 1/f$$

$$f \cong 3000 \text{ Hz}$$

$$v = 2\pi/\lambda$$
$$= 2\pi/0.11$$
$$= 57.12 \text{ m}^{-1}$$

Worked Example 7.2   Bandwidth Frequencies
Approximate the upper and lower boundary frequencies for a bandwidth with a single octave centred on 500 Hz.

Solution
Given: $f_{centre} = 500 \text{ Hz}, n = 1$
  Find: f_{upper}, f_{lower}

$$f_{upper} = 2^{n/2} f_{centre}$$
$$= 2^{1/2} \times 500$$
$$= 707.1 \text{ Hz}$$

$$f_{centre} = \sqrt{f_{upper} \times f_{lower}}$$
$$500 = \sqrt{707.1 \times f_{lower}}$$

$$f_{lower} = 353.6 \text{ Hz}$$

Worked Example 7.3   Sound Power
During a musical performance, a singer generates 0.001 W of acoustic power.
  Determine this sound power level in decibels.

Solution
Given: $\dot{W} = 0.001 \text{ W}$

Find: $L_{\dot{W}}$

$$L_{\dot{W}} = 10 \log \left(\frac{\dot{W}}{1 \times 10^{-12}} \right)$$

$$= 10 \log \left(\frac{0.001}{1 \times 10^{-12}} \right) = 10 \log \left(\frac{0.001}{1 \times 10^{-12}} \right)$$

$$= 90 \text{ dB}$$

Worked Example 7.4 Sound Intensity

An entertainment centre in a bedroom produces a sound pressure of 0.063 Pa. If the air density in the room is 1.2 kg/m^3 and the velocity of sound is 343 m/s, express this pressure as a sound intensity (W/m^2).

Solution

Given: $P = 0.063$ Pa, $\rho = 1.2 \text{ kg/m}^3$, $\overline{V} = 343$ m/s

Find: I

$$I = \frac{P^2}{\rho \overline{V}}$$

$$= \frac{0.063^2}{1.2 \times 343}$$

$$= 9.642 \times 10^{-6} \text{ W/m}^2$$

Worked Example 7.5 Sound Pressure

During a quiet conversation with a friend less than 1 m away, a listener experiences a sound pressure of 0.002 Pa.

Determine this sound pressure level in decibels.

Solution

Given: $P = 0.002$ Pa

Find: L_p

$$L_p = 20 \log \left(\frac{P}{2 \times 10^{-5}} \right)$$

$$= 20 \log \left(\frac{0.002}{2 \times 10^{-5}} \right)$$

$$= 40 \text{ dB}$$

Worked Example 7.6 Sound Pressure/Power Relationship

A machine radiates sound in a spherical field. If the sound pressure level is 80 dB at a distance of 4 m, estimate the sound power level (dB) of the machine.

Take the local atmospheric conditions as 28 °C, 100 000 Pa, and the specific gas constant and specific heat capacity ratio for air as 287 J/kg K and 1.4, respectively.

Solution

Given: $L_p = 80$ dB, $r = 4$ m, $T = 28\,°\text{C} = 301$ K, $P = 1 \times 10^5$ Pa

$$R = 287 \text{ J/kg K}, \quad \gamma = 1.4$$

Find: L_w

From the perfect gas law:

$$\rho = P/RT$$
$$= 1 \times 10^5 / 287 \times 301$$
$$= 1.158 \text{ kg/m}^3$$

Calculated the velocity of sound:

$$\overline{V} = \sqrt{\gamma RT}$$
$$= \sqrt{1.4 \times 287 \times 301}$$
$$= 347.77 \text{ m/s}$$

Calculated the area of the spherical field:

$$A = 4\pi r^2 = 4\pi \times 4^2 = 201.1 \text{ m}^2$$

Then

$$L_W = L_p - 10 \log \left(\frac{\rho \overline{V}}{400} \right) + 10 \log A$$
$$= 80 - 10 \log \left(\frac{1.158 \times 347.77}{400} \right) + 10 \log (201.1)$$
$$= 80 - 0.029 + 23$$
$$= 103 \text{ dB}$$

Worked Example 7.7   Sound Combination

The background sound level at the help desk of a University library is 40 dB (A).

The area contains four photocopiers each having a sound pressure level of 56 dB (A). Determine the resulting total sound level when all four are in use.

Solution

Given: $L_{\text{photo copier}} = 56 \text{ dB}(A)$, $L_{\text{background}} = 40 \text{ dB}(A)$

Find: L_{total}

$$L_{\text{total}} = 10 \log \left(10^{L_{\text{background}}/10} + \left[4 \times 10^{L_{\text{photocpoier}}/10} \right] \right)$$
$$= 10 \log \left(10^{40/10} + 10^{56/10} + 10^{56/10} + 10^{56/10} + 10^{56/10} \right)$$
$$= 62 \text{ dB} (A)$$

Worked Example 7.8   Sound Field

An alarm on the side of a property produces power of 110 dB.

Determine the sound pressure level at the boundary of an adjacent building 70 m distance from the alarm.

Assume a hemispherical radiation field.

Solution

Given: $L_W = 110\,\text{dB}, \quad r = 70\,\text{m}$

Find: L_p

$$L_p = L_W - 20\log r - 8$$
$$= 110 - 20\log(70) - 8$$
$$= 65\,\text{dB}$$

Worked Example 7.9   Sound Directionality

A sound source operating in a spherical field has a sound power level of 96 dB. At a given direction offset from the main axis and 10 m from the source, the sound pressure level is 70 dB.

Determine the directivity index (dB) and the directivity factor at the position of sound pressure measurement.

Solution

Given: $L_W = 96\,\text{dB}, \quad r = 10\,\text{m}, \quad L_p = 70\,\text{dB}$

Find: DI_θ, DF_θ

$$L_{p,\text{ave}} = L_W - 20\log r - 11$$
$$= 96 - 20\log(10) - 11$$
$$= 65\,\text{dB}$$

Therefore,

$$DI_\theta = L_{p,\theta} - L_{p,\text{ave}}$$
$$= 70 - 65$$
$$= 5\,\text{dB}$$

and

$$DF_\theta = 10\,\log DI_\theta$$
$$= 10\,\log(5)$$
$$= 6.99$$

Worked Example 7.10   Noise Standard

The sound pressure level in a noisy environment varies over eight hours as shown below.

Sound level (dBA)	70	63	68	75
Duration (h)	2	3	2	1

Determine the eight-hour equivalent continuous sound pressure level (dBA).

Solution

Given: L_{1-4}, t_{1-4}

Find: $L_{eq,T}$

$$L_{eq,T} = 10 \log \left[\frac{1}{T} \left(t_1 \times 10^{L_1/10} + t_2 \times 10^{L_2/10} + \cdots \cdots t_n \times 10^{L_n/10} \right) \right]$$

$$= 10 \log \left[\frac{1}{8} \left(2 \times 10^{70/10} + 3 \times 10^{63/10} + 2 \times 10^{68/10} + 1 \times 10^{75/10} \right) \right]$$

$$= 69.4 \, dB \, (A)$$

7.10 Tutorial Problems

7.1 A sound has a frequency of 1000 Hz and a wavelength of 0.32 m.
Determine its period (s^{-1}) and velocity (m/s).
[Answers: $1 \times 10^{-3} \, s^{-1}$, 320 m/s]

7.2 An octave is centred on a frequency of 250 Hz.
Determine the upper and lower banding frequencies (Hz).
[Answers: 354, 177 Hz]

7.3 The sound intensity in an office is $5 \times 10^{-7} \, W/m^2$.
Express this as a sound intensity level (dB).
[Answer: 57 dB]

7.4 A wave front with a sound intensity level of 55 dB is passing through an area of 10 m^2.
Determine the resulting sound power level (dB).
[Answer: 65 dB]

7.5 A radio generates 0.005 W of power.
Express this as a sound power level (dB).
[Answer: 97 dB]

7.6 A power tool has a sound pressure of 6 Pa.
If the density of air is 1.18 kg/m^3 and the velocity of sound is 330 m/s, determine its sound intensity level (W/m^2) and sound pressure level (dB).
[Answers: 0.092 W/m^2, 110 dB]

7.7 On the arrival of a train, the sound pressure in an underground railway platform is determined to be 2 Pa.
Express this as a sound pressure level (dB).
[Answer: 100 dB]

7.8 A room has surfaces with values of absorption coefficient at 500 Hz shown in Figure P7.8. Estimate its reverberation time (s).

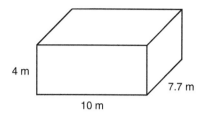

Surface absorption coefficient

Floor: 0.04

Walls: 0.1

Ceiling: 0.02

Figure P7.8 Data for tutorial Problem 7.8.

Determine the new reverberation time if the floor is covered with a material having an absorption coefficient of 0.5.
[Answer: 2.62, 0.91 s]

7.9 A building alarm has a sound power of 1 W when operating in a spherical field. Determine the sound pressure level (dB) at a distance of 50 m from the source.
[Answer: 78 dB]

7.10 The sound pressure level in a noisy environment varies over eight hours as shown below:

L_p (dB)	68	72	75	65
Duration (h)	1	3	3	1

Determine the eight-hour equivalent sound pressure level (dB)
[Answer: 73 dB]

8

Visual Factors

8.1 Overview

Visible light is the name given to a relatively small section of wavelengths in the middle of the electromagnetic spectrum ranging from 380 to 760 nm. This is important from an anthropocentric view (and those of many animals) because it happens to be the region to which the human eye has become sensitised.

Until man obtained the wherewithal to create and maintain fire, reflected sunlight from the moon and starlight had to suffice as sources of illumination after sunset. Candles (essentially fat-based) were an improvement on campfires and gas lighting a further advancement, but it is with the invention of artificial electric lighting that a good standard of both indoor and outdoor controllable illumination was made possible during the night.

Enhanced safety and amenity have facilitated the development of a 24 hours lifestyle bringing many benefits to working practices as well as enabling the playing of sport and many other leisure activities around the clock.

Learning Outcomes

- To be familiar with the structure and function of the human eye
- To be able to differentiate between parameters associated with light sources and receivers.
- To understand the basic laws of illumination
- To be familiar with lamp types and the impact of luminaire design
- To understand the need for avoiding the occurrence of unwanted light

8.2 The Human Eye

The human eye may be thought of as a three-layered approximate spheroid with a bulge at the front at the interface with the external environment. See Figure 8.1.

The outer opaque layer or covering is called the **sclerotic coat**. This layer forms the 'white' of the eye except at front and centre where there is a transparent region called the cornea, facilitating the entry of light. As well as giving some protection to other parts of the eye, the cornea also aids with the convergence of light into the eye.

Building Services Engineering: Smart and Sustainable Design for Health and Wellbeing, First Edition.
Tarik Al-Shemmeri and Neil Packer.
© 2021 John Wiley & Sons Ltd. Published 2021 by John Wiley & Sons Ltd.

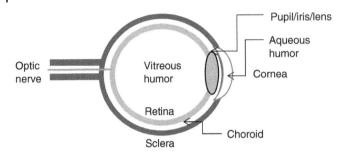

Figure 8.1 The human eye.

The middle layer or ***choroid coat*** performs a reflection dampening function inside the eye. Its function and structure changes at the front of the eye again where it extends into a variable area orifice controlling the amount of light entering the eye. This choroid coat extension is termed the ***iris*** (the coloured part of the eye), and the orifice is the ***pupil***.

The inner layer of the eye is termed the ***retina***. This is the photon-sensitive stratum containing light-sensitive receptors called ***rods*** and ***cones***. Rods provide coarse definition vision but are very sensitive and are able to operate in a low light intensity environment, e.g. at night. Cones are found in three wavelength-sensitive varieties and thus in combination provide colour vision but require relatively higher light levels to operate. Further signal processing of cone/rod output is carried out by ***interneurons*** before passing data along the ***optic nerve*** to the brain.

At the front of the eye, the retina is discontinued becoming instead a collection of muscle and tendons attached to the ***lens*** which carries out a focussing function.

The volumes behind and in front of the lens are not empty spaces. The volume of viscous liquid to the rear of the lens (the posterior chamber) is termed the ***vitreous humour*** and that in front (the anterior chamber) is the waterier ***aqueous humour***.

8.3   Light Sources and Receivers

A visible light source is an object producing electromagnetic radiation over a particular range of frequencies at which the human eye is sensitive.

The visible spectrum ranges from, at one end, violet light (wavelength 400 nm, frequency 7.5×10^{14} Hz,) to red light (wavelength 700 nm, frequency 4×10^{14} Hz,) at the other. See Figure 8.2.

All of these colours in combination are known as 'white' light whilst an absence of any of these wavelengths is known as 'darkness'.

Light can be described in terms of numbers of photons of a particular wavelength or range of wavelengths either *emitted* from a surface, *in transit*, or *arriving* at a surface.

It is more practical in studies of illumination to differentiate between these three aspects, and each is provided with its own distinct terminology and unit.

Consider a light source and a surface at some distance from the source as shown in Figure 8.3.

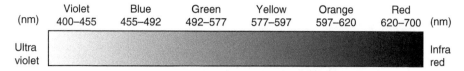

Figure 8.2 The visible spectrum.

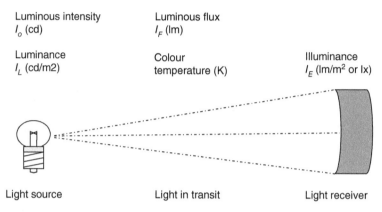

Figure 8.3 Light sources and receiver parameters.

There are five parameters of interest:

- Luminous intensity (I_o, cd)
- Luminance (I_L, cd/m^2)
- Luminous flux (I_F, lm)
- Colour temperature
- Illuminance (I_E, lm/m^2 or lx)

The photonic emission rate of a light source is described by its **luminous intensity (I_o)** measured in *candela* (cd). This unit is wavelength dependent based on the response of the human eye.

One candela is defined as being equivalent to approximately 1.464×10^{-3} W/sr emitted at a frequency of 540×10^{12} Hz, in a nominal direction.

To establish magnitude, a camera flash has a luminous intensity of approximately 1×10^6 cd, while a 4 W light-emitting diode (LED) might have a luminous intensity of a few hundred candela.

The **luminance (I_L)** measured in cd/m^2 is another term used to quantify the amount of light leaving per unit area (A) of surface in a particular direction, as seen by the viewer, i.e.

$$I_L = I_o/A \tag{8.1}$$

In everyday language, the term luminance is closely linked to *brightness*.

High-quality televisions, typically, have a luminance in excess of 500 cd/m^2. Computer monitors have a luminance of a few hundred cd/m^2. For the purposes of comparison, at its highest point, the luminance of the sun is of the order of 1×10^9 cd/m^2.

For a spherical source, the area seen by the viewer will be a circle and that for a cylinder (along the long axis) will be a rectangle.

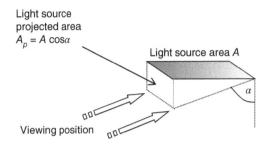

Figure 8.4 Projected area for a large area light source.

For a light source viewed at an angle other than 90° to the source surface, the projected surface area (A_p) must be used in luminance calculations.

For example, in Figure 8.4, with a viewing angle of α, the area used to calculate the luminance of the source is reduced, i.e. $A_p = A \cos \alpha$ and

$$I_L = I_o / A_p \tag{8.2}$$

The **luminous flux** (I_F) is a term referring to light in transit or the flow of light energy and is the product of the luminous intensity of a source and the solid angle (steradians) through which it passes. Luminous flux is therefore measured in cd.sr or, more commonly, *lumen (lm)*.

The ratio of the amount of luminous flux reflected from a surface to that incident upon it is termed the **reflectance** of the surface.

For almost the totality of human history our species has been reliant on the Sun and fire, i.e. hot bodies, for illumination. The Sun's surface temperature is approximately 5500 K, and this **temperature** provides a light colour quality of that experienced at noon during a clear summer's day. Sources at a lower temperature of say 2000–3500 K provide a light quality biased towards longer red-orange wavelengths redolent of a sunset/sunrise.

The quality of artificial light is often described relative to this temperature perception, even though the lamp temperatures are much lower and the resulting wavelengths reached by non-thermal means. The resulting lamp light quality is often described using terms such as 'cool white', 'warm white', etc. More formal scales assess colour fidelity and object rendition. The Colour Rendering Index (CRI) which describes how well (0–100) a light source mimics a claimed colour temperature has been in use for quite a while. The design for standardisation goes on in the face of new lamp technologies and this is likely to be replaced by newer indices (e.g. TM30) which sample a greater range of wavelengths in comparison to CRI.

The magnitude of light power incident on a surface is described by its **illuminance** (I_E) measured in *lux* (lx or lm/m^2) of receiving surface.

Artificial lighting typically provides indoor lighting levels of 200–2000 lx depending upon application.

For comparison purposes, outdoors on a sunny day the illuminance can be of the order of 100×10^3 lx. On an overcast day, this value can be reduced by at least 90%.

8.4 Laws of Illumination

In order to provide a simplified basis for analysis, light sources are often considered to have negligible dimensions or at least have dimensions that are small compared to the distance to the illuminated surface. When this simplification is made the source is termed a ***point*** source.

Another simplification, often employed, is to assume that the luminous intensity of the light source does not vary with direction. Under these circumstances, the source is said to be radiating a ***uniform*** field. These assumptions will be made in what follows.

Let us define the luminous intensity (I_o) as the luminous flux (I_F) per steradian (w), i.e.

$$I_o = I_F/\omega \tag{8.3}$$

A point source of light radiating a uniform spherical field will emit through 4π steradians; hence, the luminous flux I_F is given by:

$$I_F = I_o 4\pi \tag{8.4}$$

The illuminance (I_E) is defined as the luminous flux per area of receiving surface, i.e.

$$I_E = I_F/A \tag{8.5}$$

For a spherical radiating field with a radius (r) the light is received at right angles to the surface and:

$$\begin{aligned} I_E &= I_F/\left(4\pi r^2\right)\sqrt{a^2 + b^2} \\ &= I_o 4\pi/\left(4\pi r^2\right) \\ &= I_o/r^2 \end{aligned} \tag{8.6}$$

In words, the illuminance from a point source radiating a spherical field drops off with the inverse square of the distance from the source.

The maximum amount of light from a source will be collected by a receiving surface at 90° to the source. For a receiving surface at some other angle to the source, the illumination will be reduced. Look at Figure 8.5. Surface AB is at 90° to the incoming luminous flux I_F and receives the maximum illumination.

If the same amount of luminous flux is, instead, intercepted by a surface such as AC, it can be seen that the flux is spread over a greater area and hence the illumination (lm/m^2) is reduced compared to the original case.

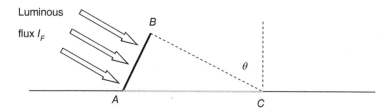

Figure 8.5 Effect of incident angle on surface illumination.

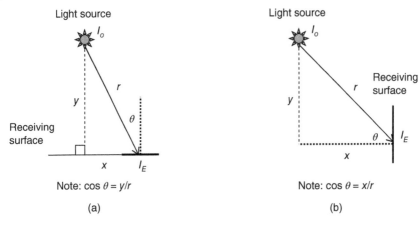

Figure 8.6 Two-dimensional point-to-point lighting geometry. (a) Horizontal receiving surface and (b) vertical receiving surface.

The amount of reduction is proportional to the cosine of the angle between the incoming flux and a 90° normal to the surface AC.

If light is striking a surface at angle other than 90°, then a correction to the simple inverse square law relationship is necessary. Consider a point on a surface receiving light at an angle θ relative to the surface normal as shown in Figure 8.6.

Dimensions x and y are the horizontal and vertical distances between source and surface.

If the distance between source and surface is, again, designated r, then the illuminance I_E is given by:

$$I_E = I_o \cos\theta / r^2 \tag{8.7}$$

where $r^2 = x^2 + y^2$ from Pythagoras's theorem.

In considering the average illuminance over a receiving *surface*, solid angle relationships for circular and rectangular geometries are useful.

The total surface area of a sphere is given by a well-known formula but how about an area less than this?

A spherical surface area of less than 4π steradians is known as a *spherical cap*.

Look at Figure 8.7. First, note the line drawn vertically downwards from the centre of the sphere to its receiving surface. Let this be a datum or reference line at 0°.

Then notice the second line and its point of contact drawn at an angle ϕ (degrees) out to the sphere surface.

Now rotate the point of contact of the second line around the surface in a horizontal plane. The surface area of the shape below the plane is an example of a spherical cap.

It can be shown that the solid angle (ω, steradians) produced between any two angles ϕ_1 and ϕ_2 is given by:

$$\omega = 2\pi \left(\cos\phi_1 - \cos\phi_2\right) \tag{8.8}$$

Since in Figure 8.7, ϕ_1 is equal to 0° then:

$$\omega = 2\pi \left(1 - \cos\phi\right) \tag{8.9}$$

Sphere centre

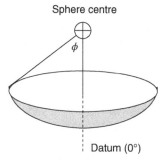

Datum (0°)

Figure 8.7 Spherical cap surface.

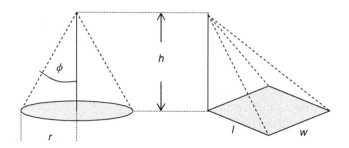

Note: tan $\phi = r/h$

Note: 1. The point source is vertically above corner of rectangle
2. tan^{-1} must be evaluated in radians

$$\omega = 2\pi \left(1 - \cos\phi\right) \qquad \omega = \tan^{-1}\left[\frac{\frac{l}{h} \times \frac{w}{h}}{\sqrt{1 + \left(\frac{l}{h}\right)^2 + \left(\frac{w}{h}\right)^2}}\right]$$

Figure 8.8 Solid angle (steradians) relationships for discs and rectangles.

Note that this solid angle is valid for the cap projection forming the circular horizontal plane and can therefore be used for estimating the average illuminance on a circular area.

Similar relationships can be derived for other receiving surface geometries.

The arrangements and derived relationships for an illuminated circle and an illuminated rectangular case are shown in Figure 8.8.

The expression for the rectangle appears complex but is really only based on three-dimensional lengths.

8.5 Lamp Types

Historically, artificial light has been produced by either heating a metal filament to such a temperature that it glows or by exciting the molecules of a gas or metallic vapour or solid to the point where a photon emission occurs. The emitted photons may not be in the visible light range in which case an intermediate substance or coating may be required for conversion. Current lamp technology is driven by the need for lighting flexibility and high-energy efficiency/low carbon impact requirements rendering some types of historical interest only.

8.5.1 Light-Emitting Diodes (LEDs)

LEDs are an established technology benefiting from more recent material development. An LED comprises a semiconductor material that will emit visible light upon the application of an electrical energy source.

Electrons traversing a p–n semiconductor junction can be encouraged to emit photonic energy.

Historically, red and green LEDs were common. Latterly blues and yellows became available. White light LEDs are now being produced by using a mixture of materials.

More recently, white light LEDs use a single wavelength producing material and layers of phosphor polymers that modify the photon wavelength to produce the range required.

In the manner by which fluorescents replaced tungsten lamps, the use of LED lamps is now commonplace in commercial and domestic applications as replacement for discharge lamps. They are quick starting, have long lives, emit little or no infrared or ultraviolet wavelengths and are inexpensive to use. The form of LED lamps has, to some extent, been dictated by the need to facilitate some aspects (e.g. easy retrofit, aesthetics) of the older lamps they seek to replace. However, as a result of their long lives, some new LED installations use integral lamp fittings.

8.5.2 Gas/Vapour Discharge Lamps

8.5.2.1 Tubular Fluorescents

These comprise a glass tube construction housing an electrode at either end. The tubes are filled with a *low pressure* ($\sim$1 Pa) mixture of argon, krypton, and a small amount of mercury. Upon the application of a potential difference the inert gases become ionised producing sufficient heat to vaporise the mercury which sustains the discharge. When excited the gas/vapour mixture produces light at mostly ultraviolet (i.e. invisible) wavelengths and so the inside of the glass tube is coated with a mixture of phosphor-based compounds which upon the receipt of UV will themselves be exited and re-radiate (fluoresce) at visible wavelengths.

Linear and compact geometries are available. Compact fluorescents simply comprise small diameter tubes in a folded arrangement.

Both types require electronic control gear.

Historically, fluorescents have been the lamp of choice for most general lighting systems in offices, shops, educational, and health applications.

The construction of these lamp types is shown in Figure 8.9.

8.5.2.2 Metal Vapour/Metal Halide Lamps

The term metal vapour represents a range of lamp types utilising a *high pressure* ($\sim$300 kPa) mercury vapour discharge in a small quartz discharge tube containing two main electrodes.

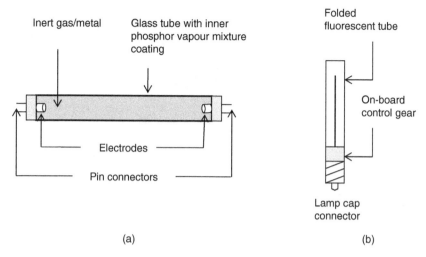

Figure 8.9 Generalised construction of fluorescent lamp types. (a) Tube fluorescent lamp and (b) compact fluorescent lamp.

The discharge is initiated by the use of a third starter electrode and the presence of an inert gas like argon.

The colour quality of the light produced by this simple arrangement is poor with wavelengths concentrated in the blue green area. The colour quality can be increased by the provision of a fluorescent phosphor powder coated outer tube.

Further improvements in colour output can be brought about by the addition of metal (e.g. thallium, indium) salts in the discharge tube as is the case in the metal halide lamp.

Lamp lives are inferior to LED and fluorescent lamps.

Metal halide lamps have municipal street lighting, sports, and architectural and aesthetic enhancement applications. Indoor industrial applications are also common.

The general arrangement of these lamps is shown in Figure 8.10a.

8.5.2.3 Sodium Lamps

Both high- and low-pressure versions are available.

In the high pressure version, the discharge takes place in sodium vapour, operating at a pressure of around 10 Pa. At this pressure, the sodium is very corrosive and so is contained inside a translucent ceramic tube which in turn is surrounded by an outer evacuated glass casing. Like fluorescent tubes and metal vapour lamps, the ionisation is initiated in an inert gas usually neon or xenon. See Figure 8.10b.

In the low pressure version operating at about 1 Pa, the discharge tube comprises a sodium resistant glass again contained in an evacuated outer glass casing. The colour rendering of the low-pressure version is inferior to that of the high pressure with yellow wavelengths predominating its output.

The colour quality of the light produced is generally not as good as other lamp types but they tend to have long lives. They have been used for municipal and commercial applications in parking, street, and road lighting. They are also suitable for industrial indoor use if colour quality is not critical. See Figure 8.10c.

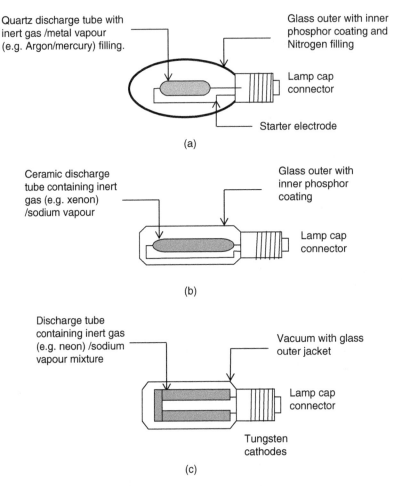

Quartz discharge tube with inert gas /metal vapour (e.g. Argon/mercury) filling.

Glass outer with inner phosphor coating and Nitrogen filling

Lamp cap connector

Starter electrode

(a)

Ceramic discharge tube containing inert gas (e.g. xenon) /sodium vapour

Glass outer with inner phosphor coating

Lamp cap connector

(b)

Discharge tube containing inert gas (e.g. neon) /sodium vapour mixture

Vacuum with glass outer jacket

Lamp cap connector

Tungsten cathodes

(c)

Figure 8.10 General construction of metal vapour/sodium lamp types. (a) Metal vapour lamp, (b) high pressure sodium vapour lamp, and (c) low pressure sodium vapour lamp.

8.5.3 Incandescent Lamps

8.5.3.1 Tungsten Filament

These comprise a tungsten filament enclosed by a glass bulb. The space between the filament and glass is filled with an inert gas to help slow the deterioration (boiling off) of the filament. Due to their ease of installation and low cost their deployment was widespread, particularly in the domestic sector. Their use and availability are now in decline.

8.5.3.2 Tungsten Halogen Lamps

These are similar in operation to tungsten filament with the difference that the filament is contained in a quartz tube containing a halogen gas, e.g. an iodide compound. The presence of the gas facilitates the recycling of tungsten back on to the filament. They may be single

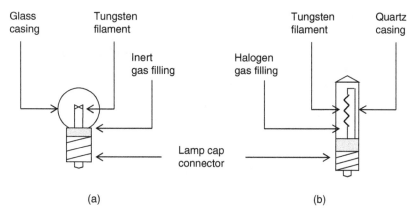

Figure 8.11 Examples of incandescent lamps. (a) Tungsten filament lamp and (b) single-ended tungsten halogen filament lamp.

or double ended and are commonly used in display settings. They are more expensive than simple tungsten filament lamps. Low voltage versions are available.

The general construction of the above incandescent lamp types is shown in Figure 8.11.

Their availability is becoming much less common in areas such as the European Union where more cost-effective alternatives, such as LEDs, are popular.

8.5.4 Luminous Efficacy

In an ideal world, a lamp would produce only visible light. In practice, lamps may also emit other wavelengths including, in some cases, a good deal of heat. The term **luminous efficacy** (lm/W) is used to compare light sources based on the amount of luminous flux produced per unit input of electrical power. A comparison of typical lamp efficacies is given in Table 8.1.

Table 8.1 Typical luminous efficacy ranges.

Lamp technology	Luminous efficacy (lm/W)
Low-pressure sodium	150–200
LED	90–150
High-pressure sodium	50–150
Metal halide	70–110
Fluorescent	50–100
Compact fluorescent	40–80
Tungsten halogen	15–25
Tungsten	5–15

8.6 Luminaires and Directional Control

Lighting systems are usually more than bare lamps. Lamps are instead contained within a fitting or **luminaire** having a range of functions, e.g. facilitating the installation, housing any attendant control gear, or being either unobtrusive or prominent dependent on application. However, after the provision of electrical, mechanical, and thermal safety, perhaps the most important function of a luminaire is the distribution and conditioning of the output from the light source.

From this point of view, luminaries may *reflect, refract,* or *diffuse* the luminous output from a lamp. Some luminaries are designed for a combination of operations.

8.6.1 Reflection

Reflection may be defined as the re-directing of a light beam due to its impact with a surface. In this case, both incoming and re-directed light beams travel in the same medium. The resulting effect to a viewer is dependent on the nature of the reflecting surface. If the surface is highly polished or *specular*, then the incoming and reflected beams are in the same plane and the angle (i) made between an incoming light beam and a line drawn normal to the surface is equal to the angle (r) between the reflected beam and the normal. See Figure 8.12.

Elliptical, parabolic, and spherical surfaces are all employed in luminaries utilising reflection as a means of directional light control.

8.6.2 Refraction

The phenomenon of refraction takes place when a light beam travels from one medium into another, e.g. from glass to air. Due to the different velocities in the two media, the beam is redirected (or 'bent') during its passage across the interface. See Figure 8.13.

The effect that a particular medium has on the velocity of light through it is expressed by the refractive index (n).

This is a ratio referenced to the speed of light in a vacuum.

The refractive index for a material is dependent on wavelength. At mid-range wavelengths of the visible spectrum, air has a refractive index of approximately unity. Typically, glass refractive indices vary from 1 to 2 depending on composition.

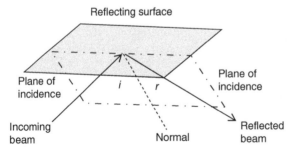

Figure 8.12 Specular reflection.

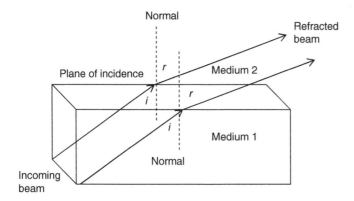

Figure 8.13 Refraction.

Look again at Figure 8.13. Let the angle of incidence (i) be that angle made between the beam in medium 1 and a normal drawn at the interface and the angle of refraction (r) be the angle made between the refracted beam in medium 2 and the normal. The amount of re-direction obtained by refraction can be calculated from an expression known as Snell's Law, thus

$$\frac{\sin i}{\sin r} = \frac{n_{medium\ 2}}{n_{medium\ 1}} \tag{8.10}$$

where n is the refractive index.

In general terms, if $n_{medium\ 1} > n_{medium\ 2}$, then the refracted beam is 'bent' away from the normal. If $n_{medium\ 2} > n_{medium\ 1}$, then the refracted beam is 'bent' towards the normal.

Lenticular and prismatic shapes are commonly employed in luminaries utilising refraction as a means of light control.

8.6.3 Diffusion

A surface or medium that scatters or alters the distribution of light in a way more complex than is indicated above is said to be exhibiting diffusion. Here reflection and refraction may both play a part but the result usually produces a general rather than coherent effect. This may be used to conceal or disguise the lamp outline or form.

8.6.4 Directional Performance Ratios

Draw an imaginary horizontal plane through the centre of a lamp, the effect of the luminaire can be quantified by some simple light output (lumen) ratios thus:

- Light output

$$\text{Light output ratio (LOR)} = \frac{\text{Total light output of luminaire}}{\text{Lamp light output}}$$

$$= \frac{I_{F,lum}}{I_{F,lamp}} \tag{8.11}$$

$$\text{Downward light output ratio (DLOR)} = \frac{\text{Downward light output of luminaire}}{\text{Lamp light output}}$$

$$= \frac{I_{F,\text{lum,down}}}{I_{F,\text{lamp}}} \tag{8.12}$$

$$\text{Upward light output ratio (ULOR)} = \frac{\text{Upward light output of luminaire}}{\text{Lamp light output}}$$

$$= \frac{I_{F,\text{lum,up}}}{I_{F,\text{lamp}}} \tag{8.13}$$

And

$$\text{LOR} = \text{DLOR} + \text{ULOR} \tag{8.14}$$

- Flux fraction
  Another term used to classify luminaires according to the percentage of downward luminous flux (DFF) directly reaching the working plane, leading to descriptions of luminaires as 'direct' (high DFF, 90–100%), 'generally diffusing' (40–60%), and 'indirect' (low DFF, 0–100%).
- Maintenance factors
  (i) *Lamp maintenance factor* (L_aMF): A ratio to account for the loss of lamp light output with age.
  (ii) *Luminaire maintenance factor* (L_uMF): A ratio to account for the loss of light output associated with the fitting's state of cleanliness.
  (iii) *Room maintenance factor* (RMF): Strictly speaking not related to the lamp/luminaire combination but affecting resulting light levels due reduced to reflectivity, again, resulting from a poor state of cleanliness.
  (iv) *Light loss factor* (LLF): A parameter accounting for the above thus:

$$\text{LLF} = L_a\text{MF} \times L_u\text{MF} \times \text{RMF} \tag{8.15}$$

8.6.5 Lumen Method

The lumen method of lighting design provides a simple alternative to the *point-to point* method of predicting resulting light levels, described earlier, by using output performance ratios.

The utilisation factor (UF) is a manufacturer supplied parameter that takes into account light output, flux distribution, as well as room proportions, room surface reflectance (typically 0.2–0.7), and the ratio of spacing to mounting height (H_{mounting}, (m)) of a proposed fitting.

The effect of room proportion (length (L,m) and width (W,m)) on utilisation factor is characterised by a parameter called the room index (RI).

$$RI = \frac{LW}{H_{\text{mounting}}(L+W)} \tag{8.16}$$

The utilisation factor can be used to calculate the required number of luminaires (N) to provide a desired average illuminance (I_L, lx) on a working plane thus

$$N = \frac{I_L A}{n\, I_F \, \mathrm{UF}\, \mathrm{LLF}}$$

(8.17)

where n is the number of lamps per fitting, I_F is the lamp output (lumen), and A is the area (m^2) of the working plane. The method is best applied to ceiling-mounted, grid arranged lighting schemes.

The actual fitting layout is directed by a manufacturer-set parameter called the (lamp) spacing–height mounting ratio (SHR) designed to ensure illumination uniformity. Designers seek to provide a luminaire layout with an SHR below that recommended by the manufacturer.

In the case of linear light sources, both axial (between luminaire centres in each row) and transverse (between rows) space mounting ratios will require calculation ensuring that the following rule is observed:

$$\mathrm{SHR}_{ax} \times \mathrm{SHR}_{trans} \leq \left(\mathrm{SHR}_{nominal}\right)^2$$

(8.18)

where SHR_{ax} is the spacing in rows and SHR_{trans} is the spacing between rows.

8.6.6 Glare

The most recognisable human factor associated with too much light is **glare**.

Glare is a phenomenon that can impair vision (disability glare) or in extreme cases even cause pain (discomfort glare).

Disability glare results when the eye is swamped with so much light that it loses it sensitivity to small differences in intensity, i.e. to detail.

Discomfort glare is thought to result from a rapid contraction of the pupil in response to the over stimulus.

Factors affecting glare include background or ambient luminance, number of glare sources, source size, and luminance as well as position (angle) relative to the viewer. General background to task light level ratios in excess of 10 : 1 are a problem.

Glare can be avoided and, generally, is the result of poor lighting design and/or installation.

A good lighting installation will not only consider source intensity and resulting light levels but also light distribution and so significant thought must be given to source location and mounting height. A number of glare indices and ratings are in use. Glare ratings run from values of 10 (unnoticeable) to 90 (unbearable) with a value of 50 being barely acceptable.

The term *light pollution* is used to describe the situation where the production and directional delivery of light is inappropriate or 'out of place'.

8.7 Worked Examples

Worked Example 8.1   Luminance
A computer monitor has a luminous intensity of 25 cd.
   The dimensions of screen are 400 mm × 250 mm.
   Determine its luminance (cd/m²)

Solution
Given: $I_o = 25$ cd, $A = 0.4 \times 0.25 = 0.1\,s^2$
   Find: I_L

$$I_L = I_o/A$$
$$= 25/0.1$$
$$= 250\,\text{cd}/\text{m}^2$$

Worked Example 8.2   Luminous Intensity of a Light Source
The luminance of a light source shown in Figure E8.2 is 400 000 cd/m².

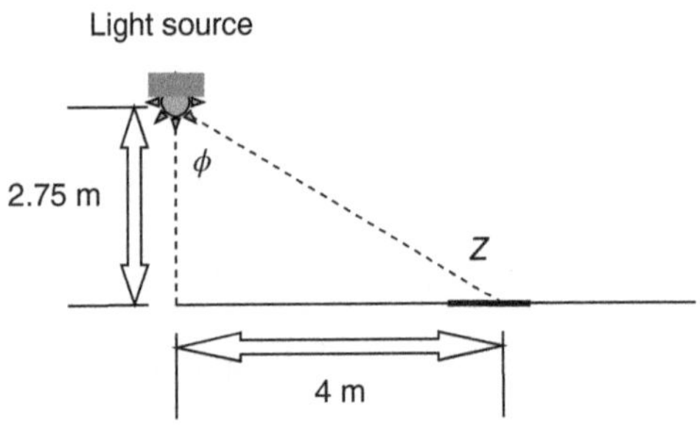

Figure E8.2   Layout for Worked Example 8.2.

   If the flat, circular source has a 7.5 cm diameter, determine the viewing angle (degrees), projected area (m²), and luminous intensity (candela) as viewed from the surface (Z).

Solution
Given: $d_{source} = 0.075$, $I_L = 400\,000$ cd/m²
   Find: I_o
   Source surface area (A_{source}):

$$A_{source} = \pi d_{source}^2/4$$
$$= \pi 0.075^2/4 = 4.418 \times 10^{-3}\,\text{m}^2$$

Viewing angle:

$$\tan\phi = 4/2.75 = 1.454$$
$$\phi = 55.49°$$
$$\cos\phi = 0.566$$

Source projected surface area (A_p):

$$A_p = A_{source} \cos \alpha = 4.418 \times 10^{-3} \times 0.566 = 2.502 \times 10^{-3} \text{ m}^2$$

Then:

$$I_L = I_0/A_p$$

$$400\,000 = I_0/2.502 \times 10^{-3}$$

$$I_0 = 1001 \text{ cd}$$

Worked Example 8.3   Illuminance

The *spherical* light source shown in Figure E8.3 uniformly emits 1500 lm.

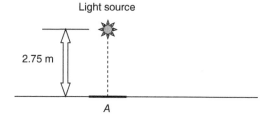

Figure E8.3   Layout for Worked Example 8.3.

Determine the luminous intensity of the source and the illuminance on a horizontal surface (A) at a distance of 2.75 m and normal to the source.

Solution

Given: $I_F = 1500$ lm, $r = 2.75$ m

Find: I_0, I_E

A spherical source with a uniform emission radiates through 4π radians.

$$I_0 = I_F/4\pi$$

$$= 1500/4\pi$$

$$= 119.35 \text{ cd}$$

And

$$I_E = I_0/r^2$$

$$= 119.35/2.75^2$$

$$= 15.78 \text{ lx}$$

Worked Example 8.4   Average and Specific Illuminance

A light source having a uniform luminous intensity of 10 000 cd is mounted 10 m above ground as shown in Figure E8.4.

If the area of illumination on the ground can be regarded as circular with a 50 m diameter, determine:

(a) the average illuminance (lux) at ground level
(b) the illuminance (lux) at a point on the ground at the periphery of the illuminated area.

Light source

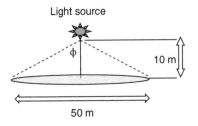

10 m

50 m

Light source

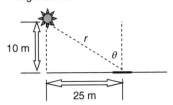

10 m

r

θ

25 m

Figure E8.4 Layout for Worked Example 8.4. (a) Average illuminance conditions. (b) Peripheral illuminance conditions.

Solution

Given: $h = 10\,\text{m}$, $I_o = 10 \times 10^3\,\text{cd}$, $d_s = 50\,\text{m}$, i. e. $r = 25\,\text{m}$

    Find: $I_{E,\,\text{ave}}$, $I_{E\,@\,25\text{m}}$

(a) *Average illuminance:*

    Source solid angle

$$\tan \phi = r/h = 25/10 = 2.5 \ \text{i.e.} \ \ \phi = 68.198° \ \text{and} \ \cos \phi = 0.371$$

    Then

$$\omega = 2\pi\,(1 - \cos \phi) = 2\pi\,(1 - 0.371) = 3.95 \ \text{sr}$$

    And

$$I_F = I_o\omega = 10\,000 \times 3.95 = 39\,500 \ \text{lm}$$

    Illuminated surface area:

$$A = \pi d_s^2/4 = \pi \times 50^2/4 = 1963.5\,\text{m}^2$$

    Then

$$I_{E,\text{ave}} = I_F/A = 39\,500/1963.5 = 20.12\,\text{lx}$$

(b) *Illuminance at periphery:*

    Using the inverse square relationship

$$r^2 = x^2 + y^2 = 25^2 + 10^2 = 725$$

$$r = 26.925\,\text{m}$$

$$\cos \theta = 10/26.925 = 0.371$$

$$I_{E,@25\,\text{m}} = \frac{I_o \cos \theta}{r^2} = \frac{10\,000 \times 0.371}{26.925^2} = 5.12\,\text{lx}$$

Worked Example 8.5   Spherical Cap

A point is 2.75 m above a surface as shown in Figure E8.5. Determine the solid angle (steradians) subtended at the point if the surface is a disc of radius 1.5 m.

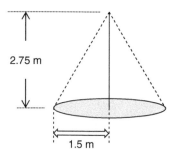

Figure E8.5   Layout for Worked Example 8.5.

Solution

Given: $R = 1.5$ m, $h = 2.75$ m

Find: ω

$$\tan \phi = R/h$$
$$= 1.5/2.75$$
$$= 0.545$$
$$\phi = 28.6^\circ$$
$$\text{Then } \omega = 2\pi \left(1 - \cos \phi\right)$$
$$= 2\pi \left(1 - \cos 28.6\right)$$
$$= 0.766 \text{ sr}$$

Worked Example 8.6   Refraction at a Glass/Air Interface

A light beam travelling in a glass exits into air. See Figure E8.6.

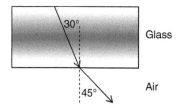

Figure E8.6   Layout for Worked Example 8.6.

If the incident and refracted angles at the interface are 30° and 45°, respectively, estimate the refractive index of the glass.

Take the refractive index of air as unity.

Solution

Given: $i = 30°$, $r = 45°$, $n_{air} = 1$

Find:

n_{glass}

$$\frac{\sin i}{\sin r} = \frac{n_{medium\ 2}}{n_{medium\ 1}}$$

$$\frac{\sin 30}{\sin 45} = \frac{1}{n_{glass}}$$

$$\frac{0.5}{0.707} = \frac{1}{n_{glass}}$$

$n_{glass} = 1.414$

Worked Example 8.7   Light Output Ratios

A *point source* of light uniformly emits 1300 cd. The source is backed by a reflector surface such that the luminaire has the geometry shown in Figure E8.7.

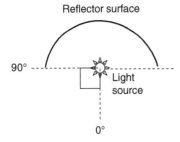

Figure E8.7   Layout for Worked Example 8.7.

If the reflectance of the surface is 0.85, determine the average intensity (cd) of the directed beam and the light output ratio (LOR).

Solution

Given: $I_o = 1300$ cd, reflectance $= 0.85$

Find: $I_{o,\ ave}$, LOR

Lamp luminous flux:

$$I_{F,lamp} = I_o \omega = 1300 \times 4\pi = 16336 \text{ lm}$$

The solid angle subtended at the lamp by the reflector is:

$$\omega = 2\pi \left(\cos 0° - \cos 90° \right)$$

$$= 2\pi \left(1 - 0 \right) = 2\pi \cong 6.283 \text{ sr}, \quad \text{i.e.a hemisphere}$$

Luminous flux emitted in lower hemisphere:

$$I_{F,down} = I_o \omega = 1300 \times 2\pi = 8168 \text{ lm}$$

Luminous flux leaving reflector:

$$I_{F,\text{ref}} = I_o \omega \times \text{reflectance} = 8168 \text{ lm}$$
$$= 1300 \times 2\pi \times 0.85$$
$$= 6943 \text{ lm}$$

Total luminous flux leaving in lower hemisphere:

$$I_{F,\text{lum}} = I_{F,\text{lower}} + I_{F,\text{ref}} = 8168 + 6943 = 15111 \text{ lm}$$

Average luminous intensity:

$$I_{o,\text{ave}} = \frac{I_{F,\text{lum}}}{\omega_{\text{lower}}} = \frac{15111}{6.283} = 2405 \text{ cd}$$

$$\text{LOR} = \frac{I_{F,\text{lum}}}{I_{F,\text{lamp}}} = \frac{15111}{16336} = 0.925 = 92.5\%$$

Worked Example 8.8   Lumen Method

A room has dimensions of 4×6 m. It is to be illuminated to a level of 250 lx with lamps each having an output 5000 lm. The utilisation factor for the fittings is 0.5 and their light loss factor is 0.92.

Determine the number of fittings required to light the room.

Solution

Given: $I_F = 5000$ lm,   UF $= 0.5$,   LLF $= 0.92$, $I_L = 250$ lx, $n = 1$

Find: N

$$A = 4 \times 6 = 24 \text{ m}$$

$$N = \frac{I_L A}{n I_F \text{ UF LLF}}$$

$$= \frac{250 \times 24}{1 \times 5000 \times 0.5 \times 0.92}$$

$$= 2.6 \cong 3$$

Worked Example 8.9   Room Index

A space has a floor with dimensions of 18 m $(L) \times 12$ m (W).

It is illuminated by luminaires 3 m above the working plane.

Determine the room index for the space.

Solution

Given: $L = 18$ m, $W = 12$ m, $H_{\text{mounting}} = 3$ m

Find: RI

$$\text{RI} = \frac{LW}{H_{\text{mounting}} (L + W)}$$

$$= \frac{18 \times 12}{3 (18 + 12)}$$

$$= 2.4$$

Worked Example 8.10   Space Height Mounting Ratio

A linear luminaire manufacturer suggests a $SHR_{nominal}$ for their fittings of 1.5.

If the fitting mounting height is 2.7 m, determine whether the plan layout in the room, shown in Figure E8.10, complies with their suggestion.

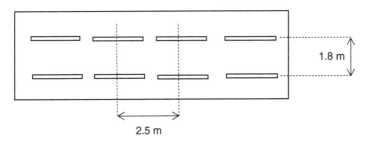

Figure E8.10   Layout for Worked Example 8.10.

Solution

Given: $SHR_{nominal} = 1.5$, $S_{trans} = 1.8$ m, $S_{ax} = 2.5$ m, $H_{mounting} = 2.7$ m

Find: Compliance of layout with $SHR_{nominal}$

$$SHR_{trans} = S_{trans}/H_{mounting} = 1.8/2.7 = 0.666$$

$$SHR_{ax} = S_{ax}/H_{mounting} = 2.5/2.7 = 0.925$$

$$SHR_{ax} \times SHR_{trans} \leq (SHR_{nominal})^2$$

$$0.925 \times 0.666 \leq (1.5)^2$$

$$0.616 \leq 2.25 \quad \text{so design complies.}$$

8.8   Tutorial Problems

8.1   A cinema screen is 10 m high and has an aspect ratio of 2.3 : 1. If its luminance is 48 cd/m^2, determine its luminous intensity (candela).
[Answer: 11040 cd]

8.2   A light source is mounted 3 m above a surface to be illuminated.
The area of the source is 1800 × 500 mm.
Determine the viewing angle (degrees) and light source projected area (m^2) for a viewer 3.5 m from a point directly beneath the source.
[Answers: 49.4°, 0.586 m^2]

8.3   A light source has a luminous intensity of 1000 cd.
Assuming point source and uniform conditions, determine the illuminance (lux) on a surface 3 m from the source.
[Answer: 111 lx]

8.4   A point source of light is 3 m above a surface.
Determine the solid angle (steradians) subtended at the point if the surface is a disc of radius 2 m.
[Answer: 1.055 sr]

8.5 A light source has a luminous flux of 3700 lm emitted through an angle of $2\pi r$. The source is centred 3.75 m above a horizontal receiving surface of diameter 10 m. Determine the average illuminance (lux) on the receiving surface and the illuminance (lux) at its outer perimeter.
[Answers: 47.11 lx, 9.05 lx]

8.6 A light beam exits from a transparent plastic medium into air.
If the interface angle *in the plastic* at the exit point is 25° and the plastic refractive index is 1.5, determine the refracted angle relative to the interface normal. Take the refractive index of air as unity.
[Answer: 39.3°]

8.7 A luminaire manufacturer supplies the following data for one of his products:
Lamp light output: 700 lm
Downward light output of luminaire: 0.85
Upwards light output of luminaire: 0.09
Determine the LOR and the total light output of the luminaire.
[Answers: 0.94, 658 lm]

8.8 A room has floor dimensions of 4×5 m. It is to be illuminated by 10 lamps each having an output of 800 lm. The utilisation factor for the fittings is 0.65, and their light loss factor is 0.94.
Determine the resulting light level (lux).
[Answer: 244 lx]

8.9 Determine the Room Index for the space shown in Figure P8.9.

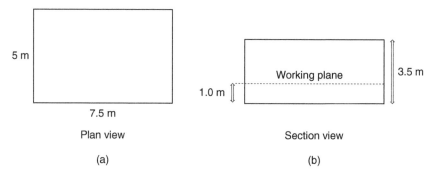

Figure P8.9 Layout for tutorial Problem 8.9. (a) Plan view. (b) Section view.

[Answer: 1.2]

8.10 A room has an overall height of 3.25 m. The working plane with respect to the lighting system is 0.8 m above ground level. If the luminaire manufacturer specifies a space height mounting ratio of 1.25, determine the spacing between fittings.
[Answer: 3.06 m]

9

Cleaning the Air

9.1 Overview

The term particulate matter (PM) or *particulates* describes solid or liquid particles released into the atmosphere from either man-made or natural sources.

At low levels in the atmosphere, particulates are harmful to human health whilst at higher atmospheric levels, they may alter the earth's heat balance due to selective absorption and reflection of electromagnetic radiation.

Natural particulate sources include volcanoes and sea spray.

Significant man-made sources include solid fuel, e.g. coal or biomass-fired electricity generation plant and diesel engines.

This section will introduce some of the significant parameters associated with particle dynamics and particle capture and will go on to apply these to common particulate collections systems such as filters and electrostatic precipitators (ESPs).

The uncontrolled introduction of external gaseous pollutants or their internally generated equivalents, into living and working spaces, is coming under increasing scrutiny. This section will also discuss their characteristics and alleviation.

Learning Outcomes

- To understand the nature of concentration and exposure.
- To be familiar with basic particle collection efficiency and associated performance definitions.
- To be familiar with basic particle forces and associated dimensionless parameters.
- To be familiar with particulate collection devices such as gravity settlers, cyclones, ESP's, etc. and use the mass conservation law to model performance.
- To understand the problems associated with ambient gaseous and vapour pollutants and their filtration.

Building Services Engineering: Smart and Sustainable Design for Health and Wellbeing, First Edition.
Tarik Al-Shemmeri and Neil Packer.
© 2021 John Wiley & Sons Ltd. Published 2021 by John Wiley & Sons Ltd.

9.2 Concentration and Exposure

9.2.1 Concentration Conversions

The basic unit for expressing pollutant concentration in the environment is weight (*of particulate*) per unit volume (*of environment*) measured in microgrammes per cubic metre ($\mu g/m^3$); however, pollutant loading concentrations are often expressed using the fractional approach described in Chapter 1, e.g. parts per million by weight (ppm_w).

A further complication is introduced when considering whether this relates to a volume fraction or mass fraction. In general, solid pollutant concentrations are usually expressed in mass terms and gas concentrations in volume terms.

To differentiate, a gaseous concentration is best expressed as 'ppmv' or 'ppbv'.

The SI equivalent (kg/m^3) of a fractional gas concentration (ppbv) can be estimated with knowledge of the specific gas constant (R, J/kg K), pressure (P, Pa), and temperature (T, K) of the gas as follows:

$$c(kg/m^3) = c(ppbv) \times 10^{-9} \frac{P}{RT} \tag{9.1}$$

The specific gas constant R can be calculated from knowledge of the molecular mass (M, kg/kmol) of the gas of interest as follows:

$$R = R_0/M \tag{9.2}$$

where R_0 is the universal gas constant (J/kmol K).

9.2.2 Pollutant Exposure

The effects of pollutant exposure are quite commonly related with time and risk. Air quality regulations/recommendations attempt to quantify the excess or relative to background probability of suffering a particular adverse medical effect associated with a specified lifetime's exposure level. So, for example, at $20\,mg/m^3$ the excess lifetime risk of an unwanted medical condition associated with exposure to a particular pollutant might be 1/10 000 whilst at $2\,mg/m^3$ the risk might be reduced to 1/1 000 000.

As a result, *acceptable* levels of pollutant are often quoted as a set (arithmetic mean) level for a set maximum time, e.g. $100\,mg/m^3$ for 20 minutes, $50\,mg/m^3$ for 1 hour, etc.

Regulations seldom differentiate 'safe' level and 'dangerous' levels.

Some pollutants are considered to have no safe level of exposure.

9.3 Particulate Pollution

9.3.1 Nature of Particulates

PM is characterised by size. Although not necessarily spherical, they are commonly specified in terms of an equivalent mean diameter (d_p, m). Due to the small scale, the diameter is often quoted using the sub-multiple micrometre (μm) or 'micron'.

A typical size range for PM is 0.01–100 μm.

The size of raindrops, for scale comparison, is typically of the order of 1000 μm and bacteria are in the range 0.4–400 μm.

Particulates having a diameter of <1 μm generally do not perceive the surrounding fluid as a flow continuum but are affected by collisions with *individual* fluid molecules.

Usually, particulate matter of a given size is abbreviated to PM_d, where d is a diameter in microns.

For example, PM_{10} relates to particulates having a diameter of 10 μm or less.

Particulates having a diameter of between 1 and 10 μm form stable suspensions and are commonly termed *smoke*. The term *dust* is often applied to the smallest particles of diameter 0.01–0.1 μm.

The human upper air tract is reasonably efficient in filtering out particulates having a diameter **in excess** of 10 μm (PM_{10}).

If particulates having a diameter of **less than** 10 μm enter the lungs, they have a tendency to remain resident in the alveoli or air sacs with the smallest able to enter the bloodstream.

Particulate-polluted gas streams are rarely homogeneous, containing a range of particle sizes. Particulate size distributions are normally characterised by either arithmetic or geometric, median or mean diameter.

The results of particulate measurements are commonly illustrated by plotting frequency distribution or cumulative distribution curves.

The World Health Organisation time-based concentration guideline limits for exposure to particulates (WHO 2018) in the ambient (outdoor) environment are:

- $PM_{2.5}$: 10 μg/m³ – annual mean, 25 μg/m³ – 24 hour mean
- PM_{10}: 20 μg/m³ – annual mean, 50 μg/m³ – 24 hour mean

The International WELL Building Institute concentration guideline limits for exposure to particulates (IWBI 2018) in the indoor environment are:

- $PM_{2.5}$: <15 μg/m³ (general spaces), <35 μg/m³ (commercial kitchens)
- PM_{10}: <50 μg/m³ – annual mean

9.3.2 Stokes Law and Terminal Velocity

When a particulate moves through a gas, the fluid viscosity produces a force, known as **drag**, on the particulate (d_p, m) in the opposite direction to the relative velocity($\overline{V_r}$).

The **drag force** (F_D) on a spherical particle is given by:

$$F_D = C_D \left(\frac{\rho_g}{2}\right)\left(\frac{\pi d_p^2}{4}\right)\overline{V_r^2} \tag{9.3}$$

where C_D is the drag coefficient.

Consider a spherical particle settling through a fluid **at rest** under the influence of gravity. The particle will accelerate until the frictional drag force + buoyancy force equals the gravity force.

At this point, the particle continues to fall at a constant velocity known as the **terminal settling velocity** ($\overline{V}_{ts}$), see Figure 9.1.

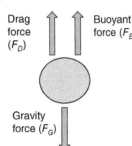

Drag
force
(F_D)

Buoyant
force (F_B)

Figure 9.1 Particle forces on body falling under the effects of gravity.

Gravity
force (F_G)

Using Newton's 2nd law, the resultant force (F_R) on the particle is given by:

$$F_R = F_G - F_B - F_D \tag{9.4}$$

where

$$F_R = \rho_p V_p \left(\frac{d\overline{V}_p}{dt} \right) \tag{9.5}$$

$$F_G = \rho_p V_p g \tag{9.6}$$

$$F_B = \rho_g V_p g \tag{9.7}$$

$$F_D = C_D \left(\frac{\rho_g}{2} \right) \left(\frac{\pi d_p^2}{4} \right) \overline{V}_r^2 \tag{9.8}$$

After substituting Eqs. (9.5)–(9.8) into (9.4) gives:

$$\rho_p V_p \left(\frac{d\overline{V}_p}{dt} \right) = \rho_p V_p g - \rho_g V_p g - C_D \left(\frac{\rho_g}{2} \right) \left(\frac{\pi d_p^2}{4} \right) \overline{V}_r^2 \tag{9.9}$$

Dividing through by $\rho_p V_p$ and noting that the particle area in the final term becomes:

$$\frac{A_p}{\rho_p V_p} = \frac{(\pi d_p^2/4)}{(\pi d_p^3/6)\rho_p} = \frac{3}{2\rho_p d_p} \tag{9.10}$$

Rearranging brings:

$$\frac{d\overline{V}_p}{dt} = \left[\frac{\rho_p - \rho_g}{\rho_p} \right] g - \left[\frac{3C_D \rho_g \overline{V}_p^2}{4\rho_p d_p} \right] \tag{9.11}$$

At terminal settling velocity:

$$\frac{d\overline{V}_p}{dt} = 0, \quad \text{i.e.acceleration is zero}$$

Solving for ($\overline{V}_p$), i.e. the terminal settling velocity($\overline{V}_{ts}$):

$$\overline{V}_{ts} = \sqrt{\left[\frac{4g(\rho_p - \rho_g)d_p}{3C_D \rho_g} \right]} \tag{9.12}$$

The drag coefficient C_D is a dimensionless number dependent on the shape or form of the particle and the friction between the fluid and the surface (or skin) of the particle.

For **spherical** particles, the approximate graphical relationship with Reynolds number is shown in Figure 9.2:

Values for C_D are found empirically and are usually correlated with Reynolds number (Re) in the following form:

$$C_D = b/Re^n \tag{9.13}$$

Correlations for a range of Re values are given in Table 9.1.

Assuming laminar (or Stokes flow) conditions:

$$C_D = 24/Re = \frac{24\mu_g}{\rho_g \overline{V}_{ts} d_p} \tag{9.14}$$

Substituting into the Eq. (9.12) gives:

$$\overline{V}_{ts} = \left[\frac{(\rho_p - \rho_g)gd_p^2}{18\mu_g} \right] \tag{9.15}$$

which is known as Stokes law.

In the study of air pollution, the particulate density (ρ_p) is usually much greater than that of the fluid (ρ_g), and so the effect of the latter is often ignored.

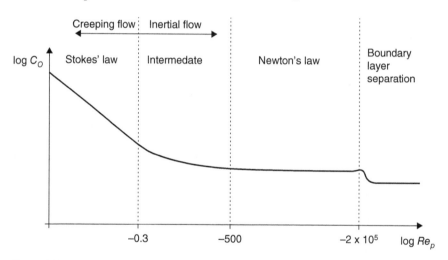

Figure 9.2 Generalised drag coefficient–Reynolds number relationship for spherical bodies. Source: Modified from Rhodes (2008).

Table 9.1 C_D – Re empirical parameters.

Region	Stokes law	Intermediate	Newton's law
Re range	<0.3	$0.3 < Re < 500$	$500 < Re < 2 \times 10^5$
C_D	24/Re	$(24/Re)(1+0.15Re^{0.687})$	∼0.44

Source: Rhodes (2008).

The Stokes number (Stk), a dimensionless parameter commonly used in particle dynamics, is defined as:

$$\text{Stk} = \left[\frac{d_p^2 \rho_p}{18\mu_g}\right]\left[\frac{\overline{V}}{d_{sys}}\right] = [\tau]\left[\frac{\overline{V}}{d_{sys}}\right] \tag{9.16}$$

where the ratio $(\overline{V}/d_{sys})$ is a velocity/length characteristic of the system, and τ may be regarded as the time constant with respect to the attainment of terminal velocity.

Stokes law analyses tend to become unreliable for particle sizes below 1 μm, and calculation of the drag force requires the use of a modification called the Cunningham correction factor (C_c) to account for the collision distance between the smallest particles and gas molecules:

$$C_c = 1 + Kn[\alpha + \beta e^{-\gamma/Kn}] \tag{9.17}$$

where Kn is the Knudsen number (λ/d_p), λ is the mean distance between collisions and α, β, and γ are empirically derived parameters

Then:

$$F_D = \left[C_D\left(\frac{\rho_g}{2}\right)\left(\frac{\pi d_p^2}{4}\right)\overline{V}_r^2\right]/C_c \tag{9.18}$$

9.4 Principles of Particulate Collection

9.4.1 Collection Surfaces

The efficiency (η_c) of the removal process for a collecting surface is defined as the mass flow rate with which the particles are removed by contact with the collecting surface, divided by the mass flow rate of particles approaching the surface in a stream tube of cross-sectional area equal to the collecting surface.

For a collecting surface comprising **cylindrical fibres** of length L_{fb} and diameter d_{fb}, see Figure 9.3.

Using the aforementioned definition:

$$\eta_c = \frac{\text{mass removal rate}}{\overline{V}_r(c)d_{fb}L_{fb}} \tag{9.19}$$

where c is the concentration of particles upstream of the collector.

For a collecting surface comprising **spherical drops**, e.g. a particulate washer or scrubber, see Figure 9.4.

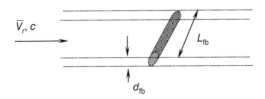

Figure 9.3 Cylindrical collector geometry.

Figure 9.4   Spherical collector geometry.

Using the aforementioned definition:

$$\eta_c = \frac{\text{mass removal rate}}{\overline{V}_r c(\pi d_{sph}^2/4)} \tag{9.20}$$

In both of the cases, $(\overline{V}_r)$ is the particle velocity *relative to the collecting surface*.

9.4.2   Collection Devices

Consider the generic pollution control device shown in Figure 9.5.

The mass flow rate of pollutant upstream of the device is $\dot{V}_0 c_0$ and that leaving is $\dot{V}_1 c_1$. Using the mass-based approach to collection efficiency, the following terms can be defined:

(1) Collection efficiency:

$$\eta_c = \frac{\dot{V}_0 c_0 - \dot{V}_1 c_1}{\dot{V}_0 c_0} = 1 - \frac{\dot{V}_1 c_1}{\dot{V}_0 c_0} \tag{9.21}$$

For equal inlet and outlet volume flow rates, this can be simplified to:

$$\eta_c = 1 - (c_1/c_0) \tag{9.22}$$

(2) Penetration:

$$p = 1 - \eta_c = \frac{\dot{V}_1 c_1}{\dot{V}_0 c_0} \tag{9.23}$$

(3) Decontamination factor:

$$DF = \frac{1}{\text{penetration}} = \frac{\dot{V}_0 c_0}{\dot{V}_1 c_1} \tag{9.24}$$

Particulate control devices are often used in combination.

For N such devices of collection efficiencies $(\eta_1 \dots \eta_N)$ **in series**, see Figure 9.6, it can be shown that:

$$\eta_{overall} = 1 - [(1 - \eta_1)(1 - \eta_2) \dots (1 - \eta_N)] \tag{9.25}$$

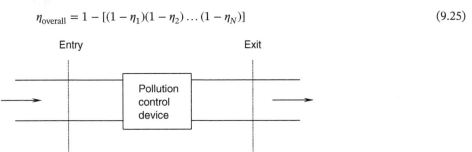

Figure 9.5   Collector operating parameters.

Figure 9.6   Schematic of collectors in series.

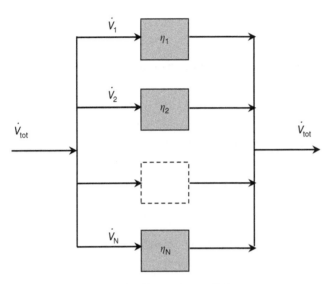

Figure 9.7   Schematic of collectors in parallel.

For N such collection devices *in parallel*, see Figure 9.7, it can be shown that:

$$\eta_{\text{overall}} = 1 - \left[(1 - \eta_1) \left(\frac{\dot{V}_1}{\dot{V}_{\text{tot}}} \right) + (1 - \eta_2) \left(\frac{\dot{V}_2}{\dot{V}_{\text{tot}}} \right) + \dots + (1 - \eta_N) \left(\frac{\dot{V}_N}{\dot{V}_{\text{tot}}} \right) \right] \quad (9.26)$$

9.4.3   Fractional Collection Efficiency

As mentioned earlier, particulate pollutant flow streams often comprise a range of particulate sizes. The *fractional efficiency* ($\eta_{c,d}$) is defined as the fraction of particles *of a given size* (d) collected compared to that (of the same size) entering a pollution control device.

The *total collection efficiency* (η_{tot}) of a collection device is given by:

$$\eta_{\text{tot}} = \sum (\eta_{c,d} \text{mf}) \quad (9.27)$$

where mf = particulate mass fraction of specific size particle.

Another term in common usage is the *cut diameter* defined as the diameter of particle for which the total collection efficiency curve has a value of 0.5, i.e. 50%.

9.5   Control Technologies

Particulate control technologies rely on a range of physical phenomena, i.e. gravity, centrifugal, electrostatic, and impaction forces, to capture and remove particulates from a flow stream. Generally, the smaller the particulate size, the greater the force magnitude required for its removal. Gravity and centrifugal forces are generally not large enough to collect the smallest particles but will be discussed here as they introduce principles, such as laminar and well-mixed particulate behaviour, important to more efficient devices.

9.5.1 Gravity Settlers

At its simplest, a gravity settler (or sedimentation chamber) is an expansion in the cross-sectional area of a duct carrying a particulate-laden fluid stream. The expansion results in a reduction in flow velocity. The change in cross section must be sufficient to allow the particles to attain their terminal velocity. Gravity settlers are not commonly used to clean air entering a building. However, their consideration gives us a deep insight into the principles of more common air cleaning devices and is a good place to start.

For the gravity settler shown in Figure 9.8 with dimensions ($L \times W \times H$):

$$A = WH \tag{9.28}$$

$$\overline{V}_{ave} = \dot{V}/WH \tag{9.29}$$

Two gas–particle interaction models are in common usage to predict particle collection efficiency. The following assumptions apply to both:

- $\overline{V}_{x,g} = \overline{V}_{x,p} = \overline{V}_{ave}$
- $\overline{V}_{y,p} = \overline{V}_{ts}$
- Stokes flow
- If a particle falls to the floor, it cannot be re-entrained

9.5.1.1 Model 1: Unmixed Flow Model
This assumes there is no mixing mechanism to redistribute *yet-to-be* collected particles across the volume. See Figure 9.9.

In time t, at a distance L downstream of the inlet, all the particles of the same diameter have fallen the same distance and the uppermost particles have fallen a distance ($H - y$)

Let

$$\overline{V}_{ave} = L/t \tag{9.30}$$

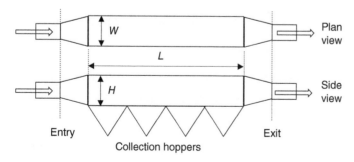

Figure 9.8 Gravity settler overall dimensions.

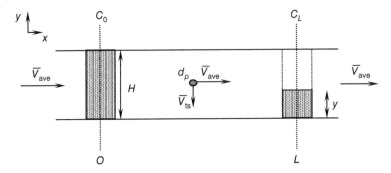

Figure 9.9 Gravity settler 'unmixed or laminar' model parameters.

and

$$\overline{V}_{ts} = \frac{H - y}{t} \tag{9.31}$$

Rearranging (9.31) and substituting from (9.30)

$$H - y = t\overline{V}_{ts} = (L/\overline{V}_{ave})\overline{V}_{ts} \tag{9.32}$$

Let the average concentration at point L be c_L:

$$c_L = (y/H)c_0, \quad \text{i.e.} \quad y = (c_L/c_0)H \tag{9.33}$$

Rearranging and substituting from Eq. (9.32):

$$c_L/c_0 = 1 - (L/H)(\overline{V}_{ts}/\overline{V}_{ave}) \tag{9.34}$$

Hence, from Eq. (9.22):

$$\eta_c = (L/H)(\overline{V}_{ts}/\overline{V}_{ave}) \tag{9.35}$$

Substituting for $\overline{V}_{ts}$ from Eq. (9.15) and ignoring gas density:

$$\eta_c = \frac{Lgd_p^2\rho_p}{18H\overline{V}_{ave}\mu_g} \tag{9.36}$$

9.5.1.2 Model 2: Well-Mixed Flow Model

This assumes that as particles of a certain size fall to the bottom, a mixing mechanism redistributes the remaining particles throughout any given volume.

Consider a length of settler dx having a uniform concentration c throughout its volume Adx. See Figure 9.10.

Performing a mass balance:

> mass flow rate of particles entering the volume = mass flow rate of particles leaving the volume + rate of particle collection in the volume

i.e.

$$c\overline{V}_{ave}A = [c + dc]\overline{V}_{ave}A + c\overline{V}_{ts}W \cdot dx \tag{9.37}$$

where $W \cdot dx$ is the collection area at the bottom of the settler.

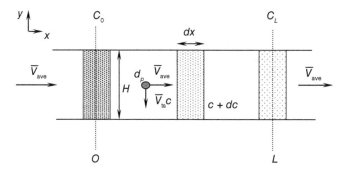

Figure 9.10 Gravity settler 'well mixed' model parameters.

Remembering that: $A = WH$, divide through by $\overline{V}_{ave} A$
Rearranging and separating variables:

$$c - c\frac{\overline{V}_{ts}}{\overline{V}_{ave}H} \cdot dx = c + dc$$

$$c\left(1 - \frac{\overline{V}_{ts}}{\overline{V}_{ave}H} \cdot dx\right) = c + dc$$

$$-\frac{\overline{V}_{ts}}{\overline{V}_{ave}H} \cdot dx = \frac{c + dc}{c} - 1$$

$$-\frac{\overline{V}_{ts}}{\overline{V}_{ave}H} \cdot dx = \frac{c}{c} + \frac{dc}{c} - 1 = \frac{dc}{c}$$

Integrating between $c = c_0$ and c_L, and $x = 0$ and L:

$$\int_{c_0}^{c_L} \left(\frac{1}{c}\right) \cdot dc = \int_0^L -(\overline{V}_{ts}/\overline{V}_{ave}H) \cdot dx$$

[Remembering that: $\int \frac{1}{x}dx = \ln x$]

$$\ln c_L - \ln c_0 = -(\overline{V}_{ts}/\overline{V}_{ave})(L/H)$$

[Remembering that: $\ln x - \ln y = \ln(x/y)$]
Rearranging: $c_L/c_0 = \exp[-(\overline{V}_{ts}/\overline{V}_{ave})(L/H)]$
Using the definition of (9.22):

$$\eta_c = 1 - \exp[-(\overline{V}_{ts}/\overline{V}_{ave})(L/H)]$$

Substituting from (9.15):

$$\eta_c = 1 - e^{\left(-\frac{Lgd_p^2\rho_p}{H\overline{V}_{ave}18\mu_{gas}}\right)} \tag{9.38}$$

The well-mixed and unmixed (or laminar) collection efficiency models are related by:

$$\eta_{well\ mixed} = 1 - e^{(-\eta_{unmixed\ flow})} \tag{9.39}$$

The generalised form of the collection efficiency–particle diameter relationship is shown in Figure 9.11.

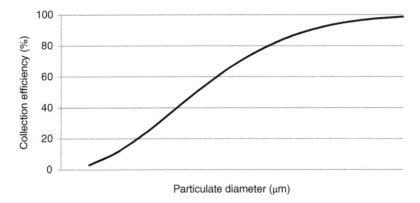

Figure 9.11 Generalised collection efficiency–particle size relationship for well-mixed model assumptions.

9.5.2 Centrifugal Separators or Cyclones

Due to the relatively weak gravitational force, gravity settlers are only suitable for the collection of large particles. In order to be effective for smaller particles, the gravitational force must be replaced by a stronger force.

In the cyclone, the collection force is centrifugal in nature.

Consider a particle in a fluid stream flowing at a radius r in a curved duct of height H. See Figure 9.12.

Assume:

- Solid body rotation
- The tangential velocities of the gas and particle are equal, i.e. $\overline{V}_g = \overline{V}_p$
- The gas has no radial velocity component
- In the radial direction, the centrifugal force is balanced by the viscous drag force

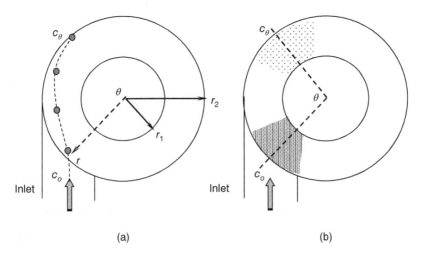

Figure 9.12 Curved duct collection principles. (a) Particle collection path and model parameters. (b) Well-mixed model, i.e. redistribution of uncaptured particles across duct.

- Well-mixed conditions, i.e. redistribution of any remaining particles resulting in concentration varying with angle or curvature not radius

Let the fluid velocity be a function of radius r, i.e. $\overline{V}_g = C_1 r$
From continuity considerations, the volume flow rate is given by:

$$\dot{V} = \int_{r_1}^{r_2} \overline{V}_g H \cdot dr = C_1 H(r_2^2 - r_1^2)/2 \tag{9.40}$$

Hence,

$$C_1 = 2\dot{V}/H(r_2^2 - r_1^2) \tag{9.41}$$

and

$$\overline{V}_g = [2\dot{V}/H(r_2^2 - r_1^2)]r \tag{9.42}$$

Define the average tangential velocity as $\overline{V}_{g,ave} = \dot{V}/H(r_2 - r_1)$ and an average Stokes number thus:

$$\text{Stk}_{ave} = \frac{\tau \overline{V}_{g,ave}}{r_2} = \frac{\tau \dot{V}}{r_2 H(r_2 - r_1)} \tag{9.43}$$

Consider the forces acting on a particle at the cyclone outer wall. See Figure 9.13:

- A centrifugal force (F_C) acting in the direction of the outer wall
- A retarding viscous drag force (F_D) opposing the motion

Using Newton's 2nd law, the resultant force (F_R) is given by:

$$F_R = F_C - F_D \tag{9.44}$$

where

$$F_R = \rho_p V_p \left(\frac{d\overline{V}_p}{dt} \right)$$

$$F_C = \rho_p \left(\frac{\pi d_p^3}{6} \right) \left(\frac{\overline{V}_g^2}{r_2} \right)$$

$$F_D = C_D A_p \rho_g \overline{V}_p^2 / 2$$

After substitution:

$$\rho_p V_p d\overline{V}_p/dt = \rho_p \left(\frac{\pi d_p^3}{6} \right) \left(\frac{\overline{V}_g^2}{r_2} \right) - C_D A_p \rho_g \overline{V}_p^2 / 2$$

Centrifugal
force (F_C)

Drag force
(F_D)

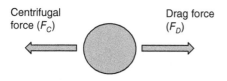

Figure 9.13 Particle forces in a cyclone.

Make the following assumptions:

Stokes flow, i.e. $C_D = 24/Re = 24\mu_g/\rho_g \overline{V}_{ts}d_p$

$$A_p = \pi d_p^2/4$$

After substitution and rearrangement:

$$\overline{V}_{ts} = \left(\frac{\rho_p d_p^2}{18\mu_g}\right)\left(\frac{\overline{V}_g^2}{r_2}\right) = \frac{\tau \overline{V}_g^2}{r_2} \tag{9.45}$$

A mass balance over an angle θ yields:

Mass flow rate of particles entering volume = mass flow rate of particles leaving volume + rate of particle collection in volume

$$c\dot{V} = (c + dc)\dot{V} + c\overline{V}_{ts}Ad\theta$$

Substituting from (9.45):

$$c\dot{V} = (c + dc)\dot{V} + c(\tau\overline{V}_g^2/r_2)r_2H \cdot d\theta$$

Substituting from (9.42) gives:

$$c\dot{V} = (c + dc)\dot{V} + c[\tau 4\dot{V}^2 r_2^2/H^2(r_2^2 - r_1^2)(r_2^2 - r_1^2)r_2](r_2H) \cdot d\theta$$

Dividing through by $\dot{V}$ and rearranging:

$$c - c[\tau 4\dot{V}r_2^3/H(r_2^2 - r_1^2)(r_2^2 - r_1^2)] \times (r_2) \cdot d\theta = c + dc$$

$$c(1 - [\tau 4\dot{V}r_2^3/H(r_2^2 - r_1^2)(r_2^2 - r_1^2)] \times (r_2) \cdot d\theta) = c + dc$$

$$(-[\tau 4\dot{V}r_2^3/H(r_2^2 - r_1^2)(r_2^2 - r_1^2)] \times (r_2) \cdot d\theta) = \frac{c + dc}{c} - 1$$

$$(-[\tau 4\dot{V}r_2^3/H(r_2^2 - r_1^2)(r_2^2 - r_1^2)] \times (r_2) \cdot d\theta) = \frac{c}{c} + \frac{dc}{c} - 1 = \frac{dc}{c}$$

Let

$$C_2 = \frac{4r_2^3}{(r_2 + r_1)(r_2^2 - r_1^2)} \tag{9.46}$$

Giving:

$$\frac{dc}{c} = -(\text{Stk}_{ave}C_2 \cdot d\theta)$$

Integrating between 0 and θ

$$\int_{c_0}^{c_\theta}\left(\frac{1}{c}\right) \cdot dc = \int_0^\theta (-[\text{Stk}_{ave}C_2 \cdot d\theta)$$

[Remembering that: $\int \frac{1}{x}dx = \ln x$]

$$\ln c_\theta - \ln c_0 = -\text{Stk}_{ave}C_2\theta$$

[Remembering that: $\ln x - \ln y = \ln(x/y)$]

Then rearranging:

$$c_\theta / c_0 = \exp[-\text{Stk}_{\text{ave}} C_2 \theta]$$

[Remembering that in general: $\eta_c = 1 - (c_{\text{exit}}/c_{\text{entry}})$]

$$\eta_c = 1 - \exp[-\text{Stk}_{\text{ave}} C_2 \theta]$$
$$= 1 - \exp[-\text{Stk}_{\text{ave}} C_2 2\pi N] \tag{9.47}$$

where N is the number of (*descending*) turns made by the fluid stream.

The equation derived earlier indicates the relationship between collection efficiency, particle size, and cyclone dimensions.

The construction and relative dimensions of a typical top entry/exit cyclone are shown in Figure 9.14.

Shortly after the inlet, the particulate-contaminated gas is forced to turn in the main body of the cyclone.

Due to their inertia, the particles, however, collide with the walls and fall to the bottom of the cone section for collection.

As shown, the gas flow follows a double helical pattern, the outer helix descending, the inner helix rising to the outlet. Only the outer helix contributes to collection.

An empirical expression commonly used for the number of descending turns N is:

$$N = (1/H)(2H_1 + H_2) \tag{9.48}$$

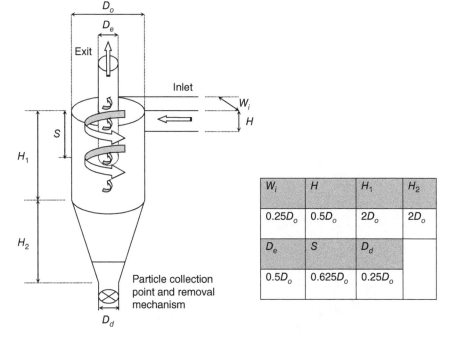

W_i	H	H_1	H_2
$0.25D_o$	$0.5D_o$	$2D_o$	$2D_o$
D_e	S	D_d	
$0.5D_o$	$0.625D_o$	$0.25D_o$	

Figure 9.14 Typical or 'classical' cyclone dimensions based on the cylinder diameter (D_o).

In order to maintain high collection efficiency across a range of particle size $\overline{V}_g$ could be increased. This, however, would incur additional power consumption at the system fan.

Alternatively, W_i could be reduced but to maintain classical cyclone dimensions, this would reduce its overall size and hence flow rate.

One solution to this problem is the multi-cyclone consisting of many small cyclones in parallel.

9.5.3 Electrostatic Precipitators (ESPs)

Gravitational and centrifugal forces are too weak to provide good collection efficiencies for particles having a diameter of less than 5 µm.

In this case, one alternative would be to use electrostatic forces to drive particles to a collecting surface.

In the ESP shown in Figure 9.15, a wire hangs between two plates. One end of the wire is connected through a rectifier to a high voltage supply setting up an electric field between wire (the charging electrode) and plate (the collecting electrode).

The other end of the wire is weighted but hangs freely.

Electronic charges leave the wire (the '**coronal discharge**') on their journey to the grounded or otherwise charged plates. See Figure 9.16.

Particles carried in the gas flow collect the negative charge as a result of their proximity to the wires and are then driven by electrostatic attraction to the plates.

Tubular ESP geometries are also in use employing cylindrical collector surrounding a central charging electrode.

An electrical field EF (V/m) is said to exist when electrical potential (V) varies from one point to another in space.

The **permittivity** is a measure of how well a material supports an electric field.

The charge associated with an electron is 1.6×10^{-19} C.

The total charge collected by a particle passing through an electric field will depend on the number of electrons collected and will be a multiple of the aforementioned.

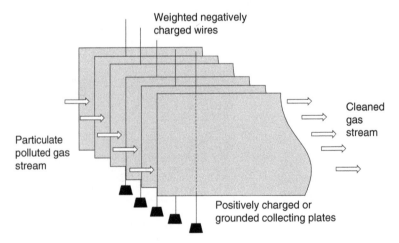

Figure 9.15 Simple plate and wire electrostatic precipitator arrangement.

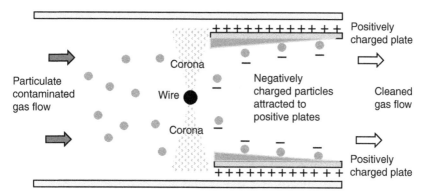

Figure 9.16 Schematic of coronal discharge and particle collection in a single channel.

An often-used relationship for estimating the charge q (C) accumulated by a particle d_p (m) being charged in an electric field EF_o (V/m) is:

$$q = 3\pi \left(\frac{\varepsilon}{\varepsilon + 2}\right) \varepsilon_o d_p^2 EF_o \tag{9.49}$$

where ε is a material property known as the **dielectric constant** (or relative permittivity), and ε_o is the permittivity of free space (or vacuum) taken as $(8.85 \times 10^{-12}$ C/V m$)$.

The electrostatic force (F_E) on a particle is then:

$$F_E = qEF_l \tag{9.50}$$

where EF_l is the local field strength where the particles are collected, i.e. causing the force. Note that EF_l may not be equal to EF_o. For example, the particle may acquire its charge in an area of high electric field strength, say close to the wire electrode.

However, for practical purposes let $EF_o = EF_l = EF$.

Then substituting into the aforementioned:

$$F_E = 3\pi \left(\frac{\varepsilon}{\varepsilon + 2}\right) \varepsilon_o d_p^2 EF^2 \tag{9.51}$$

Consider the forces acting on a particle finding itself in an electric field, see Figure 9.17: They are:

- An electrostatic attraction force (F_E) in the direction of the voltage potential difference
- A retarding viscous drag force (F_D) opposing the attraction

Using Newton's 2nd law, the resultant force (F_R) is given by:

$$F_R = F_E - F_D \tag{9.52}$$

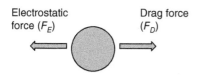

Figure 9.17 Particle forces in an ESP.

where

$$F_R = \rho_p V_p \left(\frac{d\overline{V}_p}{dt}\right)$$

$$F_E = qEF$$

$$F_D = C_D A_p \rho_g \overline{V}_p^2/2$$

After substitution:

$$\rho_p V_p d\overline{V}_p/dt = qEF - C_D A_p \rho_g \overline{V}_p^2/2$$

Making the following assumptions:

Stokes flow, i.e. $C_D = 24/Re = 24\mu_g/\rho_g\overline{V}_p d_p$

$$A_p = \pi d_p^2/4$$

$$qEF = 3\pi \left(\frac{\varepsilon}{\varepsilon+2}\right)\varepsilon_o d_p^2 EF^2$$

At zero acceleration: i.e. $\frac{d\overline{V}_p}{dt} = 0$

$$0 = 3\pi \left(\frac{\varepsilon}{\varepsilon+2}\right)\varepsilon_o d_p^2 EF^2 - \left(\frac{24\mu_g}{\rho_g\overline{V}_p d_p}\right)\left(\frac{\pi d_p^2}{4}\right)\rho_g(\overline{V}_p^2/2)$$

Solving for $\overline{V}_p$ the particle terminal settling velocity $\overline{V}_{ts}$ is given by:

$$\overline{V}_{ts} = \left[d_p \varepsilon_o EF^2 \left(\frac{\varepsilon}{\varepsilon+2}\right)\right]/\mu_g \tag{9.53}$$

In ESP technology, the terminal settling velocity is more commonly termed the particle ***drift or migration velocity***.

Inspection of the aforementioned reveals several important points:

(i) $\overline{V}_{ts} \propto EF^2$, so raising the field strength by increasing the voltage or decreasing the wire to plate distance will significantly increase drift velocities.

Unfortunately, this will also increase the incidence of sparking to the point where particle build-up on the plate is disrupted.

(ii) $\overline{V}_{ts} \propto d_p$, and so compared to a cyclone, larger drift velocities would be anticipated at any given particle size.

(iii) $\overline{V}_{ts}$ is independent of $\overline{V}_{ave}$, hence the residence time of the particle and its probability of capture can be increased by providing a large area.

In terms of collection efficiency analysis, an ESP is similar to that of the well-mixed gravity settler model.

Let an ESP have a plate of width (i.e. height) W, length (in the flow direction) L, and a wire-to-plate distance of d. Consider a section dx having a uniform concentration c throughout its volume Adx. See Figure 9.18.

Performing a mass balance:

Mass flow rate of particles entering the volume = mass flow rate of particles

leaving the volume + rate of particle collection in the volume

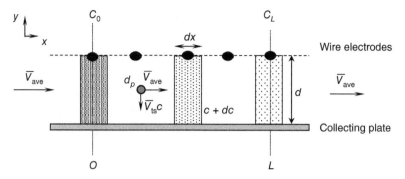

Figure 9.18 Wire and single-plate ESP parameters.

i.e.

$$c\overline{V}_{ave}A = [c + dc]\overline{V}_{ave}A + c\overline{V}_{ts}W \cdot dx$$

where $W \cdot dx$ is the plate collection area.

[Remembering that flow area $A = WD$, divide through by $\overline{V}_{ave}A$]

Rearranging and separating variables:

$$c - c\frac{\overline{V}_{ts}}{\overline{V}_{ave}d} \cdot dx = c + dc$$

$$c\left(1 - \frac{\overline{V}_{ts}}{\overline{V}_{ave}d} \cdot dx\right) = c + dc$$

$$-\frac{\overline{V}_{ts}}{\overline{V}_{ave}d} \cdot dx = \frac{c + dc}{c} - 1$$

$$-\frac{\overline{V}_{ts}}{\overline{V}_{ave}d} \cdot dx = \frac{c}{c} + \frac{dc}{c} - 1 = \frac{dc}{c}$$

Integrating between $c = c_0$ and c_L, and $x = 0$ and L:

$$\int_{c_0}^{c_L}\left(\frac{1}{c}\right) \cdot dc = \int_0^L -(\overline{V}_{ts}/\overline{V}_{ave}d) \cdot dx$$

[Remembering that: $\int \frac{1}{x}dx = \ln x$]

$$\ln c_L - \ln c_0 = -(\overline{V}_{ts}/\overline{V}_{ave})(L/d)$$

[Remembering that: $\ln x - \ln y = \ln(x/y)$]

Then rearranging:

$$c_L/c_0 = \exp[-(\overline{V}_{ts}/\overline{V}_{ave})(L/d)]$$

[Remembering that in general: $\eta_c = 1 - (c_{exit}/c_{entry})$]

$$\eta_c = 1 - \exp[-(\overline{V}_{ts}/\overline{V}_{ave})(L/d)]$$

At this point, the earlier derived expression for terminal velocity could be substituted but instead most ESP analyses take the following approach.

Substituting $\overline{V}_{ave} = \dot{V}/Wd$ and remembering that $A_{plate} = WL$:

$$\eta_c = 1 - \exp\left(-\frac{\overline{V}_{ts}A_{plate}}{\dot{V}}\right) \tag{9.54}$$

However, practical ESPs have collection plates on both sides of the wire electrodes. So, total plate collecting area $A_{total\ plate} = 2 \times A_{plate}$ and total volume flow rate $\dot{V}_{total} = 2\ \dot{V}$. Hence,

$$\eta_c = 1 - \exp\left(-\frac{\overline{V}_{ts}A_{total\ plate}}{\dot{V}_{total}}\right) \tag{9.55}$$

ESP performance is less than that predicted by theory, due to several factors including re-entrainment of trapped particles during removal via plate rapping and problems with the electrical resistivity of collected particles.

For example, low-resistivity particles will result in small voltage gradients in the collected particles or 'cake' formed on the plate and hence reduced electrostatic forces holding the collected particles to the plate. Again, re-entrainment could result.

Alternatively, high-resistivity particles can produce low voltage gradients at the wire resulting in poor charging of the particles or very high voltage gradients in the cake leading to thermal energy generation and minor gas explosions potentially blowing the cake off the plate.

These problems can be overcome by conditioning the gas stream with sulfur trioxide (SO_3) prior to the ESP or separating the charging and collection zones.

Ozone production can also be a problem associated with the field generation, and a specialist catalyst may be needed to reduce resulting O_3 concentrations.

9.5.4 Fabric Filters

Filters function by dividing the flow to the point where particles can be captured by **interception**, **impaction**, and **diffusion** on fibres and **sieving** by already collected particles. See Figure 9.19.

(A) *Impaction*: This mechanism is based on the inertia of the particle and its inability to flow around the fibre with the gas flow path (streamline).

(B) *Interception*: The flow path of the particle is such that the fibre is directly in its path.

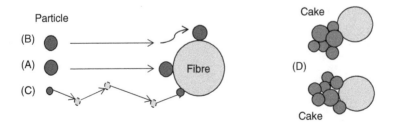

Figure 9.19 Fibre collection mechanisms.

(C) *Diffusion*: A phenomenon experienced by smaller particles resulting from Brownian motion, i.e. gas molecule bombardment.

(D) *Sieving*: Particles trapped by a filter surface may bridge over the gaps between fibres with the resulting cake then sieving particles of a size smaller than the bare filter.

Empirically it can be shown that for **impaction** on a **single cylindrical fibre**, the collection efficiency (η_{fb}) is given by:

$$\eta_{fb} = [\text{Stk}/(\text{Stk} + 0.85)]^{2.2} \tag{9.56}$$

where

$$\text{Stk} = [d_p^2 \rho_p / 18\mu_g][\overline{V}_r / d_{fb}] \tag{9.57}$$

[Note that the filter *face velocity* is commonly taken as representative of the relative system velocity ($\overline{V}_r$) in the calculation of Stokes number]

The total volume occupied by the filter comprises the fibre volume and void volume (v).

The percentage or fraction of the overall filter volume that is composed of solid fibres is termed the **fibre solid fraction** (f_f).

If the filter solid is regarded as being composed of one long length of fibre (L_{fb}) *per unit volume of filter*, then:

$$f_f = \frac{\pi d_{fb}^2 L_{fb}}{4} \tag{9.58}$$

Rearranging:

$$L_{fb} = \left(\frac{4}{\pi}\right)\left(\frac{f_f}{d_{fb}^2}\right) \tag{9.59}$$

The percentage or fraction of the overall filter volume that is free space is commonly termed the *voidage* or *porosity*, i.e.

$$\text{porosity} = 1 - f_f \tag{9.60}$$

From continuity considerations:

$$\overline{V}_{face} A_{face} = \overline{V}_v A_v \tag{9.61}$$

where $\overline{V}_v$ is the void velocity and A_v is the void area.

In terms of flow, the porosity may also be regarded as the fraction of total cross-sectional (*face*) area of the filter that is free space or void, i.e.

$$A_{face} = \frac{A_v}{1 - f_f} \tag{9.62}$$

Consider a particulate-laden airstream passing through the filter of thickness H as shown in Figure 9.20.

A mass balance for the particles across section dx gives:

Mass flow rate of particles entering filter = mass flow rate of particles leaving filter + mass collection rate in filter

$$c\overline{V}_{face} A_{face} = (c + dc)\overline{V}_{face} A_{face} + (c\overline{V}_v d_{fb} L_{fb} \eta_{fb} A_{face} \cdot dx)$$

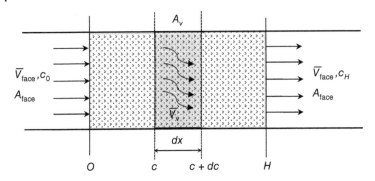

Figure 9.20 Fabric filter parameters.

Substituting from the continuity equation:

$$c \overline{V}_v A_v = (c + dc) \overline{V}_v A_v + (c \overline{V}_v d_{fb} L_{fb} \eta_{fb} A_{face} \cdot dx)$$

Substituting for A_{face} (9.58), L_{fb} (9.59):

$$c = (c + dc) + \left(\frac{4}{\pi}\right) \left(\frac{f_f}{1 - f_f}\right) c \eta_{fb} \left(\frac{1}{d_{fb}}\right) \cdot dx$$

Rearranging and separating variables:

$$c - c \left(\frac{f_f}{1 - f_f}\right) \left(\frac{4}{\pi}\right) \left(\frac{\eta_{fb}}{d_{fb}}\right) \cdot dx = c + dc$$

$$c \left(1 - \left(\frac{f_f}{1 - f_f}\right) \left(\frac{4}{\pi}\right) \left(\frac{\eta_{fb}}{d_{fb}}\right) \cdot dx\right) = c + dc$$

$$- \left(\frac{f_f}{1 - f_f}\right) \left(\frac{4}{\pi}\right) \left(\frac{\eta_{fb}}{d_{fb}}\right) \cdot dx = \frac{c + dc}{c} - 1$$

$$- \left(\frac{f_f}{1 - f_f}\right) \left(\frac{4}{\pi}\right) \left(\frac{\eta_{fb}}{d_{fb}}\right) \cdot dx = \frac{c}{c} + \frac{dc}{c} - 1 = \frac{dc}{c}$$

Integrating between $c = c_0$ and c_H, and $x = 0$ and H:

$$\int_{c_0}^{c_H} \left(\frac{1}{c}\right) \cdot dc = \int_0^H - \left\{ \left(\frac{f_f}{1 - f_f}\right) \left(\frac{4}{\pi}\right) \left(\frac{\eta_{fb}}{d_{fb}}\right) \right\} \cdot dx$$

[Remembering that: $\int \frac{1}{x} dx = \ln x$]

$$\ln c_H - \ln c_0 = - \left(\frac{f_f}{1 - f_f}\right) \left(\frac{4}{\pi}\right) \left(\frac{\eta_{fb}}{d_{fb}}\right) H$$

[Remembering that: $\ln x - \ln y = \ln(x/y)$]
Then rearranging:

$$c_H / c_0 = \exp \left\{ - \left(\frac{f_f}{1 - f_f}\right) \left(\frac{4}{\pi}\right) \left(\frac{\eta_{fb}}{d_{fb}}\right) H \right\}$$

[Remembering that in general: $\eta_c = 1 - (c_{exit}/c_{entry})$]

$$\eta_c = 1 - \exp\left[-\left(\frac{f_f}{1-f_f}\right)\left(\frac{4}{\pi}\right)\left(\frac{\eta_{fb}}{d_{fb}}\right)H\right] \tag{9.63}$$

Most building fabric filters have a framed panel or pleated (parallel bag) cartridge geometry, aligned normal to the airflow. The fibres themselves can be cellulose, glass fibre, or other synthetic materials, e.g. polyester, acrylic, and polyamide.

In an effort to enhance collection throughout the depth of a filter and dust handling capacity, fibres may be impregnated with a synthetic oil at the cost of lower efficiency.

9.6 Non-particulate Pollutants

9.6.1 Oxides of Nitrogen (NO_x, NO, NO_2)

Oxidation of nitrogen, such as occurs in some combustion processes, produces progressively nitric oxide (NO) and then nitrogen dioxide (NO_2).

In short:

$$0.5N_2 + 0.5O_2 \rightleftharpoons NO$$

$$NO + 0.5O_2 \rightleftharpoons NO_2$$

At high levels of concentration, NO_2 is a lung irritant causing inflammation of the airways and bronchitis-like symptoms. The World Health Organisation time-based concentration guideline limits for exposure to the gas (WHO 2018) in the ambient (outdoor) environment are:

- $40\,\mu g/m^3$ – annual mean
- $200\,\mu g/m^3$ – 1-hour mean

The International WELL Building Institute standards (IWBI 2019) suggest a guideline for NO_2 in operational commercial kitchens of less than 100 ppbv.

9.6.2 Ozone (O_3)

Ozone (O_3) is a blue-coloured, malodorous gas. Ozone occurs throughout the thickness of the stratosphere with the peak concentration occurring typically at elevations of about 24 km. Ninety percent of atmospheric ozone is to be found in the stratosphere and is essential to the well-being of life on earth. Ozone creation and destruction are attributable to the ultraviolet (UV) component of solar radiation.

However, at ground level, ozone is considered a lung irritant.

NO_x in combination with hydrocarbon compounds (HC) plays a part in the formation of low-level ozone (O_3).

$$NO_x + HC + sunlight + O_2 \rightarrow O_3 + photochemical\ smog$$

Ozone can also be generated in the presence of strong electrostatic fields as can be found, for example, in photocopiers and ESPs.

The World Health Organisation time-based concentration guideline limit for exposure to the gas (2018) in the ambient (outdoor) environment is a mean of 100 μg/m^3 – 8-hour mean.

The International WELL Building Institute standards (IWBI 2019) suggest an internal space general guideline for O$_3$ of less than 51 ppbv.

9.6.3 Volatile Organic Compounds (VOCs)

The term volatile organic compound (VOC) covers a large range of hydrocarbon-based species commonly performing roles in adhesives, cleaning agents, office equipment (printers, copiers computers), paints, and solvents. Examples include acetic acid, benzaldehyde, formaldehyde, methylene chloride, phenol, and toluene. Most are generated within buildings, but some maybe carried in from the external environment by a ventilation system. Assessments of their presence tend to aggregate all species under the term *total volatile organic compounds* (TVOCs).

Unacceptable VOC concentrations in air commonly result in allergic reactions, skin, throat, and eye irritation, and respiratory problems. High-level concentrations may cause headaches and nausea. Some agents are suspected carcinogens (including lung cancer and leukaemia), and some species, for example, benzene, have no recommended safe exposure levels.

The International WELL Building Institute standards (IWBI 2019) suggest an internal space general guideline for TVOCs of less than 500 μg/m^3 and less than 27 ppbv for formaldehyde. A guideline of less than 81 ppbv is suggested for operational commercial kitchens.

9.6.4 Radon

Radon is a colourless, chemically inert gas which is, however, radioactive. Radon is commonly emitted by some igneous rocks and can build up in spaces either composed of granite or founded on granite. Prolonged inhalation to radon 'particles' can result in lung damage and cancer. Their presence is usually described in terms of the amount of (α) radiation present per cubic metres of space (Becquerels/m^3).

According to the World Health Organisation (WHO 2018), the radon concentrations associated with an excess lifetime risk of lung cancer are 1 per 100 and 1 per 1000 at concentrations of 1670 Bq/m^3 and 167 Bq/m^3 respectively for lifelong non-smokers.

The International WELL Building Institute standards (IWBI 2019) suggest an internal space general guideline for Radon of less than 0.148 Bq/l in the lowest occupied level of a building.

The recommended response to a building suffering from this problem is to ventilate the space to limit concentration build-up.

9.6.5 Carbon Monoxide (CO)

Carbon monoxide is the result of inefficient or incomplete combustion. It is oxygen-hungry and reacts with any oxygen it encounters. Thus, inhaling CO results in the reduction of oxygen for biological processes, interfering with the haemoglobin cycle.

Elevated CO levels can result in reduced activity levels.

The World Health Organisation time-based concentration guideline limits for exposure to the gas (WHO 2018) are:

- $100\,mg/m^3$ – 15-minute mean
- $35\,mg/m^3$ – 1-hour mean
- $10\,mg/m^3$ – 8-hour mean
- $7\,mg/m^3$ – 24-hour mean

The International WELL Building Institute standards (IWBI 2019) suggest an internal space general guideline for CO of less than 9 ppmv and less than 35 ppmv in operational commercial kitchens.

9.6.6 Micro-organisms

This is a category encompassing bacteria, viruses, pollen, moulds, and fungi. Their dimensions may be of the order of 0.1 µm. Their presence at high levels is implicated in the occurrence of respiratory discomfort and allergic reaction. Their survival rate in the air is dependent on a range of factors such as temperature, humidity as well as the presence of other pollutants to varying degrees. Micro-organisms may find their way into a building via the ventilation system or be introduced by the occupants and their activities. If poorly maintained, the building and its building services, e.g. its air-conditioning humidifiers and coolers may provide a third source of contamination.

9.7 Principles of Non-particulate Collection

9.7.1 Adsorption

Adsorption is a surface process resulting from unbalanced atomic forces at the surface of a material (the absorbent) attracting the molecules of the gas/vapour (the adsorbate).

Adsorption is dependent on surface area, temperature, and pressure. Activated by acids or salts, carbon filters are highly microporous, can have a surface area in excess of $1000\,m^2/g$ and are commonly used for the removal of a range of gaseous/vapour pollutants. Some metal oxides can also be effective. For example, manganese and copper oxides are able to provide ozone filtration.

9.7.2 Ultraviolet Technologies

UV radiation is electromagnetic energy with a wavelength in the range of approximately 100–400 nm. This range has three sub-classifications UV-A, UV-B, and UV-C. UV-C has a wavelength of approximately 100–280 nm and has been proposed for micro-organism filtration and air cleaning as UV can disrupt DNA. Ultraviolet germicidal irradiation (UVGI) has been used in medical facilities such as hospitals and clinics and found application in prisons and shelters. Configuration types include room upper air, in-duct (room recirculation) and those dedicated to high-risk plant areas. The use of UV may result in ozone generation and hence a need for catalytic removal downstream of the device.

Two systems are in use:

(i) Direct use

The technology is based on the fluorescent lamp. For the purposes of illumination, the discharge inside the lamp generates UV which is converted to visible light by the lamp tube coating. Without the coating, the UV is available for disinfection. UV systems have been shown to have some success against bacteria and moulds but less so in the case of viruses. However, studies are incomplete. Its effectiveness will depend on the exposure time and the lamp intensity or dosage.

(ii) Photo-catalysis

Energetic UV radiation is used to knock out electrons from a titanium dioxide substrate (TiO_2). The electrons go on to ionise local water vapour creating OH^- radicals that can break hydrocarbon bonds in VOCs to form carbon dioxide and water.

9.7.3 Plasma Cleaning

An ionisation process, similar to electrostatic precipitation, aims to cause particulate clustering, DNA disruption, and oxidation. Current technologies claim to be suitable for in-room treatment as well as supply, exhaust, and recirculation flows. They can be operated in series with a fabric filter.

9.8 Pressure Drop Considerations

Increasing the contact area between an air flow and a surface or making a flow change direction will result in friction manifesting itself as a static pressure drop. Thus, from the perspective of the fan, any clean-up device will just be another pressure loss to overcome.

The pressure drops associated with clean-up devices are usually correlated with flow rate or the mass of particles collected (sometimes called the 'dust load'). Like duct fittings, the relationships can often be modelled by a positive slope square law.

Hence, the method chosen for cleaning an air flow will have consequences for fan size. Consider the methods described earlier:

- Gravity settlers
  The device relies on an increase in cross-sectional area, resulting in increasing contact area and hence friction. However, the reduced velocity may result in an increased static regain resulting in a low pressure drop. Particles should be removed out of the airstream by hoppers and so should contribute to friction. Unfortunately, as demonstrated earlier, gravity settlers are associated with a low collection efficiency.
- Cyclones
  A device based on making the air flow to perform repeated spirals and a 180° vertical change in direction must expect to incur a pressure drop in excess of that in a gravity settler. Its capture mechanism is also based on a high flow velocity, again unhelpful in terms of pressure loss. Its compensation is a higher collection efficiency. If particles do not aggregate on the cyclone cylinder, its pressure drop will not, in theory, increase significantly over time.

- Electrostatic precipitator

  As particles aggregate on a collector plate, it is likely that a small pressure loss increase would be incurred, and if not cleaned regularly, the flow channel cross-sectional area could be reduced. However, with good maintenance, the pressure loss should not increase significantly with time. In any event, the plates should be kept clean to maintain optimum electric field conditions. The compensation for a higher maintenance demand is a higher collection efficiency than the cyclone.
- Fabric filter

  Particulate collection can occur on the upstream face and the interstices of the filter. Particle agglomeration on the front face of the filter is sometimes termed 'cake'. As the particulates remain in the flow path, they cause significant increase in pressure drop with time and maintenance and/or replacement will be required. Nevertheless, filters remain the most widely used air cleaning technology.
- UV technologies

  UV technologies are low pressure drop by comparison with some other devices. Some general arrangements are shown in Figure 9.21.

  In-duct devices are generally tubular in form and may be parallel or in-line (a) or cross-duct (b) with respect to air flow direction. Upper room systems (c) commonly have dedicated fans. Being surface-based, subsequent catalytic ozone removal may increase the pressure drop significantly.

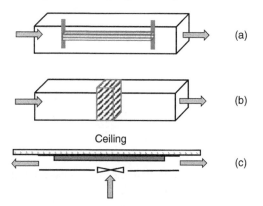

Figure 9.21 General UVGI arrangements. (a) In-duct, in-line tubes, (b) in-duct, crossflow tubes, (c) shrouded tube, dedicated fan.

9.9 Worked Examples

Worked Example 9.1 Concentration Conversion

A carbon monoxide (CO) concentration is reported to be 4000 ppbv.

If the reporting temperature and pressure are 20 °C and 1.013 bar respectively, use the data supplied to express this concentration in mg/m^3.

Data

Molecular mass of carbon: 12 kg/kmol
Molecular mass of oxygen: 16 kg/kmol
Universal gas constant: 8.314 kJ/kmol K

Solution

Given: $c = 4000\,\text{ppbv}$, $T = 20°C = 293\,\text{K}$, $P = 1.013\,\text{bar} = 1.013 \times 10^5\,\text{Pa}$, $M_{CO} = 12 + 16 = 28\,\text{kg/kmol}$

Find: c

$$R = R_o/M = 1000 \times 8.314/28 = 296.6\,\text{J/kg K}$$

$$c(\text{kg/m}^3) = c(\text{ppbv}) \times 10^{-9}\frac{P}{RT} = 4000 \times 10^{-9} \times \frac{1.013 \times 10^5}{296.6 \times 293}$$

$$= 4.66 \times 10^{-6}\,\text{kg/m}^3$$

$$= 4.66\,\text{mg/m}^3$$

Worked Example 9.2 Stokes Law

A particulate having a diameter of 25 µm falls in air.

Assuming Stokes flow, use the data supplied to determine its terminal velocity (m/s).

Data

Particle density: $1200\,\text{kg/m}^3$, air dynamic viscosity: $1.8 \times 10^{-5}\,\text{kg/ms}$, air density: $1.2\,\text{kg/m}^3$.

Solution

Given: $\rho_p = 1200\,\text{kg/m}^3$, $\rho_g = 1.2\,\text{kg/m}^3$, $\mu_g = 1.8 \times 10^{-5}\,\text{kg/ms}$, $d_p = 25 \times 10^{-6}\,\text{m}$

Find: $\overline{V}_{ts}$

$$\overline{V}_{ts} = \left[\frac{(\rho_p - \rho_g)g d_p^2}{18\mu_g}\right]$$

$$= \left[\frac{(1200 - 1.2) \times 9.81 \times (25 \times 10^{-6})^2}{18 \times 1.8 \times 10^{-5}}\right]$$

$$= 0.023\,\text{m/s}$$

Worked Example 9.3 Fractional Efficiency

A pollution control device has the fractional efficiency profile shown in Table E9.3.

Determine its total collection efficiency (%) when attempting to clean up a gas stream contaminated with particulates having the weight fractions indicated.

Table E9.3 Performance data for Worked Example 9.3.

Particle size (µm)	Fractional efficiency (%)	Weight fraction (%)
50	99	10
30	95	30
10	85	35
5	60	15
1	55	10

Solution

Given: Range of d_p, $\eta_{c,d}$, mf
  Find: η_{tot}

$$\eta_{tot} = \sum(\eta_{c,d}\text{mf})$$
$$= (0.99 \times 0.1) + (0.95 \times 0.3) + (0.85 \times 0.35) + (0.6 \times 0.15) + (0.55 \times 0.1)$$
$$= 0.8265 = 82.65\%$$

Worked Example 9.4 Overall Collection Efficiency and Particulate Concentration
A polluted air stream has its particulate concentration reduced from 175 to 10 µg/m³ by a
collection device shown in Figure E9.4.

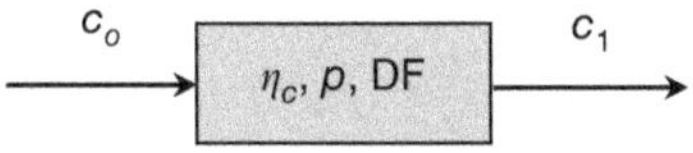

c_o → [η_c, p, DF] → c_1

Figure E9.4 Layout for Worked Example 9.4.

  Assuming the volume flow rate remains unchanged, determine the collection efficiency,
penetration, and decontamination factor for the collection device.

Solution

Given: $c_o = 175\,\mu g/m^3$, $c_1 = 10\,\mu g/m^3$
  Find: η_c, p, DF

$$\eta_c = 1 - (c_1/c_o)$$
$$= 1 - (10/175)$$
$$= 94.3\%$$

  Penetration $p = 1 - \eta_c = 1 - 0.943 = 0.057$

  Decontamination factor $\text{DF} = \dfrac{1}{\text{penetration}} = \dfrac{1}{0.057} = 17.54$

Worked Example 9.5 Particulate Collectors in Combination
A filter has a collection efficiency of 80%. Determine the overall collection efficiency (%) of
two such filters used in series as shown in Figure E9.5.

Solution

Given: $\eta_{1-2} = 0.8$

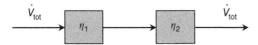

Figure E9.5 Layout for Worked Example 9.5.

Find: η_{series}

$$\eta_{overall} = 1 - [(1 - \eta_1)(1 - \eta_2) \cdots (1 - \eta_N)]$$
$$= 1 - [(1 - 0.8)(1 - 0.8)]$$
$$= 0.96 = 96\%$$

Worked Example 9.6   Collector Arrangements and Cut Diameter

A collection device comprises a single row of 400 μm diameter fibres on 500 μm centres as shown in Figure E9.6.

The air face velocity is 0.5 m/s, using the data supplied estimate the cut diameter (i.e. at 50% collection efficiency) for the device.

Data

Dynamic viscosity of air: 1.8×10^{-5} kg/ms
Particle density: 1100 kg/m³

Solution

Given: $\rho_p = 1100$ kg/m³, $\mu_g = 1.8 \times 10^{-5}$ kg/ms, $\overline{V}_{face} = 0.5$ m/s, $d_{fb} = 400 \times 10^{-6}$ m, d_{fb-fb} spacing $= 100 \times 10^{-6}$ m
  Find: $d_{p,\,cut}$

Fraction blocked per unit length of row :

%blocked $= 400/(400 + 100) = 0.8 = 80\%$

$\eta_{cut} = $ %blocked $\times \eta_{fb,cut}$

$0.5 = 0.8 \times \eta_{fb,cut}$,   i.e. $\eta_{fb,cut} = 0.625$

Then

$$\eta_{fb,cut} = [Stk/(Stk + 0.85)]^{2.2}$$
$$0.625 = [Stk/(Stk + 0.85)]^{2.2}, \quad \text{i.e. } Stk = 3.6$$

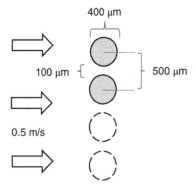

Figure E9.6   Conditions for Worked Example 9.6.

Then

$$Stk = \frac{d_p^2 \rho_p}{18\mu_g} \times \frac{\overline{V}_{face}}{d_{fb}}$$

$$3.6 = \frac{d_{p,cut}^2 \times 1100}{18 \times 1.8 \times 10^{-5}} \times \frac{0.5}{400 \times 10^{-6}}$$

$$d_{p,cut} = 28.99 \times 10^{-6} \text{ m} \cong 29 \text{ μm}$$

Worked Example 9.7 Cyclone Separator Performance

An air stream containing particles having a density of 1200 kg/m^3 is passed through a cyclone separator having an overall diameter of 1 m at a rate of $0.8 \text{ m}^3/\text{s}$.

If the dynamic viscosity of the air is $1.8 \times 10^{-5} \text{ kg/ms}$, determine:

(a) The number of effective turns (N) of the air stream.
(b) The collection efficiency for 10 μm particles.

Solution

Given: $\rho_p = 1200 \text{ kg/m}^3$, $D_0 = 1 \text{ m}$, $\mu_g = 1.8 \times 10^{-5} \text{ kg/ms}$, $d_p = 10 \times 10^{-6} \text{ m}$, $\dot{V} = 0.8 \text{ m}^3/\text{s}$
Find: N, $\eta_{c, 10 \mu m}$
From 'classical' cyclone dimensions (see Figure E9.14):

$$D_o = 1 \text{ m}, \ D_e = 0.5 \text{ m}, \ r_2 = 0.5 \text{ m}, \ r_1 = 0.25 \text{ m}, \ H = 0.5 \text{ m}, \ H_1 = 2 \text{ m}, \ H_2 = 2 \text{ m}$$

$$C_2 = \frac{4r_2^3}{(r_2 + r_1)(r_2^2 - r_1^2)}$$

$$= 4 \times 0.5^3 / [(0.5 + 0.25) \times (0.5^2 - 0.25^2)] = 3.55$$

(a)

$$N = (1/H)(2H_1 + H_2)$$

$$= (1/0.5)((2 \times 2) + 2) = 12 \text{ turns}$$

(b)

$$Stk_{ave} = \left(\frac{\rho_p d_p^2}{18\mu_g} \right) \left(\frac{\dot{V}}{r_2 H(r_2 - r_1)} \right)$$

$$= [1200 \times (10 \times 10^{-6})^2 / 18 \times 1.8 \times 10^{-5}] \times [0.8/(0.5 \times 0.5(0.25))]$$

$$= 0.00474$$

$$\eta_c = 1 - \exp[-Stk_{ave} C_2 2\pi N]$$

$$= 1 - \exp[-0.00474 \times 3.55 \times 2 \times \pi \times 12] = 0.719, \text{ i.e. } 71.9\%$$

Worked Example 9.8 Electrostatic Precipitator Performance

A $0.8 \text{ m}^3/\text{s}$ gas stream containing 10 μm diameter particles passes through an ESP of electric field strength 150 kV/m.

If the collection plate surface area is 25 m^2, using a well-mixed model determine:

(a) The particle drift velocity (m/s)
(b) The collection efficiency (%)

Take the gas stream dynamic viscosity as 1.8×10^{-5} kg/ms and the particle dielectric constant as 6.

Solution

Given: $\mu_g = 1.8 \times 10^{-5}$ kg/ms, $\varepsilon = 6$, EF $= 150 \times 10^3$ V/m, $d_p = 10 \times 10^{-6}$ m, $\dot{V} = 0.8$ m^3/s, $A_{\text{total plate}} = 25$ m^2, $\varepsilon_o = 8.85 \times 10^{-12}$ C/V m

Find: $\overline{V}_{ts}$, $\eta_{c,2.5\ \mu m}$

(a)

$$\overline{V}_{ts} = \left[d_p \varepsilon_o \text{EF}^2 \left(\frac{\varepsilon}{\varepsilon + 2} \right) \right] / \mu_g$$

$$= \left[10 \times 10^{-6} \times 8.85 \times 10^{-12} \times (150 \times 10^3)^2 \left(\frac{6}{6 + 2} \right) \right] / 1.8 \times 10^{-5}$$

$$= 0.083 \ \text{m/s}$$

(b)

$$\eta_c = 1 - \exp \left(-\frac{\overline{V}_{ts} A_{\text{total plate}}}{\dot{V}_{\text{total}}} \right)$$

$$= 1 - \exp \left(-\frac{0.083 \times 25}{0.8} \right)$$

$$= 0.9252 = 92.52\%$$

Worked Example 9.9   ESP Performance and Load Variation

An ESP is treating a particle-laden gas stream with a collection efficiency of 98%.

If all other operating parameters remain constant, determine the new collection efficiency (%) if the air flow rate is increased by 25%.

Solution

Given: $\eta_{\text{existing}} = 0.98$, $\dot{V}_{\text{new}} = 1.25 \times \dot{V}_{\text{existing}}$

Find: η_{new}

Existing

$$\eta_{c,\text{existing}} = 1 - \exp \left(-\frac{\overline{V}_{ts} A_{\text{total plate}}}{\dot{V}_{\text{total}}} \right)$$

$$0.98 = 1 - \exp \left(-\frac{\overline{V}_{ts} A_{\text{total plate}}}{\dot{V}_{\text{total}}} \right)$$

$$\exp \left(-\frac{\overline{V}_{ts} A_{\text{total plate}}}{\dot{V}_{\text{total}}} \right) = 0.02$$

$$\left(-\frac{\overline{V}_{ts} A_{\text{total plate}}}{\dot{V}_{\text{total}}} \right) = -3.912$$

New

$$\eta_{c,new} = 1 - \exp\left(-\frac{\overline{V}_{ts} A_{total\ plate}}{\dot{V}_{total} \times 1.25}\right)$$

$$= 1 - \exp\left(-\frac{\overline{V}_{ts} A_{total\ plate}}{\dot{V}_{total} \times 1.25}\right)$$

$$= 1 - \exp\left(\frac{-3.912}{1.25}\right)$$

$$= 0.9563 = 95.63\%$$

Worked Example 9.10 Fabric Filter Performance

A manufacturer proposes to produce filters having the following specification for the collection of 1 μm particles in an air stream:

Air viscosity: 1.8×10^{-5} kg/ms	Particle density: 1200 kg/m^3
Filter thickness: 10 mm	Filter face velocity: 0.5 m/s
Filter fibre solid fraction: 5%	Cylindrical fibre diameter: 10 μm

Determine their collection efficiency (%).

Solution

Given: $\rho_p = 1200$ kg/m^3, $\mu_g = 1.8 \times 10^{-5}$ kg/ms, $H = 0.01$ m, $\overline{V} = 0.5$ m/s, $d_p = 1 \times 10^{-6}$ m, $d_f = 10 \times 10^{-6}$ m, $f_f = 0.05$

Find: $\eta_{c,\ 10\,\mu m}$

$$\text{Stk} = [d_p^2 \rho_p / 18\mu_g][\overline{V}_r / d_{fb}]$$

$$= [(1 \times 10^{-6})^2 \times 1200 / 18 \times 1.8 \times 10^{-5}] \times [0.5 / 10 \times 10^{-6}]$$

$$= 0.185$$

$$\eta_{fb} = [\text{Stk} / (\text{Stk} + 0.85)]^{2.2}$$

$$= [0.185 / (0.185 + 0.85)]^{2.2}$$

$$= 0.0226$$

$$\eta_c = 1 - \exp\left[-\left(\frac{f_f}{1 - f_f}\right) \times \left(\frac{4}{\pi}\right) \times \left(\frac{\eta_{fb}}{d_{fb}}\right) \times H\right]$$

$$= 1 - \exp\left[-\left(\frac{0.05}{1 - 0.05}\right) \times \left(\frac{4}{\pi}\right) \times \left(\frac{0.0226}{10 \times 10^{-6}}\right) \times 0.01\right]$$

$$= 0.7813 = 78.13\%$$

9.10 Tutorial Problems

9.1 A nitrogen dioxide (NO_2) emission concentration is reported to be 20 ppbv. If the reporting temperature and pressure are 25 °C and 1.013 25 bar respectively, use the data below to express this concentration in $\mu g/m^3$.

Data

Molecular mass of nitrogen: 14 kg/kmol
Molecular mass of oxygen: 16 kg/kmol
Universal gas constant: 8.314 kJ/kmol K
[Answer: 38 $\mu g/m^3$]

9.2 A particulate having a diameter of 10 μm settles in air. Assuming Stokes flow, use the data supplied to determine its terminal velocity (m/s).

Data

Particle density: 1100 kg/m^3, air dynamic viscosity: 1.8×10^{-5} kg/ms, air density: 1.2 kg/m^3.
[Answer: 0.0033 m/s]

9.3 Using the correlation parameters supplied, determine the Cunningham correction factor for 10, 1, and 0.1 μm particles.

α: 1.142, β: 0.558, γ: 0.999, λ: 0.0673 μm
[Answers: 1.007 7, 1.0769, 1.8537]

9.4 A pollution control device has the fractional efficiency profile shown in Table P9.4. Determine its total collection efficiency (%) when attempting to clean up a gas stream contaminated with particulates having the weight fractions indicated.

Table P9.4 Performance data for tutorial Problem 9.4.

Particle size (μm)	Fractional efficiency (%)	Weight fraction (%)
50	99	10
25	92	30
10	80	45
2	70	10
1	60	5

[Answer: 83.5%]

9.5 A polluted air stream has its particulate concentration reduced from 240 to 40 $\mu g/m^3$ by a collection device. Assuming the volume flow rate remains unchanged, determine the collection efficiency (%), penetration and decontamination factor for the collection device.
[Answers: 83.3%, 0.167, 6]

9.6 A pollution control device has a collection efficiency of 85%.
Determine the overall collection efficiency (%) of three such filters used in series.
[Answer: 99.66%]

9.7 Two pollution control devices are connected in parallel.
The collection efficiencies and flow rate split between collectors are shown in Table P9.7.
Determine the overall collection efficiency (%) for the combination.

Table P9.7 Performance data for tutorial Problem 9.7.

Collector	Collection efficiency (%)	Flow rate split (%)
1	88	56
2	85	44

[Answer: 86.7%]

9.8 An inexperienced designer believes that a section of ductwork of height 1 m and length 3 m can act as a pre-filter and capture 50% of 50 μm diameter particles entering an air handling plant. If the velocity in the settling space is 0.75 m/s, verify using a well-mixed model, whether this is correct.
Take $\rho_p = 1100 \, \text{kg/m}^3$ and $\mu_g = 1.8 \times 10^{-5} \, \text{kg/ms}$
[Answer: Incorrect, 28.33%]

9.9 A 0.5 m³/s gas stream containing 5 μm diameter particles passes through an ESP of electric field strength 200 kV/m.
If the plate surface area is 20 m², using a well-mixed model determine:
(a) The particle drift velocity (m/s)
(b) The collection efficiency (%)
(c) The area required for 99% collection efficiency (m²)
Take the gas stream dynamic viscosity as 1.8×10^{-5} kg/ms and the particle dielectric constant as 6.
[Answers: 0.0738 m/s, 94.77%, 31.2 m²]

9.10 A manufacturer proposes to produce air filters having the following specification for the collection of particles with a density of 1100 kg/m³ and a mean diameter of 1 μm:

Air viscosity: 1.8×10^{-5} kg/ms	
Filter thickness: 10 mm	Filter face velocity: 0.4 m/s
Filter fibre solid fraction: 5%	Cylindrical fibre diameter: 10 μm

(a) Show that this design is likely to have a poor collection efficiency.
(b) If the remainder of the specification remains unchanged, determine the thickness (mm) required to increase the efficiency to *at least* 99%

[Answers: 57.487%, 54 mm]

10

Solar Energy Applications

10.1 Overview

Historically, the interaction between insolation and an inhabited structure has been regarded as a problem. Solar gain in the summer is something to be engineered away with the use of refrigeration. Solar gain in the winter can help offset a building's heating requirement but cannot be relied on and will not figure in heating plant sizing.

However, solar energy must now, through harvesting, be regarded as an important opportunity to reduce a building's carbon footprint. Harvesting can take place on a building's interior surfaces, external surfaces, or its surrounding ground to condition a space, heat water, generate electricity, or aid with space illumination.

Learning Outcomes

- To be familiar with solar collection technologies for thermal processes in buildings
- To be familiar with solar collection technologies for electricity generation in buildings
- To be familiar with the potential for near-surface ground energy recovery
- To gain a basic knowledge of both solar and electrical energy storage technologies
- To understand the provision of daylighting in space illumination

10.2 Solar Thermal Collector Technologies

Solar thermal energy can be used to provide hot water for heating and catering/washing purposes. A range of technologies are available.

10.2.1 Flat-Plate Glazed Collectors

These comprise a roof mounted box with transparent top cover and insulated sides and rear. The box contains an absorber plate with passages on its underside through which passes the heat transfer fluid, which is usually a water/glycol mixture. See Figure 10.1. Although thermal efficiency is relatively low, temperatures of up to 55 °C are attainable, making these types of collectors popular for washing/catering purposes.

Building Services Engineering: Smart and Sustainable Design for Health and Wellbeing, First Edition.
Tarik Al-Shemmeri and Neil Packer.
© 2021 John Wiley & Sons Ltd. Published 2021 by John Wiley & Sons Ltd.

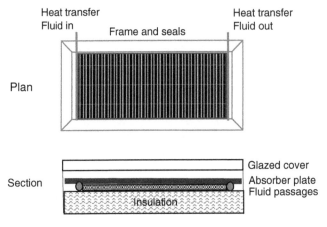

Figure 10.1 Flat-plate glazed collector.

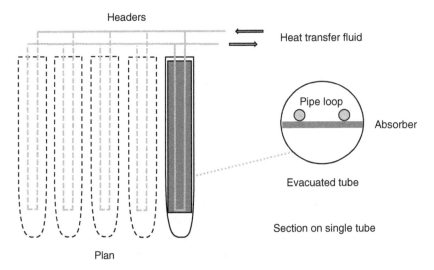

Figure 10.2 Evacuated tube collector arrangement (flow through type).

10.2.2 Evacuated Tube Collectors

In this type of collector, the passages or pipes containing the heat transfer fluid are housed co-axially in evacuated (<0.01 bar) glass tubes that reduce the convective heat loss and greatly improve the collection efficiency. Tube arrays are connected via flow and return headers. A generalised layout for an array of flat absorber, single-wall tubes is shown in Figure 10.2.

Double-wall, cylindrical absorber designs are also in use.

As an alternative to the flow through type, evacuated tube *heat pipe* collectors are also available. Here the evacuated tube houses a closed-end pipe containing a liquid with a low evaporation temperature. The liquid is evaporated by the insolation causing the vapour

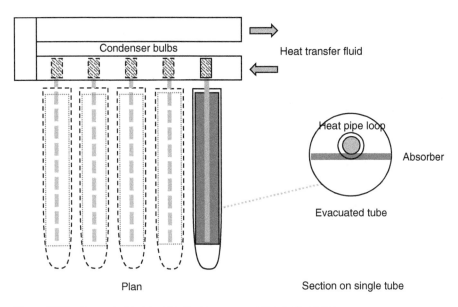

Figure 10.3 Evacuated tube collector arrangement (heat pipe type).

to rise to the header end of the tube where it condenses, releasing its energy to a flow of water/glycol via heat exchanger fins.

As a result of the heat pipe collector's condensate gravity return requirement, they cannot be installed horizontally. A generalised layout for an array of flat absorber, single-wall tubes is shown in Figure 10.3

Double-wall, concentric, cylindrical absorber designs are also available.

Evacuated tube models can provide temperatures of up to 90 °C with a thermal efficiency higher than that of a flat plate while occupying less space. At this temperature, the energy can be available for space heating support in addition to hot water provision. This advantage comes at a higher cost relative to flat-plate collectors.

10.2.3 Solar Thermal Collector Efficiency (η_c)

The losses from a flat-plate solar collector can be divided into two categories: optical losses and heat losses. See Figure 10.4:

- *Optical losses*: This factor accounts for reflection from the transparent outer surface and the less-than-perfect absorptivity of the absorber surface.
  If the solar irradiation (I_{irr}, W/m^2) absorbed by the absorber plate (absorptivity, α) passes through the cover plate (transmissivity, τ), then:

$$\dot{Q}_{\text{optical loss}} = I_{irr} A_c \left(1 - \alpha\tau\right) \tag{10.1}$$

  Where A_c (m^2) is the collection area.
- *Heat losses*: This factor accounts for convection and radiation heat losses from collector surfaces (top, back, and edges) based on the temperature difference between the average

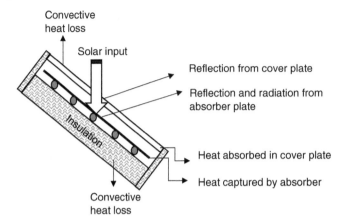

Figure 10.4 Collector losses associated with thermal collectors.

absorber temperature and the ambient temperature ($T_{\text{ave}}-T_{\text{amb}}$):

$$\dot{Q}_{\text{heat loss}} = U_{\text{comb}}A_c\left(T_{\text{ave}} - T_{\text{amb}}\right) \tag{10.2}$$

Where U_{comb} is the combined (convective and radiative) heat loss coefficient (W/m² °C).

The surface areas associated with this loss are commonly expressed with reference to the collection area.

Using Eqs. (10.1) and (10.2), the useful thermal power $\left(\dot{Q}_{\text{th}}, W\right)$ captured by a solar thermal collector is given by:

$$
\begin{aligned}
\dot{Q}_{\text{th}} &= I_{\text{irr}}A_c - \dot{Q}_{\text{losses}}\\
&= \left[I_{\text{irr}}A_c - \left(\dot{Q}_{\text{optical loss}} + \dot{Q}_{\text{thermal loss}}\right)\right]\\
&= I_{\text{irr}}A_c - \left[I_{\text{irr}}\left(1 - \alpha\tau\right) + U_{\text{comb}}A_c\left(T_{\text{ave}} - T_{\text{amb}}\right)\right]\\
&= I_{\text{irr}}A_c\alpha\tau - U_{\text{comb}}A_c\left(T_{\text{ave}} - T_{\text{amb}}\right)
\end{aligned} \tag{10.3}
$$

This is the energy captured by the heat transfer fluid. Hence:

$$\dot{Q}_{\text{th}} = \dot{m}C_p\left(T_{f,\text{out}} - T_{f,\text{in}}\right) \tag{10.4}$$

where $\dot{m}$ is the mass flow of heat transfer fluid (kg/s) and C_p its specific heat capacity.

The collector efficiency is calculated using:

$$\eta_c = \dot{Q}_{\text{th}}/I_{\text{irr}} \tag{10.5}$$

$$= \frac{I_{\text{irr}}A_c\alpha\tau - U_{\text{comb}}A_c\left(T_{\text{ave}} - T_{\text{amb}}\right)}{I_{\text{irr}}A_c} \tag{10.6}$$

$$= \alpha\tau - \left[U_{\text{comb}}\left(T_{\text{ave}} - T_{\text{amb}}\right)/I_{\text{irr}}\right] \tag{10.7}$$

Collector efficiency is more commonly and usefully expressed utilising the fluid inlet temperature (T_{in}) instead of the absorber average temperature (T_{ave}).

This approach addresses the energy transfer between the absorber and collector heat exchanger more directly.

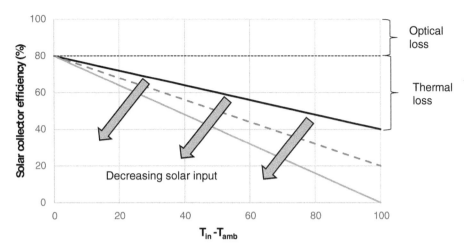

Figure 10.5 Indicative collection efficiency relationship for solar thermal hot water collector.

The **Hottel–Whillier–Bliss** efficiency expression utilises this improvement:

$$\eta_c = F_R \left[\alpha\tau - \left[U_{comb} \left(T_{in} - T_{amb} \right) / I_{irr} \right] \right] \tag{10.8}$$

Where F_R is the *collector heat removal factor*, a parameter that is determined by the absorber geometry, its thermal properties, and its heat transfer effectiveness with the heat transfer fluid. Eq. (10.8) is a straight line of intercept $F_R \, \alpha\tau$ and slope of $F_R U_{comb}$.

A plot of η_c vs. $U_{comb}(T_{in} - T_{amb})/I_{irr}$ is shown in Figure 10.5.

Claims of thermal efficiencies up to 70% for evacuated tube varieties and 60% for flat-plate collectors are commonly made, but annual operating efficiencies ~20% lower than these figures are more feasible.

10.2.4 Solar Thermal Air Heaters

A very simple concept for raising the temperature of ambient air before distributing to an occupied space.

Two varieties are in use.

- Glazed flat plate

  This is an arrangement and a thermodynamic analysis very similar to the solar thermal hot water collector comprising a box with a transparent front cover, absorber plate, ambient air inlet, heated air outlet ducting, and fan. The thermal efficiency can be estimated by an algorithm similar to that producing Eq. (10.5), i.e.

$$\eta_c = \dot{Q}_{th}/I_{irr} = \dot{m}C_{p,air} \left(T_{out, \ air} - T_{amb} \right) / I_{irr} \tag{10.9}$$

- Solar (or Trombe) wall collectors

  These employ a building's opaque facades for the purpose of collecting solar energy. Essentially, a solar wall collector comprises an external wall, a glass cover, and an air gap. Solar energy is transmitted through the glass cover raising the temperature of the wall. Air between the glass and the wall's external surface is warmed and rises to enter

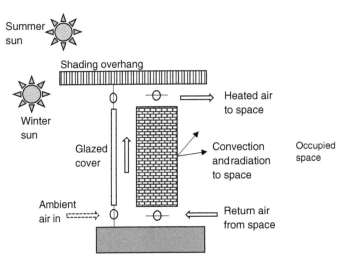

Figure 10.6 Schematic of solar wall collector (in full recirculation mode).

the building space at a high level with room air re-entering the air gap at a low level to be reheated. Flow of fresh/recirculated air is facilitated via dampers as shown in Figure 10.6. As a result of the raised wall temperature, additional heat will also, after a time lag, be convected and radiated to the occupied space via the wall internal surface.

10.3 Solar Electricity

Conversion of solar energy to (DC) electricity requires the capture of electromagnetic or photonic energy in sunlight. Electromagnetic energy is absorbed and emitted in discrete packages known as **quanta** (*pl*).

The potential for capture ($E_{photonic}$, J/photon) depends on the frequency (f, Hz) or wavelength (λ, m) of the light and is described by:

$$E_{photonic} = hf = hc/\lambda \tag{10.10}$$

where h is known as Planck's constant having a value of 6.626×10^{-34} Js and c is the speed of light $= 3 \times 10^8$ m/s.

Electromagnetic energy is also often expressed in **electron volts** (eV) where:

$$1 \text{ eV} = 1.602 \times 10^{-19} J.$$

Solar irradiation at the Earth's surface comprises a range of wavelengths from approximately 300 nm to 3 µm. The visible light component of this is in the range from 400 to 700 nm.

Photoelectricity is concerned with an atom's electron orbits.

The outermost band (or atomic orbit) that an electron can occupy and still be closely associated with its atom is called the **valence band**.

An electron receiving energy can jump to higher **conduction band** (or orbit) where it may roam through a substance essentially producing an electric current.

The gap between these bands is called the ***forbidden band,*** and the amount of energy required to jump the gap is material-dependent. In photovoltaic devices (PVs), photons of solar radiation are used to raise an electron from the valence band to the conduction band. Photons having less energy than the material band gap energy will not raise any electrons, and photons with more energy will dislodge electrons but expend their *surplus energy* in heat.

Clearly good candidates for producing a useful current would be materials with a large availability of electrons and a low forbidden band gap energy tuned to the wavelength of the incoming solar radiation.

Electron availability and energetics can be used to classify materials into *insulators* (full valence bands, high band gap energies), *conductors* (relatively empty valence bands, low band gap energies), and *semi-conductors* (relatively full valence bands, low band gap energies). For semi-conductors, band gap energies tend to be <3 eV, e.g. silicon at 1.11 eV.

10.3.1 Photovoltaic (PV) Cells

First-generation solar PV cells were manufactured from single-crystal silicon. These forms are black and homogeneous in appearance. A section through a typical first-generation silicon PV cell is shown in Figure 10.7.

The cell is a sandwich comprising a transparent glass cover/anti-reflective combination and two photoactive layers of silicon seeded with other elements (e.g. boron, phosphorous) to set up an electron field potential difference and electrodes.

Larger blue/black, multifaceted, polycrystalline forms cast from a silicon mould are also available. In both cases the operational solar energy conversion efficiency is typically <15%.

Silicon-based PVs are heavy and their orientation and support require significant thought. They can, however, be reasonably easily integrated into atria, facades, and roofs. Attempts have also been made to incorporate them into the building fabric, e.g. roof tiles. Once installed they have the advantages of low maintenance and little (local) environmental impact. After electronic inversion to AC, their output can be used locally or exported to the electricity grid. Their output is also suitable for storage via batteries. With a rapid fall in cell production costs, the electrical, i.e. non-generating components of a first-generation PV system are now starting to dominate total cost of an installation.

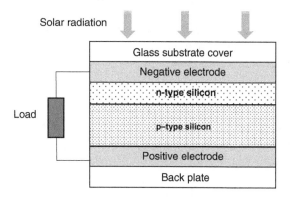

Figure 10.7 General arrangement of first-generation PV cell.

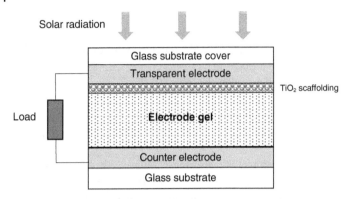

Figure 10.8 General arrangement of DSSC.

Second-generation thin-film PVs currently (2019) account for about 10% of the solar PV market. Their main attractions over first-generation cells are their mechanical flexibility and low mass. Unlike first-generation cells, requiring manufacturing processes such as melting, ingot formation, and sawing/wafering, second-generation cells are fabricated by sequential layer deposition. However, they are made, in part, with less common materials, some of which can be rather unattractive, e.g. cadmium, tellurium, arsenic, etc.

Third-generation dye-sensitised solar cells (DSSC) are actually a subset of thin-film devices. They differ from second-generation thin-film devices in that they are photochemical in nature. They essentially comprise two transparent glass substrates, two electrodes, and an intervening electrode gel containing a photosensitive dye as shown in Figure 10.8. The upper electrode is transparent to solar radiation exposing the dye to solar energy. The upper electrode also has a localised nanoparticle TiO_2 scaffolding to enhance surface area and charge transport while the lower (or counter) electrode (typically Pt) is catalytic in nature.

The dye chemistry is crucial to the range of absorbed radiation. The process within the cell is often described as being analogous to photosynthesis. Their best solar conversion efficiency (typically <10%) is poorer than that of first- and second-generation devices, but their material requirement and energy of manufacture are much lower. Although of lower performance than some manufactured recipes, natural dyes (anthocyanins) are preferred on grounds of cost and low toxicity.

The latest research into third-generation cells includes the use of new materials such as perovskite, the substitution of glass layers with polymers, and the use of different types of cells in tandem to increase the photonic energy capture envelope.

10.3.2 PV Cell Shading

Although receiving diffuse solar input, clearly a PV cell in full shade is not generating at anything like full capacity but even partial array shading requires consideration.

To achieve useful currents and voltages, PV arrays comprise series/parallel connection of individual PV cells.

The cell with the lowest illumination determines performance as it effectively becomes a load rather than energy source.

The problem of shading may be alleviated (but not overcome) by the installation of suitable electronics (bypass diodes) in PV electrical circuit; however, local shading potential must be a concern.

Shading is classified into four types:

- *Building systems shading*: The result of direct shadows from building services such as communications equipment flues, lightning conductors, etc.
- *Location shading*: The result of solar obscuration from neighbouring buildings and trees.
- *Self-shading:* An overlapping phenomenon occurring in parallel rack-mounted systems. Clearly it makes sense to maximise the energy generation from any allocated surface area. However, in the case of angled PVs on horizontal surfaces, thought needs to be given to their packing density as this can cause problems at some sun angles.
- *Temporary shading*: Dependent on the local meteorological, biological, and anthropological environment, e.g. bird droppings, leaf litter, particulate pollution, snow. The problem is alleviated by tilting the array (10°–15°).

10.3.3 PV Energy Production

The annual energy output (E_{array}, kWh/m^2) from a PV array can be roughly estimated by taking into account some operational efficiency and rating parameters as follows:

$$E_{array} = I(C_1/100)(C_2/100)C_3C_4 \qquad (10.11)$$

where

$I = $ Location annual solar irradiation (kWh/m^2)
$C_1 = $ PV module efficiency (%) – dependent on type of cells used
$C_2 = $ Power system efficiency (%) – inverter and other electrical losses – typically 85%
$C_3 = $ Shading and soiling de-rating – typically 0.95
$C_4 = $ Heat loss de-rating – typically 0.97.

10.4 Ground-Based Energy Sources

The core of the Earth is thought to be at a temperature of approximately 5000 °C. Hot spots having temperatures sufficient for steam generation and electrical power production can be found at reasonably shallow depths in some places around the globe, and this resource is best termed *geothermal energy*. However, in general, with a few exceptions, the temperatures at the *near surface*, i.e. in rock outcrops and soils, are a result of the storage of solar irradiation, fluctuations in local meteorological conditions, and vegetation cover. Using this energy via a heat exchanger or a heat pump can provide useful heating or cooling for conditioned air.

At a depth of 1–2 cm, soil temperatures are closely allied to local air temperatures to within 1–2 °C. Soil temperatures at a given depth are not, however, static and respond to changes in prevailing conditions.

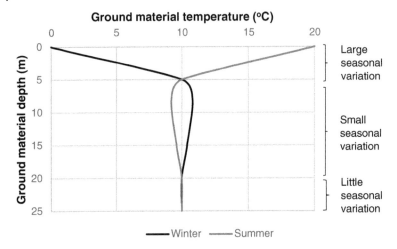

Figure 10.9 Generalised seasonal ground material temperature variations.

Surface temperatures can demonstrate large *hourly* changes. In general, heat moves downward during the day and upward during the night. Typically, *diurnal* changes of this nature can be detected down to a depth of about 50 cm.

In addition to variation in energy input at the surface, resulting ground temperature will depend on soil/rock properties such as thermal conductivity, heat capacity, and thermal diffusivity. These in turn will depend on the surface material texture, structure state of consolidation, and air/water content. Soil/rock thermal conductivities can vary by an order of magnitude. Typical values can be in the range 0.3–3 W/m² K. Typical specific heat capacities can vary from 750 to 1500 J/kg K and densities from 1500 to 2700 kg/m³.

Annual ground material temperature cycles may be detected over a greater depth.

A generalised seasonal temperature variation within 25 m of the surface at a temperate, mid-latitude location is shown in Figure 10.9.

The aforementioned diagram is for illustrative purposes only. It is important to remember that the actual temperature gradients will depend on local ground conditions. In addition, they are not static. As a result of three-dimensional transient effects, temperature gradients can be complex displaying points of inflexion as heat flow directions and rates vary within the soil/rock body.

Clearly it would be advantageous from a design and operational point of view to carry out energy exchange with a heat sink/source of known temperature, i.e. at a depth below the level of hourly/diurnal/annual fluctuation where the temperature swing is minimal. For the case illustrated earlier, material annual temperature swings are small at depths greater than 20 m. A ground-coupled heat exchanger located at a depth of less than 5 m would, over the year, be required to perform under a wide range of temperature conditions.

Care must be taken with ground temperature computer modelling which is best supported by on-site measurements and actual temperatures in the excavated or disturbed ground.

10.4.1 Direct Ground Heating/Cooling

This is a relatively simple way of ground coupling as a means to pre-heating /pre-cooling a building's air supply with the use of underground labyrinths, pathways, or tubes buried to a depth of typically 5 m. With good insulation, passages may be directly beneath the building or alternatively located in the building's environs. Velocities sufficient to promote turbulent flow in the passages are important for good heat transfer. As a result of burial, maintenance and condensate removal require serious consideration at the design stage.

10.4.2 Ground Source Heat Pumps (GSHPs)

The term heat pump was introduced in Section 6.3 as a device identical in operation and components to that of a refrigerator but with the difference that its prime function would be to provide heat. Heat can be extracted from ambient air. However, in temperate climates the air temperature in winter, when heat is required, could drop below freezing. Remembering Eq. (6.10), large temperature differences between heat source and sink result in low coefficient of performance (COP) values and therefore air operation could result in high running costs as well as the potential for freezing up of the evaporator. Better performance could be expected using a source 10–15 °C higher, like the near-surface ground.

If the heat pump is able to operate in reverse mode, i.e. the roles of the evaporator and condenser are reversed, then the system could be employed to cool its building by using the ground as a heat sink.

A heat pump's evaporator is not, normally, suitable for directly embedding in the ground. Instead water/glycol/corrosion inhibitor ground-coupled circuits (loops/coils) are commonly deployed which then exchange their heat with the evaporator via a liquid pump.

Closed ground loops can be vertical or horizontal. Horizontal loops require the provision of a trench of up to 1–2 m deep. Depending on ground water content, heat extraction rates of 8–40 W/m^2 are used for design purposes.

Vertical loops employ boreholes drilled into bedrock of typically ~100–200 m depth and 100–150 mm in diameter, housing a U-tube or spiral pipe fixed in place with a high thermal conductivity grout. Typically, heat extraction rates of 25–100 W/m are used for design purposes. Dry, unconsolidated rock tends to produce lower returns than strata with an associated groundwater flow.

It should be noted that adjacent ground and surface water bodies such as lakes and rivers can also be used as heat pump sources. In this case, open-loop collectors may be considered. However, issues with extraction licenses as well as water quality will need consideration.

10.5 Energy Storage

When the supply of energy is intermittent and energy demand is inflexible, storing energy is the only route to avoid loss of service.

10.5.1 Thermal Energy Storage

Thermal energy can be stored by, for example, increasing the pressure of a fluid, by chemical reaction, by increasing the temperature, or changing the phase of a substance. The last two options are the most common in building applications. Storage media are then limited to solids and liquids.

For a given volume the mass will be determined by the material density (kg/m³).

Perhaps the most important thermodynamic material parameter for a store is its energy capacity calculated on a mass (J/kg) or volume (J/m³) basis which will determine its physical size for a given storage capacity.

Other important storage medium parameters include thermal conductivity (k, W/m K) volumetric change (%), and store losses (W/m²).

Store material density, specific heat capacity, latent heat, and thermal conductivity, in combination, will set limits on store charging/discharging rates. The form of heat exchange surface will also require careful consideration.

Non-toxicity, chemical stability, and low cost are also, of course, desirable.

10.5.1.1 Sensible Heat Storage

Temperature-related energy change is often termed *sensible energy*.

For a substance undergoing an increase in temperature *without any phase* change, the amount of energy stored can be expressed using the specific heat capacity of the storage medium (C_p, J/kg K). Typical specific heat capacities for construction materials are of the order of 1000 J/kg K. Soil can have a specific heat capacity of around 2000 J/kg K depending on water content. Water, itself, is an excellent thermal storage medium having a specific heat capacity of approximately 4200 J/kg K. However, specific heat capacity is not a constant as it is temperature-dependent.

The energy associated with a sensible energy change is given by:

$$Q_{\text{sensible}} = m_{\text{store}} C_p \Delta T \tag{10.12}$$

Where ΔT is the difference between the store final and initial temperature.

When the thermal energy carrying fluid and storage media are both water, then for sensible heat stores, *tank storage* is common with heat exchange facilitated by a relatively simple coil arrangement.

Water-based systems in temperate climates usually operate with a back-up source of heat such as gas or electricity as shown in Figure 10.10. Here the solar thermal collector pre-heats the incoming cold water via a heat exchanger. The back-up source of heat tops up the store when there is insufficient solar energy to reach the desired hot water temperature. Typically, the store's hot water is used for washing and catering. More complex 'Combi' systems are available that supplement both hot water and heating demand.

For control, a minimum of *two* temperature sensors are required:

One to measure the temperature at the hottest part of the solar circuit, i.e. the flow from the collector, and a second to measure the temperature in the store at the height of the solar circuit heat exchanger or the store solar circuit exit.

The temperature signals are compared and a decision is made by a controller, via a relay, on solar circuit pump switching.

A *third* sensor is often installed in the upper area of the store to measure the store take-off temperature, enabling switch off of the solar system if maximum store temperature is attained.

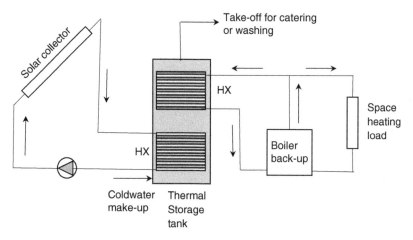

Figure 10.10 Typical solar thermal sensible heat storage arrangements.

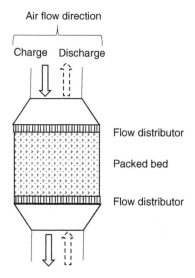

Figure 10.11 Packed bed particulate energy store.

Heat meters utilise solar circuit flow and return temperature and volumetric flow rates to calculate the energy yield and CO_2 saved by the solar circuit.

Solar air heating systems employ loosely packed particulate bed technologies for thermal storage. See Figure 10.11. Charging/discharging is cyclic and not contemporaneous. Flow direction through the bed is reversed during the discharge portion of the cycle.

10.5.1.2 Latent Heat Storage

Energy change associated with phase transition is often simply termed *latent energy*.

The specific latent heat (h_{pc}, J/kg) is a property that describes the heat storage capabilities of a substance undergoing a particular phase change. Phase change occurs at constant

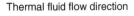

Thermal fluid flow direction

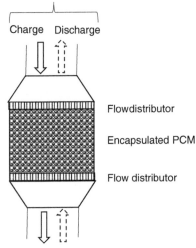

Charge   Discharge

Flowdistributor

Encapsulated PCM

Flow distributor

Figure 10.12   Encapsulated PCM energy store.

temperature. Fusion/melting (solid–liquid) is the commonly deployed transition and so the term *heat of fusion* is also associated with latent heat stores.

The energy associated with latent heat or phase change energy is given by:

$$Q_{pc} = mh_{pc} \qquad\qquad (10.13)$$

Energy stores based on phase change materials (PCMs) need careful consideration with respect to the phase change temperature (e.g. freezing/melting point), the charging temperature of the energy source, and the energy demand temperature.

For example, ice/water has an excellent latent heat value of around 330 kJ/kg, and it is a readily available material. However, its fusion phase change temperature is 0 °C, and consequently, other materials such as paraffin waxes and fatty acids, eutectics (a combination of two or more low melting point materials), salt hydrates all with higher melting points are considered more suitable.

There are a large number of commercially available PCMs with melting points covering a range of 0–120 °C and latent heat values in the range of 60–360 kJ/kg.

For storage, again, stand-alone packed beds are in use in an arrangement similar to sensible solar air heating stores described earlier. See Figure 10.12. However, in this case, the PCM material is, typically, physically separated from the thermal fluid by encapsulation in, for example, spheres.

The encapsulated phase change renders the heat transfer process more complex, taking place across a part-liquid, part-crystalline medium.

10.5.2   Electrical Energy Storage

If PV-sourced electrical energy exceeds a building's demand, then (providing an economic grid feed-in tariff is not available) it should be stored in a localised battery storage system. The electrical energy store can be used to minimise the economic impacts of time-based grid electricity tariffs by load shifting or peak shaving of the building's electrical demand.

An electrical storage system has many important characteristics in common with thermal systems e.g. gravimetric energy density or specific energy (J/kg or Wh/kg), volumetric energy density (J/m³ or Wh/m³), charge/discharge times (h), power output (kW), and cycle (or round trip) efficiency.

The term specific power (W/kg) is also in common use.

In the case of electrical storage systems, the energy capacity (or charge state) of the store is traditionally quoted in Ampere-hours (Ah) rather than Joules.

A little care is required with efficiency as there are several definitions in use based on charge, voltage, energy, etc.

The *Coulombic* efficiency ($\eta_{coulomb}$) is defined as:

$$\eta_{coulomb} = \frac{\text{Discharge } (Ah)}{\text{Charge } (Ah)} \tag{10.14}$$

The voltage efficiency ($\eta_{voltage}$) is defined as:

$$\eta_{voltage} = \frac{\text{Average Discharge voltage } (V)}{\text{Average Charge voltage } (V)} \tag{10.15}$$

The energy efficiency (η_{energy}) is defined as:

$$\eta_{energy} = \eta_{coulomb} \times \eta_{voltage} \tag{10.16}$$

Overall efficiency definitions may also refer to AC/DC conversion units.

Battery life is usually quoted in terms of the number of charge/discharge cycles. Increased temperature, chemical change (producing unwanted chemical reactions), and depth of discharge can significantly lower battery life and efficiency.

With careful design, an electrical storage system can also perform as an uninterruptible power supply or as a frequency regulator. In these cases, short response times (ms) will be important.

The primary performance characteristic for a battery system is the charge/discharge curve. This presents the available battery voltage as a function state of charge or remaining discharge capacity (Ah).

Generalised, typical charge/discharge (voltage vs. capacity) curves are shown in Figure 10.13. Looking at the discharge curve, three distinct areas and voltage conditions are discernible. They are: the full (100%) charge voltage followed by an *exponential* phase: the voltage at the end of the exponential phase and its slow decrease during the useful *operational* phase of the battery: the voltage at the beginning of a rapid voltage *decay* phase. Systems are usually cut out at low capacity (i.e. >0%) due to the precipitous loss of voltage in the decay phase.

A 'flat' (constant voltage) curve during the operational discharge phase would provide a more controllable power transfer process. In reality, however, for most configurations the voltage continues to fall during this period.

Operational curves can also be affected by temperature and other state of life factors.

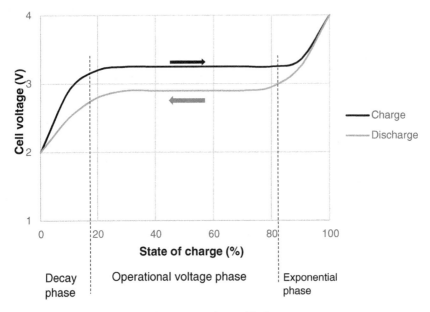

Figure 10.13   Generalised, typical battery charge/discharge curves.

10.5.3   Battery Technologies

There is an ever-expanding range of technologies available for electrical energy storage:

- Lead acid batteries
  Although relying on toxic substances (lead electrodes, aqueous sulphuric acid electrolyte), historically, the most commonly deployed technology for use in buildings and
  transport. Even though low-cost, they are now being overtaken by other technologies
  in building power supply applications. Specific energies of 30–40 Wh/kg and energy
  densities of 60–75 Wh/l are commonly reported. Single cells can provide a nominal
  voltage of 2–2.1 V. They are not well suited to continuous, heavy discharge.
- Lithium-ion batteries
  A Lithium-ion battery comprises a lithiated metal (e.g. cobalt, manganese, aluminium,
  titanium) oxide or phosphate positive electrode and a carbon negative electrode.
  Their operation depends on the electrochemical interaction of lithium ions in
  non-metallic solution. Single cells can provide a nominal maximum voltage of
  about 3.8–4 V. They are low-temperature and low-maintenance devices that provide a
  reasonably flat operational discharge curve and short charge times (1–4 hours). Gravimetric energy density is superior to lead acid batteries being in the range 100–250 Wh/kg.
  Reported energy efficiencies are of the order of 75%. A specialised protection circuit is
  required to prevent overheating which can be a problem with these devices.
  The opportunities for improved lithium-based electrical storage technologies have not yet
  been exhausted as lithium–air, lithium–metal, lithium–sulphur, and solid-state lithium
  systems are all in development which may provide better specific energy, energy density, etc.

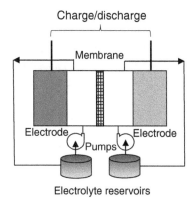

Figure 10.14 Flow battery schematic.

- Sodium sulphate (Na/S) batteries

A sodium sulphur battery comprises a molten sulphur positive electrode, a molten sodium negative electrode, and a solid ceramic alumina electrolyte. To maintain the molten electrode condition, cells must operate at high temperature ($\sim$300 °C) and will therefore require heating. Single cells can provide a nominal voltage of 2 V. They have a high capital cost and the requirement for independent heating entails a parasitic load and running cost. Gravimetric densities are high in the range 150–750 Wh/kg. Volumetric energy densities of 150 Wh/l are typical. Reported efficiencies are of the order of 75% or better. Installations in the range of 10–30 MW are operational.

- Flow batteries

Like other battery technologies, flow devices feature two electrodes. However, this is where the similarity ends as each (graphite) electrode is fed with its own individual electrolyte flow circulated from a tank via a pump. The electrodes and electrolyte pairs are separated from each other by an ion exchange membrane. See Figure 10.14.

Ionic exchange at the electrodes is based on the redox (reduction/oxidation) principle and so the electrolyte must be capable of existing in a number of oxidation states. Material varieties on the principle include vanadium, iron–chromium, and zinc–bromine. As the electrolyte acts, essentially as a fuel store, this arrangement has led to flow batteries being thought of as an analogue to fuel cells.

They are capable of fast recharge times due to the availability of fresh electrolyte and have long lives as the electrodes only degrade slowly. However, they are more complex than other battery types as they require more secondary equipment such as pumps and tanks. Single cells can provide a nominal voltage of 1–1.5 V. Specific energies are of the order of 40 Wh/kg with energy densities of 20–35 Wh/l. For vanadium cells, efficiencies in the range 65–75% are reported. Again, installations in the 10 MW range are operational.

- Ultracapacitors

Although having a wet or liquid electrolyte between graphite electrodes, ultracapacitors do not rely on an electrochemical process. Their charge storage mechanism is electrostatic in nature meaning they are not, strictly speaking, regarded as batteries. However, they are worthy of mention here as they can be employed as electrical energy storage devices capable of providing a large amount of power over a short period of time in the range of seconds to minutes and so they have the potential to complement battery systems and provide a peak lopping function. Recharge times are also short (seconds); however, discharge voltage profiles do not display any 'flat' operational phase and show, instead, a steady decline. Specific energy is of the order of 5 Wh/kg. The energy storage capacity of an ultracapacitor is dependent on its capacitance (C, F). Like batteries, working powers can be built up by series/parallel combinations of devices.

10.6 Daylighting

Daylighting, i.e. the deliberate and controlled introduction of daylight, can be used to bring colour, internal environmental variability, and a visual connection with the outside into an occupied, internal space. Given careful design and integration with the space electric lighting, daylighting can bring energy savings.

Unobstructed, daytime, external light levels can range from 5000 (overcast) to 100 000 (clear sky) lux. Normally the luminance of the sky is not uniform over its hemisphere. Assuming *overcast*, i.e. sun-independent conditions, the variation in sky luminance (L, cd/m^2) is often modelled as a function of altitude angle (θ).

A simple model for the spatial distribution of sky luminance is provided by:

$$I_{L,\,\theta} = 0.333 I_{L,\text{zenith}} (1 + 2 \sin \theta) \tag{10.17}$$

The daylight factor (DF, %) at a point on a working plane height (typically 0.85 m above floor level) in a space is the ratio of actual illuminance to the illuminance (lux) prevailing with the point of interest outside the building under an unobstructed, standard overcast sky.

Spaces having average DF's of 5% or better would be considered well provided with daylight. DFs of less than 2% are considered gloomy.

Tall windows provide the best daylighting. Daylight distribution to a space can be improved by glazing on more than one façade, the use of light shelves, diffuse glazing, or light pipes.

However, maximising the daylight factor in a space should not come at the cost of introducing unacceptable glare and overheating via a direct unobscured view of the Sun.

For a side-lit space, the average daylight factor on a working plane (0.85 m above floor level) resulting from a direct view of the sky through glazing can be calculated from:

$$\text{DF}_{\text{wall glazing}} = \frac{(\text{MF}) A_{\text{glazing}} \tau_{\text{glass}} \theta}{A_{\text{room surface total}} \left(1 - \rho_{\text{room surfaces}}^2\right)} \tag{10.18}$$

Where θ is the visible sky angle (degrees) from the mid-point of the window. See Figure 10.15.

The maintenance factor (MF, range 0–1) is dimensionless and makes an allowance for glazing cleanliness and the presence of glazing architecture such as bars and dividers. For normal exposures, typical values would be in the range 0.75–0.95.

Other empirical DF formulae are in use.

On top-lit atria, another dimensionless parameter, the Well Index (WI), is used to predict the daylight factor on the ground directly beneath the glazing. If the atria height is H (m)

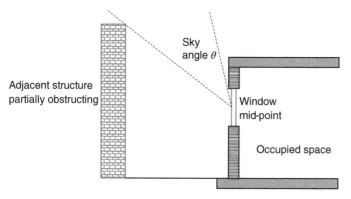

Figure 10.15 Daylight factor and visible sky angle.

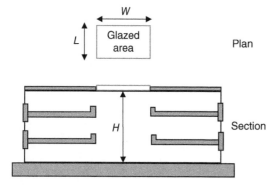

Figure 10.16 Top-lit atrium and the Well Index.

and the opening dimensions are width W (m) and length L (m), as shown in Figure 10.16, then the Well Index is given by:

$$WI = \frac{(H\,(W+L))}{(2LW)} \tag{10.19}$$

and the daylight factor at the well base is given by:

$$DF_{\text{atria base}} = 100\tau_{\text{glass}}e^{-WI} \tag{10.20}$$

DF equations can be modified to account for internally and externally generated reflections.

10.7 Worked Examples

Worked Example 10.1 Solar Thermal Collector Losses
A 3 m^2 flat-plate solar collector operates under the conditions detailed as follows.

Calculate the collector's optical and heat losses (W) and determine the collector's thermal power (W).

Operating conditions

Glass cover transmissivity: 0.9	Collector surface absorptivity: 0.92.
Solar irradiation: 500 W/m²	Combined U value: 5 W/m² K
Surface average temperature: 318 K	Ambient temperature: 285 K

Solution

Given: $A_c = 3\,\text{m}^2$, $\tau = 0.9$, $\alpha = 0.92$, $I_{irr} = 500\,\text{W/m}^2$

$$U_{comb} = 5\,\text{W/m}^2\,\text{K}, \quad T_{ave} = 318\,\text{K}, \quad T_{amb} = 285\,\text{K}$$

Find: $\dot{Q}_{th}$

$$\dot{Q}_{optical\ loss} = I_{irr}A_c\,(1 - \alpha\tau)$$
$$= 500 \times 3 \times (1 - (0.92 \times 0.9\,))$$
$$= 258\,\text{W}$$

$$\dot{Q}_{heat\ loss} = U_{comb}A_c\,(T_{ave} - T_{amb})$$
$$= 5 \times 3 \times (318 - 285)$$
$$= 495\,\text{W}$$

$$\dot{Q}_{th} = I_{irr} - \dot{Q}_{losses}$$
$$= 1500 - (258 + 495)$$
$$= 747\,\text{W}$$

Worked Example 10.2 Solar Thermal Collector Efficiency

The test results from a glazed flat-plate solar collector are as shown in Figure E10.2.

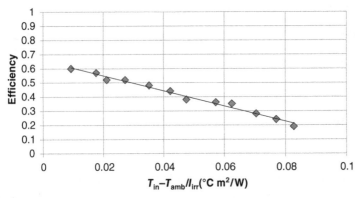

Figure E10.2 Performance data for Worked Example 10.2.

The transmissivity of the glass cover is 0.9.

The absorptivity of the absorber surface is 0.92.

Determine:

(a) The collector heat removal factor F_R
(b) The overall conductance U_{comb} (W/m² K)
(c) The collector efficiency (%) and useful thermal power (W/m²) if the solar irradiation is 450 W/m², the ambient temperature is 5 °C, and the inlet water temperature is 15 °C.

Solution

Given: $\tau = 0.9$, $\alpha = 0.92$, $I_{irr} = 450$ W/m², $T_{amb} = 5 °C$, $T_{in} = 15 °C$

Find: F_R, U_{comb}, η_c, $\dot{Q}_{th}$

From the graph, approximate a linear line of best fit for the data and estimate the intercept and gradient of the line:

$$Intercept \sim 0.66$$

$$Gradient \sim (-)5.4$$

Then from the Hottel–Whillier–Bliss equation:

(a)

$$Intercept\ value = F_R \alpha \tau$$
$$0.66 = F_R \times 0.92 \times 0.9$$
$$F_R = 0.797$$

(b)

$$Gradient\ value = F_R U_{comb}$$
$$5.4 = 0.797 \times U_{comb}$$
$$U_{comb} = 6.77\ \text{W/m}^2\text{K}$$

(c)

$$\eta_c = F_R \left[\alpha\tau - \left[U_{comb}\left(T_{in} - T_{amb}\right)/I_{irr}\right]\right]$$
$$= 0.797\left[(0.92 \times 0.9) - \left[6.77\,(15 - 5)/450\right]\right]$$
$$= 0.54 = 54\%$$

$$\eta_c = \dot{Q}_{th}/I_{irr}$$
$$0.54 = \dot{Q}_{th}/450$$
$$\dot{Q}_{th} = 243\ \text{W/m}^2$$

Worked Example 10.3   Ground Source Heat Pump for Space Heating and Hot Water Provision

A ground source heat pump (GHSP) uses R134a as a working fluid. The refrigerant enters the compressor as saturated vapour at 2 bar and on leaving the condenser it is saturated liquid at 12 bar. Both pressures are absolute. Assume that the compression process is 100%

(a)

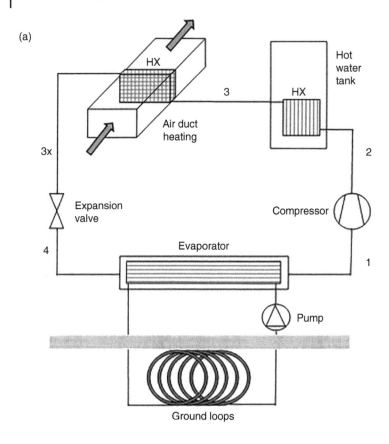

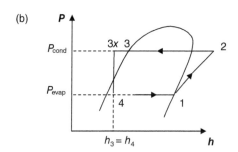

Figure E10.3 (a) GSHP plant layout (b) thermodynamic cycle for worked example 10.3

isentropic. Upon leaving the compressor, the superheated vapour is piped through a hot water heater. The refrigerant leaves the water heater in a saturated liquid condition and then goes on to lose further energy to an air duct space heater, exiting as a subcooled liquid at 20 °C. The circuit is completed via an expansion valve back to the evaporator at 2 bar. The evaporator is heated by underground heat source circulating water at constant temperature of 10 °C. The compressor is rated at 1.25 kW. See Figure E10.3a,b.

Assuming 80% overall compressor efficiency and the liquid refrigerant has a specific heat capacity of 1.4 kJ/kg K, calculate, *using the tables in Appendix B*:

(a) the mass flow of refrigerant
(b) the heat absorbed by the hot water heater and that by the space heater
(c) the actual coefficient of performance for heating.

Solution

Given : $s_1 = s_2$ (100% isentropic compression), $\eta_{comp} = 0.8$, $C_p = 1.4$ kJ/kg K

$P_1 = P_4 = 2$ bar, $P_2 = P_3 = 12$ bar (no pressure losses), $\dot{W}_{comp,act} = 1.25$ kW

Find : $\dot{m}_{ref}$, $\dot{Q}_{hot\ water}$, $\dot{Q}_{air\ heater}$, COP_{HP}

From the Table B.1 in Appendix B:

At 2 bar, saturation temperature $T_c = -10.09°C = 262.91$ K

Saturation vapour conditions : $h_1 = 241.3$ kJ/kg, $s_1 = 0.9253$ kJ/kg K $= s_2$

At 12 bar, saturation temperature, $T_h = 46.32°C$

Saturation liquid conditions : $h_3 = 115.76$ kJ/kg

Subcooled liquid conditions : $h_{3x} = h_3 - Cp\,(T_3 - T_x)$
$$= 115.76 - 1.4\,(46.32 - 20) = 78.91\ \text{kJ/kg}$$

Taking s_2 to be 0.9253 kJ/kg it can be seen, by inspection, that T_2 must have a value between 50 and 60 °C

Using interpolation on the enthalpy table at T_2, h_2 is found to be 278.44 kJ/kg

(a) *Mass of refrigerant*:

Across the compressor : $h_2 - h_1 = 278.44 - 241.3 = 37.14$ kJ/kg

$\dot{W}_{comp,act} \times \eta_{comp} = \dot{m}_{ref}\,(h_2 - h_1)$
$$1.25 \times 0.8 = \dot{m}_{ref} \times 37.14$$
$$\dot{m}_{ref} = 0.0269\ \text{kg/s}$$

(b) *Hot water and air heaters*:

$\dot{Q}_{hot\ water} = \dot{m}_{ref}\,(h_2 - h_3) = 4.38$ kW
$\dot{Q}_{air\ heater} = \dot{m}_{ref}\,(h_3 - h_{3x}) = 0.99$ kW

(c) *Coefficient of performance*:

$COP_{HP} = (\dot{Q}_{hot\ water} + \dot{Q}_{air\ heater})/\dot{W}_{comp,act}$
$$= \frac{4.38 + 0.99}{1.25}$$
$$= 4.4$$

Worked Example 10.4 PV Array Sizing

Estimate the annual energy output (E_{array}, kWh/m^2) from a PV array operating under the following parameters:

Average annual daily solar irradiation: 2.8 kWh/day m^2.
PV module efficiency = 12%
Power system electrical losses = 85%
Shading and soiling de-rating = 0.95
Heat loss de-rating = 0.97.

Solution

Given: $I = 2.8$ kWh/day m^2, $C_1 = 12\,\%$, $C_2 = 85\,\%$, $C_3 = 0.95$, $C_4 = 0.97$

Find: E_{array}

Annual solar irradiation $I = 2.8 \times 365 = 1022$ kWh/m^2

$$E_{\text{array}} = I(C_1/100)(C_2/100)C_3 C_4$$
$$= 1022(12/100)(85/100)0.95 \times 0.97$$
$$= 96 \text{ kWh/m}^2$$

Worked Example 10.5 Sensible Heat Thermal Store Sizing

Water is to be used as a sensible heat storage medium. The water temperature will be raised from 20 to 45 °C. The store is to be cylindrical in form having a diameter of 1 m.

Using the properties supplied, determine the store height (m) for an energy capacity of 200 MJ.

Water density: 995 kg/m^3
Water specific heat capacity: 4180 J/kg K

Solution

Given: $\Delta T = 45 - 20 = 25\,^\circ$C, $D_{\text{store}} = 1$ m, $\rho_{\text{water}} = 995$ kg/m^3,

$$C_{\text{pwater}} = 4180 \text{ J/kg K}, \quad Q_{\text{sensible}} = 200 \text{ MJ} = 200 \times 10^6 \text{ J}$$

Find: H_{store}

$$Q_{\text{sensible}} = m_{\text{store}} C_p \Delta T$$
$$200 \times 10^6 = m_{\text{store}} \times 4180 \times 25$$
$$m_{\text{store}} = 1913.88 \text{ kg}$$

$$m_{\text{store}} = \rho V_{\text{store}} = \rho \times \left[(\pi D_{\text{store}}^2/4) \times H_{\text{store}} \right]$$
$$1913.88 = 995 \times \left[0.7855 \times H_{\text{store}} \right]$$
$$H_{\text{store}} = 2.45 \text{ m}$$

Worked Example 10.6 Phase Change Thermal Storage

An energy store uses a PCM having the following properties:

Specific (sensible) heat capacity (solid phase) = 1950 J/kg K

Specific latent heat = 240 kJ/kg
Specific (sensible) heat capacity (liquid phase) = 3550 J/kg K
Melting point = 32 °C

Initially the PCM is at 25 °C. After being charged its final temperature is 40 °C. Determine the energy stored in the material (kJ/kg)

Solution

Given: $C_{p,\text{solid phase}} = 1950\,\text{J/kg K}$, $h_{pc} = 240\,\text{kJ/kg}$, $C_{p,\text{liquid phase}} = 3550\,\text{J/kg K}$

$$T_i = 25°\text{C}, \quad T_{\text{melt}} = 32°\text{C}, \quad T_f = 40°\text{C}$$

Find: Q_{total}

$$Q_{\text{total}} = Q_{\text{sensible}} + Q_{\text{pc}}$$
$$= Q_{\text{solid phase}} + Q_{\text{pc}} + Q_{\text{liquid phase}}$$
$$= C_{p,\text{solid phase}} \left(T_{\text{melt}} - T_i\right) + h_{pc} + C_{p,\text{liquid phase}} \left(T_f - T_{\text{melt}}\right)$$
$$= \left[1950 \times (32 - 25)/1000\right] + 240 + \left[3550 \times (40 - 32)/1000\right]$$
$$= 13.65 + 240 + 28.4$$
$$= 282.05\,\text{kJ/kg}$$

Worked Example 10.7   Battery Energy Efficiency

A battery is found to have a coulombic efficiency of 98%.

During a test the device was found to have an average charge voltage of 3.3 V and an average discharge voltage of 2.7 V.

Determine the battery energy efficiency (%).

Solution

Given: η_{coulomb}, V_{charge}, $V_{\text{discharge}}$

Find: η_{energy}

$$\eta_{\text{voltage}} = \frac{\text{Average discharge voltage } (V)}{\text{Average charge voltage } (V)}$$
$$= \frac{2.7}{3.3}$$
$$= 0.8182 = 81.82\%$$

$$\eta_{\text{energy}} = \eta_{\text{coulomb}} \times \eta_{\text{voltage}}$$
$$= 0.98 \times 0.8182$$
$$= 0.8018 = 80.18\%$$

Worked Example 10.8   Overcast Sky Luminance

On an overcast day, the zenith sky luminance is 2500 cd/m².

What is the sky luminance (cd/m²) at an altitude angle of 50°?

Solution

Given: $I_{L,\text{zenith}} = 2500\,\text{cd/m}^2$, $\theta = 50°$

Find: $I_{L,\theta}$

$$I_{L,\theta} = 0.333 \; I_{L,\text{zenith}} \; (1 + 2\sin\theta)$$
$$= 0.333 \times 2500 \times (1 + 2\sin(50))$$
$$= 2108 \; \text{cd}/\text{m}^2$$

Worked Example 10.9   Side-Lit Daylight Factor

An occupied space has glazing along one wall as shown in Figure E10.9.

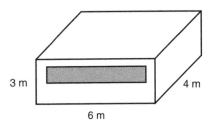

3 m

4 m

6 m

Figure E10.9   Layout for Worked Example 10.9.

The glazed area is $6\,\text{m}^2$ and the glazing transmittance is 0.7.

The maintenance factor for the glass is estimated to be 0.9.

Due to obstructions, the sky angle is 60°.

If the average interior surface reflectance is 0.5, determine the average daylight factor (%) in the room.

In the absence of electric lighting, would the room be described as 'bright' or 'gloomy'? What would be the internal light level (lux) if the external level is taken to be 5000 lx?

Solution

Given: $A_{\text{glazing}} = 6\,\text{m}^2$, $\tau_{\text{glass}} = 0.7$, MF $= 0.9$, $\theta = 60°$, $\rho = 0.5$

$$A_{\text{room surface total}} = (4 \times 6 \times 2) + (3 \times 4 \times 2) + (3 \times 6 \times 2) = 108\,\text{m}^2$$

Find: DF

$$DF_{\text{wall glazing}} = \frac{\text{MF} \times A_{\text{glazing}} \tau_{\text{glass}} \theta}{A_{\text{room surface total}} \left(1 - \rho^2_{\text{room surfaces}}\right)}$$
$$= \frac{0.9 \times 6 \times 0.7 \times 60}{108 \left(1 - 0.5^2\right)}$$
$$= 2.8\%$$

i.e. tending towards 'gloomy'. Artificial lighting would be required

The resulting light level in the room would be $0.028 \times 5000 = 140\,\text{lx}$.

Worked Example 10.10   Well Index and Atrium Daylight Factor

The rectangular glazed area in an atrium has dimensions of length 4 m and width 5 m. The glazing is 10 m above ground level and has a transmissivity of 0.9.

Determine the Well index and estimate the daylight factor at ground level at the base of the atrium.

Solution

Given: $L = 4\,\text{m}$, $W = 5\,\text{m}$, $H = 10\,\text{m}$, $\tau_{\text{glass}} = 0.9$

Find: WI, $\text{DF}_{\text{atria base}}$

$$\text{WI} = \frac{(H(W+L))}{(2LW)}$$

$$= \frac{(10 \times (5+4))}{(2 \times 4 \times 5)}$$

$$= 2.25$$

$$\text{DF}_{\text{atria base}} = 100 \times \tau_{\text{glass}} \times e^{-\text{WI}}$$

$$= 100 \times 0.9 \times e^{-2.25}$$

$$= 9.5\%$$

10.8 Tutorial Problems

10.1 A building occupied all year around housing 30 people consumes 8 l of hot water per person per day. The water requires heating from 10 to 60 °C for storage purposes.

The building owners would like 20% of the annual hot water heating demand to be supplied by a solar thermal collector.

The annual average solar irradiation is 700 kWh/m² annum and the collector has an annual overall collection efficiency of 40%.

Determine the required collector area (m²).

Take the density of water as 1000 kg/m³ and its specific heat capacity as 4200 J/kg °C. 1 kWh $= 3.6 \times 10^6$ J

[Answer: 3.65 m²]

10.2 A solar thermal collector having a collector heat removal factor of 0.92 is operating under the conditions detailed as follows.

Estimate its collection efficiency (%) using the Hottel–Whillier–Bliss expression.

Specification

Glass cover transmissivity: 0.89	Collector surface absorptivity: 0.92
Insolation 800 W/m²	Combined U value: 6 W/m² K
Collector fluid inlet temperature: 320 K	Ambient temperature: 291 K

[Answer: 55.3%]

10.3 Assuming visible light to have an average photonic energy of 3.68×10^{-19} J/photon, determine the rate of visible light photons arriving on a surface receiving 300 W of solar irradiation.

[Answer: 8.15×10^{20} photons/s]

10.4 The owners of a building having an annual electrical energy consumption of 12 000 kWh/annum wish to generate 10% of its energy demand via a PV array having the specification shown:

Specification

Average annual daily solar irradiation: 4.19 kWh/d m².

PV module efficiency = 10%

Shading and soiling de-rating = 0.95

Power system electrical losses = 85%

Heat loss de-rating = 0.97

Determine the required collector area (m²).

[Answer: 10.01 m²]

10.5 For a sensible energy store, the *volumetric energy capacity* is given by the product of a storage material density (ρ, kg/m³) and its specific heat capacity (C_p, J/kg K). The ability of a material to charge/discharge its energy in rate terms is characterised by its *thermal diffusivity* (α, m²/s) where:

$$\alpha = k/\rho C_p$$

Where k is the storage material thermal conductivity (W/m² K).

For the two materials listed in Table P10.5, compare the aforementioned parameters and describe each material's relative strengths.

Table P10.5 Thermal storage material properties for tutorial Problem 10.5.

Material @20 °C	Density (kg/m³)	Specific heat capacity (J/kg K)	Thermal conductivity (W/m K)
Concrete	2000	880	1.4
Water	998	4180	0.6

[Answer: concrete – lower volumetric heat capacity, higher thermal diffusivity, water – high volumetric heat capacity, low thermal diffusivity.]

10.6 A PCM has a latent heat value of 130 kJ/kg and a density of 900 kg/m³. The material is encapsulated in 80 mm diameter spheres.

The manufacturers suggest the volumetric packing efficiency is 0.66, i.e.0.66 m³ of PCM per m³ of total storage volume. Estimate the number of spheres required and the volume (m³) of a store able to retain 20 MJ of heat.

Ignore the sphere encapsulation material.

[Answer: 638, 0.259 m³]

10.7 In a battery cycle test, the charge delivered to the battery was 32 mAh. On discharge, the charge available was 31.4 mAh. If the voltaic efficiency for the device 80%, determine the battery energy efficiency (%).

[Answer: 78.5%]

10.8 A person at point A in Figure P10.8 views an overcast sky through a window. If the zenith sky luminance is 2000 cd/m², estimate the maximum sky luminance (cd/m²) seen by the viewer.

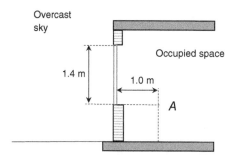

Figure P10.8 Layout for tutorial Problem 10.8.

Assume the glazing has perfect transmissivity.
[Answer: 1750 cd/m²]

10.9 The prospective owners of the new building shown in Figure P10.9, specify that the interior is to have a minimum average daylight factor of 5% via a glazed area on one of the long (10 m) walls.

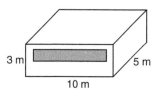

Figure P10.9 Layout for tutorial Problem 10.9.

The glazing transmittance is 0.7 and the maintenance factor for the glass is estimated to be 0.9. The sky angle is 70°.
If the average interior surface reflectance is 0.5, determine the glazing area (m²) required to meet the owners' wish.
What will be the resulting interior light level (lux) if the exterior level is 5000 lux?
[Answers: 16.2 m², 250 lux]

10.10 The *square* glazed area in an atrium has a dimension of length L (m). The glazing is 12 m above ground level and has a transmissivity of 0.9. Determine the dimension of the glazing if the daylight factor at ground level at the base of the atrium is to be 5%.
[Answer: 4.15 m]

11

Measurements and Monitoring

11.1 Overview

This chapter covers a range of internal environmental measurement devices working on different principles. In fact, too many to do justice to in a short overview. Instead, a group of sensor/meter performance factors that are common to all devices and should be considered before use or purchase will be briefly mentioned. The list is by no means exhaustive:

Accuracy (or inaccuracy): A measure of uncertainty defining the closeness of the measured value with the true value.

Calibration: A means of determining a device's accuracy by comparing its returned value at a known reference point (This may be followed by a device adjustment to minimise the inaccuracy).

Error: The difference between a measured value and the true value.

Intrusiveness error: An error resulting from the ability of the placement of the sensor/meter itself to alter the measurement field.

Precision: An indication of measurement repeatability.

Range: The values of the measured parameter to which the sensor/meter is suited to read accurately.

Resolution: The degree with which a sensor/meter can follow a continuously and rapidly varying input with a representative output.

Spatial error: An error resulting from the poor positioning of a sensor producing an unrepresentative result.

Learning Outcomes

- To be familiar with gaseous and particulate concentration measurement.
- To be familiar with a range of techniques for the measurement of air physical parameters such as pressure, temperature, and moisture content.
- To be familiar with techniques for the measurement of local illumination and sound levels.
- To be familiar with the application of a range of measurement techniques in utility (e.g. electricity, heat, gas, and water) measurement.

Building Services Engineering: Smart and Sustainable Design for Health and Wellbeing, First Edition. Tarik Al-Shemmeri and Neil Packer. © 2021 John Wiley & Sons Ltd. Published 2021 by John Wiley & Sons Ltd.

11.2 Compositional Parameters

11.2.1 Gaseous Concentration Measurement

The interaction of matter (i.e. atoms and molecules) with photons provides a means of measuring the presence of a gas.

The interaction might result in:

- The photon passing through the molecule with both unaffected.
- The photon being absorbed resulting in an increase of molecule oscillation.
- The photon being absorbed by the molecule resulting in the emission of another photon of the same wavelength (elastic scattering).
- The photon being absorbed by the molecule resulting in the emission of another photon of a different wavelength (nonelastic scattering).
- Ionisation.

The probability of any particular occurrence depends on the incoming photon wavelength and the gas species present. The degree of its occurrence depends on the path length taken by the photons and the gas concentration (number of atoms per unit volume).

These phenomena lead to a number of measurement techniques:

Absorption spectroscopic techniques rely on the attenuation of a photon source due to the presence of an intervening pollutant species.

Infrared and UV wavelengths are commonly used as photon sources.

Extractive techniques rely on a comparison between a beam passed through a gas sample and another passed through a reference cell.

Luminescence techniques rely on the production of electromagnetic radiation by an excited gas molecule.

Photoluminescence relies on an incoming photon exciting a gas molecule resulting in the emission of a photon of different wavelength. For example, sulphur dioxide molecules excited with 210 nm UV light will emit light at 350 nm when returning to their ground state.

Chemiluminescence is the emission of light from a gas molecule excited in a chemical reaction with another gas under carefully controlled conditions in a reaction chamber. For example, nitric oxide (NO) reacts with ozone (O_3) to give nitrogen dioxide (NO_2) in a reaction that generates light:

$$NO + O_3 \rightarrow NO_2 + O_2 + \text{light emission.}$$

The intensity of the light emitted is proportional to the nitric oxide concentration.

Optical methods rely on the scattering of a light (visible, infrared, or laser) beam due to an intervening gas medium, thus reducing the amount of light received at a detector.

Colorimetric methods rely on changes in the optical properties, such as reflectivity or colour, of a reagent or medium (gel) after exposure to the gas of interest. The medium is usually selected such that exposure to the gas of interest darkens its surface lowering its reflectance. The change is detected by a light source/receiver arrangement.

Electrochemical methods rely on exposing the gas to a fuel cell type arrangement comprising two electrodes and an ion conductor/electrolyte resulting in attendant oxidation/reduction reactions and the generation of a measurable current in an external circuit.

The electrode exposed to the gas is termed the *working* electrode and the other the *counter* electrode.

For example, for carbon monoxide, the half-cell reactions are:

Working electrode: $CO + H_2O \rightarrow CO_2 + 2H+ + 2e$
Counter electrode: $\frac{1}{2}O_2 + 2H+ + 2e- \rightarrow H_2O$
Overall: $CO + \frac{1}{2}O_2 \rightarrow CO_2$

Both electrodes are catalytically coated with, for example, Ag, Au, or Pt.

The magnitude of the current is proportional to the gas concentration.

Some cells operate with a third *reference* electrode for improved sensitivity.

MOS (metal oxide semiconductor) methods rely on changes in the electrical resistance of heated, mixed metal (e.g. Chromium, Titanium, and Tin) oxide semiconductors when exposed to a gas of interest. Oxygen exchange at the gas–surface interface will determine the availability of charge in the semiconductor. Hence, the sensor response (resistance increase/decrease) is dependent on the nature of the gas (oxidising/reducing) and sensor material selection. For a given combination, the change in resistance is proportional to the gas concentration.

11.2.2 Particulates Concentration Measurement

Particulate concentration measurement employs several techniques such as the optical and colorimetric methods described above. Examples include transmissivity, beta-attenuation and black smoke analysers. Gravimetric techniques such as filter weighing and tapered element oscillating microbalance (TEOM)analysers are also available.

11.3 Physical Parameters

11.3.1 Pressure Measurement Principles

Commonly, pressure transducers convert the deflection of a diaphragm, having a reference pressure on one side and the pressure of interest on the other, into an electrical signal. The output can be used to produce a reading on an indicator or used to control a process. Transducers are available covering a wide range of pressure with a fast response. Varieties include.

- *Piezoresistive strain gauge*: uses the Piezoresistive effect of bonded or formed strain gauges to detect strain due to applied pressure.
- *Capacitive*: uses a diaphragm and pressure cavity to create a variable capacitor to detect strain due to applied pressure.
- *Magnetic*: measures the displacement of a diaphragm or flexible chamber by means of changes in inductance (reluctance), LVDT, Hall effect, or by eddy current principle.
- *Piezoelectric*: uses the piezoelectric effect displayed by certain materials, such as quartz, to measure the strain upon the sensing mechanism due to pressure.
- *Optical*: uses the physical change of an optical fibre to detect strain due to applied pressure.

- *Potentiometric*: uses the motion of a wiper along a resistive mechanism to detect the strain caused by applied pressure.
- *Resonant*: uses the changes in resonant frequency in a sensing mechanism to measure stress, or changes in gas density, caused by applied pressure.

The following *mechanical* pressure measurement devices are so common that they also warrant coverage here.

- *Bourdon pressure gauge*
  In the Bourdon gauge, a curved, flattened metal tube is closed at one end and connected to the pressure source at the other end. Under pressure, the tube tends to straighten and causes a deflection of a pointer through a lever and rack and pinion amplifying system. See Figure 11.1. This gauge can be used for liquids or gases from a fraction of a bar pressure up to 10 000 bar. High-quality gauges can have an accuracy of the order of 0.5% of full scale.
- *Manometers*
  The U-tube manometer may be used to measure a pressure relative to atmospheric pressure, or the difference between two pressures. If one 'leg' is much larger in diameter than the other, a '*well*' manometer is obtained and only a single reading is required. The '*inclined*', single-leg manometer gives greater accuracy. When the manometer fluid is less dense than the fluid, the pressure of which is to be measured, an inverted manometer is used. When pressure is measured relative to atmospheric pressure, the air density is assumed to be negligible compared with that of the manometer fluid. See Figure 11.2.

11.3.2   Temperature Measurement Principles

The necessary property that a substance must possess for it to be considered for use in the measurement of temperature is that it should experience some recognisable change when its temperature is changed. Electrical property and dimensional change are most commonly employed. Moreover, this change must be repeatable without deterioration.

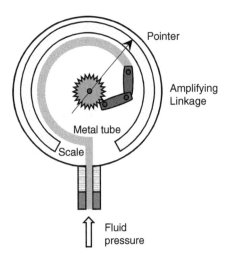

Pointer

Amplifying
Linkage

Metal tube

Scale

Fluid
pressure

Figure 11.1   Bourdon pressure gauge schematic.

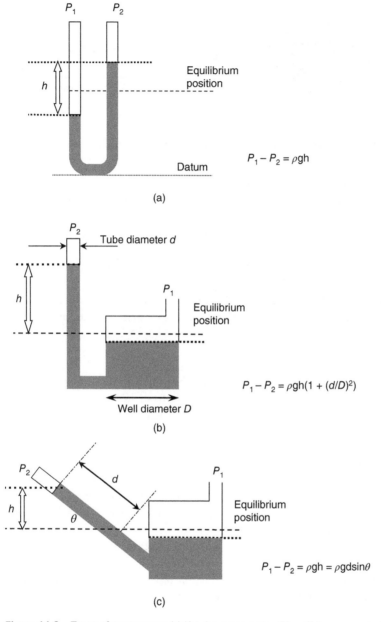

Figure 11.2   Types of manometer. (a) U-tube manometer, (b) well-type manometer, and (c) inclined manometer.

The choice of material and effect used for measurement is governed by the temperature range considered, degree of accuracy required, type of installation, and cost. Thermocouples, electrical resistance thermometers, and thermistors are described here as they generate an electrical signal which can be transmitted and used for control.

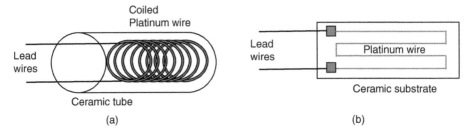

Figure 11.3 Examples of resistance thermometer sensor geometry. (a) Coiled wire sensor and (b) thin film sensor.

11.3.2.1 Resistance Thermometers

This type is based on the fact that the electrical resistance of metal increases as the temperature of the metal is raised.

A resistance thermometer consists of a coil of wire suitably wound and enclosed in a sheath.

For low temperatures (up to 100 °C), silk-covered nickel wire is used. For temperatures up to 1000 °C, platinum wire is used, wound on a piece of mica.

If the resistance of the metal at temperature T °C is R_T and its resistance at temperature 0 °C is R_0 then, in its abbreviated form, the approximate resistance–temperature relationship is given by:

$$R_T = R_0(1 + \alpha T) \tag{11.1}$$

For a platinum wire, an appropriate value for α would be 0.003 851/$^\circ$C. High-quality resistance thermometers can have an accuracy of the order of ±0.001 °C.

Sensors can be fabricated into a number of geometries depending on the application. See Figure 11.3.

11.3.2.2 Thermocouples

In the early 1830s Berlin, **Thomas Johann Seebeck** heated the junction of a circuit comprising two different metals. He detected a magnetic field close to the circuit but failed to make the link with the presence of a voltage in the circuit. Nevertheless, his work is celebrated by the term '**Seebeck effect**'.

The magnitude of the voltage is proportional to the temperature of the junction and the chosen materials.

The slope of the voltage–junction temperature relationship is termed the **Seebeck coefficient** (α_{ab}, V/K) for the combination of materials a and b.

Mathematically:

$$\alpha_{ab} \equiv \lim_{\Delta T \to 0} \left(\frac{\Delta V}{\Delta T} \right) \quad \text{or} \quad V = \int_{T_1}^{T_2} \alpha_{ab} dT \tag{11.2}$$

where $\alpha_{ab} = \alpha_a - \alpha_b$

Practical thermocouple circuits have a reference temperature junction in addition to the sensing temperature junction.

The value of the emf (mV) developed depends upon the difference in the temperature of the hot and cold junctions. A basic thermocouple schematic is shown in Figure 11.4.

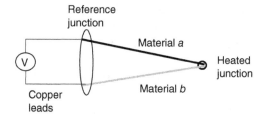

Figure 11.4 Seebeck effect.

Table 11.1 Emf values for typical Thermocouple material combinations.

Type	Combination	Typical range	Typical emf (mV@100 °C)
S	Platinum: 10%, Rhodium/Platinum	0 to 1500 °C	0.645
R	Platinum: 13% Rhodium/Platinum	0 to 1500 °C	0.647
J	Iron/Copper-Nickel	−200 to 850 °C	5.268
K	Nickel-Chromium/Nickel-Aluminium	−180 to 1100 °C	4.095
T	Copper/Copper-Nickel	−250 to 400 °C	4.277
E	Nickel-Chromium/Copper-Nickel	−200 to 850 °C	6.137

For a known reference temperature, say ice point (0 °C), then the emf will be a measure of the temperature of hot junction. This relation is usually linear, but the voltage is very small and amplification is necessary. A range of thermocouple material combinations are available. See Table 11.1.

High-quality thermocouples can have an accuracy of the order of ±0.02 °C.

11.3.2.3 Thermistor

Thermistors for temperature measurement are semiconductors displaying a large, negative slope, nonlinear resistance response to temperature change, typically expressed in the following form:

$$1/T = A + B \ln R + C(\ln R)^3 \tag{11.3}$$

where A, B, and C are material-specific constants and T is in Kelvin. There are however several ways of linearising an output signal from such a sensor. Thermistors are best suited to temperatures below 100 °C. High-quality thermistors can have an accuracy of the order of ±0.1 ° C.

Positive slope thermistors are commonly used in an electrical circuit protection role.

The following *mechanical* temperature measurement devices, utilising thermal expansion, are so common that they also warrant coverage here.

- Liquid-in-Glass Thermometer
  This method utilises the expansive properties of a liquid with temperature.
  Suitable liquids include alcohol, acetone, aniline, etc.

The liquid is contained in a very fine glass capillary tube, which terminates in an enlarged part called the bulb. Before the tube is sealed off at its upper end, the space above the liquid is exhausted of air and filled with an inert gas, such as nitrogen.

- Bi-Metal Thermometer

  This makes use of the difference in the expansion of two metal strips, robust and suitable for temperatures up to 550 °C.

  Two strips of dissimilar metal firmly attached to each other throughout their length, usually steel and copper, or in some cases steel and brass.

  This movement of the top end can be used as a method for recording temperature.

- Gas Thermometer

  Gases such as argon, hydrogen, helium, nitrogen, and oxygen, if relatively pure, display a large, uniform rate of thermal expansion over a large temperature range. If they can be regarded as perfect gases, then the variation of temperature with volume is governed by Charles Law (at constant pressure) given by:

$$\frac{V_1}{V_2} = \frac{273 + T_1}{273 + T_2} \tag{11.4}$$

A simple gas thermometer might then comprise a probe, flexible connection, a Bourdon gauge (as described in Section 11.3.1) filled with a near perfect gas. See Figure 11.5.

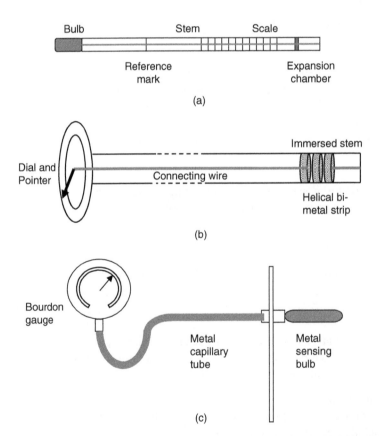

Figure 11.5 Mechanical temperature measurement devices. (a) Liquid-in-glass thermometer, (b) bi-metal thermometer, and (c) gas thermometer.

11.3.3   Humidity Measurement Principles

There are over a dozen different methods of measuring humidity. Some are direct, i.e. related to the amount of moisture present. Others use another parameter, e.g. temperature or pressure to infer the moisture. Electrical and optical methods are discussed here.

(a) Electrical property methods

Changes in material electrical property, e.g. electrical conductivity and resistance or dielectric constant, capacitance and impedance, can facilitate moisture measurement.

- Ceramics

Ceramics such as TiO_2, Al_2O_3, doped SiO_2, or lithium chloride-doped $MnWO_4$ readily adsorb water onto their surfaces producing improved proton transfer (via hydrogen bonding) that result in a decrease in electrical conductivity and an increase in dielectric constant.

- Semiconductors

Semiconductors such as SnO_2, ZnO, and In_2O_3 are able to absorb water onto their surfaces in both molecular and hydroxyl forms producing a local change in electron concentration. This results in an increase in the electrical conductivity of n-type materials and a decrease in p-type materials

- Polymers

Both resistive and capacitance changes can be effected with a suitable choice of material. In general, polymer electrical conductivity decreases with increasing moisture content. Polyelectrolytic polymers employ ammonium, sulphur, or phosphorus salts in porous thin films.

- Thermal conductivity hygrometer

In the absence of convection and radiation, the heat lost from a heated wire surrounded by gas in an enclosure takes place by conduction. The thermal conductivity of a gas is dependent on temperature and moisture content. Hence, for a constant applied voltage, the resulting temperature of the wire and its electrical resistance depends on humidity. Practical measurements are comparative requiring a reference dry air sample at the same temperature as the sensor.

- 'Dewcel' hygrometer

Another moisture measurement related to electrical resistance. The sensor comprises a sample of lithium chloride wound on a plastic former and a pair of contact wires. The lithium chloride is hygroscopic and absorbs moisture present in the air thus lowering its electrical resistance. The current required to maintain a given voltage difference across the sensor is a measure of the moisture present.

- Electrolytic hygrometer

In essence, the sensor is a cell having platinum electrodes and a phosphorous pentoxide electrolyte. Moisture in a gas sample is absorbed by the electrolyte and dissociated to produce hydrogen and oxygen ions. The resulting cell current is proportional to the moisture content.

(b) Optical methods

As described in Section 11.2.1, the absorption of specific wavelength electromagnetic radiation is dependent on the amount or concentration of intervening medium. Water vapour exhibits sufficient attenuation, between a source of either ultraviolet or infrared wavelength and detector, to enable the construction of a measurement device based on this principle. The degree of attenuation is indicative of the absolute humidity.

An alternative optical system relies on the temperature at which condensation or 'dew' forms on a cooled surface. In this arrangement, a mirrored surface comprising one side of a thermoelectric generator is irradiated with a light beam and the reflected light measured. On the occurrence of condensation, the incoming light is scattered and a reduced reflected signal is received. The temperature of the surface at this time is the Dew point, and the moisture content can be derived.

The following *mechanical* humidity measurement device is so common that it also warrants coverage here.

11.3.3.1 Wet- and Dry-Bulb Hygrometer (Relative humidity)

The evaporation of water requires an amount of energy equivalent to the latent heat (J/kg) to perform the operation. As a result of the process, the temperature of any remaining water is lowered. The rate of evaporation is dependent on the amount of moisture in the atmosphere. Using this phenomenon, it is possible to construct a humidity measuring instrument known as the wet- and dry-bulb hygrometer.

It comprises two thermometers, a tight-fitting wick and small water reservoir. See Figure 11.6. One end of the wick fits around the bulb of one thermometer (the *wet bulb*), while the other end of the wick dips into a vessel of water. The second thermometer (the dry bulb), in an adjacent position, is normal and simply exposed to the environment of interest. The action of the water evaporating from the wick of the wet-bulb thermometer cools the bulb, and its temperature, (T_{wb}), falls below that of the dry bulb (T_{db}), by an amount depending upon the relative humidity of the air.

A value for relative humidity (ϕ, %) can be estimated from:

$$\phi = \frac{P_{sat,wb} - 100KP_{atm}(T_{db} - T_{wb})}{P_{sat,db}} \qquad (11.5)$$

where $K = 0.00066(1 + 0.00115\ T_{wb})$

The saturation pressures at both dry- and wet-bulb conditions can be calculated using Eq. (4.2).

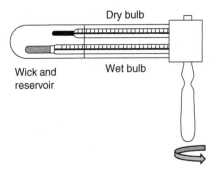

Figure 11.6   Wet- and dry-bulb hygrometer.

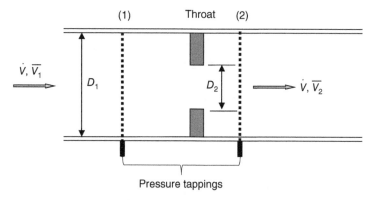

Figure 11.7 Restriction method of velocity measurement.

11.3.4 Velocity Measurement Principles

Three methods of measurement will be discussed here.

11.3.4.1 Differential Pressure Meters

These methods calculate the velocity of flow through a duct or pipe by inserting a deliberate change of cross section to flow (a restriction known as the '*throat*') thus causing a pressure drop related the speed of flow. See Figure 11.7.

Applying the Bernoulli equation and the continuity equation between positions (1) and (2), it can be shown that the velocity in the throat (2) is given by:

$$\overline{V}_2 = \sqrt{\frac{2(P_1 - P_2)}{\rho[1 - (A_2/A_1)^2]}} \qquad (11.6)$$

The resulting differential $(P_1 - P_2)$ can be measured by one of the pressure techniques described earlier.

The pressure drop is based on the ideal pressure/velocity change and takes no account of friction.

Applying the continuity equation at the restriction (2), the real volume flow rate $(\dot{V})$ is determined by:

$$\dot{V} = C_d \overline{V}_2 A_2 \qquad (11.7)$$

where C_d is a discharge coefficient that accounts for frictional loss.

For large cross-sectional air ducts, a non restriction solution is provided by the pitot-static tube or grid.

A pitot-static device comprises a slender concentric tube aligned with the flow used to measure velocity by means of a pressure difference.

The **inner tube** is open to the flow direction and experiences the **total pressure**, i.e. *the sum of velocity and static pressure.*

The **outer tube** is closed in the flow direction but has peripheral holes and thus experiences the **static pressure** only. See Figure 11.8.

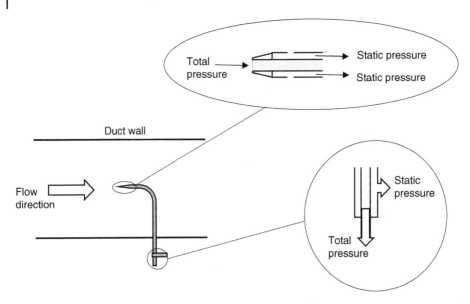

Figure 11.8 Pitot-static tube arrangement.

If the inner and outer tubes are connected across a differential pressure transducer, it can be shown that the velocity in the pipe or duct carrying the flow is given by:

$$\overline{V}_{\text{duct}} = \sqrt{\frac{2(P_{\text{total}} - P_{\text{static}})}{\rho}} \tag{11.8}$$

Using a pitot-static tube to average elementally across a large pipe or duct:

$$\overline{V}_{\text{duct}} = (1/A_{\text{total}}) \sum \overline{V}_i A_i \tag{11.9}$$

where 'i' denotes a sampling point in the cross section of the duct.

Accuracy will not only depend on the pressure transducer deployed across the pipe ends but also on the probe orientation. The probe should be aligned parallel to the flow. Misalignment will produce severe error at angles of >15 °C from the optimum. Multi-tube, averaging devices used at grilles, diffusers, heating coils, and other fittings are termed *flow grids*.

11.3.4.2 Anemometers
Various types of anemometer are used to measure the velocity, usually of air. The 'cup type', Figure 11.9a, is used for free air and has hemispherical cups on arms attached to a rotating shaft. The shape of the cups gives a greater drag on one side than the other and results in a speed of rotation approximately proportional to the air speed. Velocity is found by measuring revolutions over a fixed time.

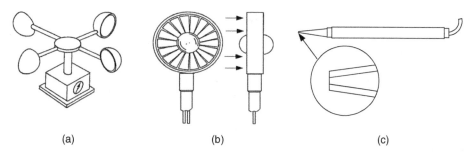

Figure 11.9 Types of anemometers. (a) Cup anemometer, (b) vane anemometer, and (c) hot-wire anemometer.

The vane anemometer, Figure 11.9b, has an axial impeller attached to a handle with extensions and an electrical pick-up which measures the revolutions. A meter with several ranges indicates the velocity.

The '*hot-wire*' anemometer, Figure 11.9c, is used where it is necessary to investigate the change in velocity over a small distance, e.g. in a boundary layer. A probe terminating in an extremely small heated wire element is situated in the fluid stream. If the fluid is at a different temperature from the wire, heating/cooling of the wire occurs changing its electrical resistance. A change in current to the wire is required to maintain its original temperature. The current change is indicative of the flow velocity. Simple hot-wire devices are only suitable for clean fluids.

11.3.4.3 Optical Methods

Optical methods use the scattering of laser light by small particles in the flow to determine flow rate. Units typically comprise two laser beams focused a short distance (D) apart in the plane of the flow path. Each beam has its own photodetector on the other side of the conduit. A flow particle crossing the first beam will cause it to scatter producing a pulse signal at its photodetector. As the same particle crosses the second beam, its detecting optics collects its scattered light and produces a second electrical pulse. By measuring the time interval (t) between these pulses, the gas velocity is calculated as:

$$\overline{V} = D/t \tag{11.10}$$

Smaller insertion type versions are available which remove the need for optical windows in the conduit.

Laser–Doppler techniques, again, use the presence of particles ($\sim$1 μm) for measurement. Here, two angled beams are both focused on a measurement volume in the flow where they interfere to produce a series of optical fringes (an alternation of light and dark areas) suitable for detection as pulses. A particle passing through the volume scatters the laser light, producing a frequency shift (the Doppler frequency, $f_{doppler}$) and a modifying the fringe pattern in a way that is related to velocity.

Both methods assume that the particle is travelling at the same velocity as the carrier fluid relies on a minimum particle concentration.

11.4 Visual and Aural Parameters

11.4.1 Light Measurement Principles

Illuminance or Lux meters detect brightness by converting photons from a light source into current via a photovoltaic semiconductor such as silicon or selenium. The magnitude of the current generated is proportional to the light intensity (lm/m^2 or lux) on a working plane. The photo sensing element is usually covered by diffusing globe to eliminate directional effects (the so-called cosine correction) and a frequency filter to mimic the spectral sensitivity of the human eye. Depending on the complexity and cost, the filter may be factory-set for a particular type of lamp or, alternatively, tuneable.

11.4.2 Sound Measurement Principles

At the heart of every sound meter is a microphone which is essentially a variable air capacitor supplied with a constant voltage. The '*squeeze and release*' effect of a passing sound wave changes its capacitance by changing the capacitor gap. This, in turn, imposes a sinusoidal voltage on top of the supply voltage. The combination voltage provides a measurable output suitable for signal conditioning. See Figure 11.10.

Microphones based on a magnetic effect are available, but they cannot match the frequency sensitivity of capacitor or 'condenser' systems.

Sound level meters measure frequency-dependent sound pressures (P_{RMS}). However, a simple pressure measurement device would not bias its sensitivity between frequencies in the way that the human ear does. Practical sound level meters are therefore equipped with weighting scales (the *A*, *B*, *C*, and *D* scales) dependant on requirement.

The '*A*' scale attempts to reproduce the sensitivity of the human ear by applying the following corrections (dB), Table 11.2, to unprocessed pressure levels collected by the meter:

After applying the corrections, the overall sound pressure level is calculated by deploying the law of sound addition. See Eq. (7.17).

The '*B*' scale is related to pure tone sound. The '*C*' rating is often approximated as linear due to its lack of weighting at mid-range frequencies, whilst the '*D*' weighting is used for aircraft sound measurement.

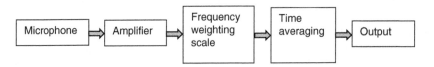

Figure 11.10 Sound meter signal processing.

Table 11.2 Frequency-dependent '*A*' weighting corrections.

Frequency (Hz)	31.5	63	125	250	500	1000	2000	4000	8000	16 000
'*A*' weighting correction (dB)	−39.4	−26.2	−16.1	−8.6	−3.2	0	1.2	1	−1.1	−6.6

The use of a particular scale in an assessment is indicated by appending the bracketed letter to the unit, e.g. dB(*A*).

Sound meters usually require averaging (sampling) time setting depending on the nature of the noise. More sophisticated devices have the capability to measure time varying noise to provide equivalent continuous values.

11.5 Utility Measurement and Metering

This section describes the measurement of utilities such as electricity, gas, water, and heat consumption.

11.5.1 Electricity Metering

Strictly speaking, energy is measured in Joules and power in Watts. However, in terms of electrical energy, these are both small magnitude expressions and in commercial terms, multiples of these are used, e.g. kW, MW, etc.

Further, for electrical energy consumption, the Joule is substituted with the *Watt-hour* and its multiples, e.g. kWh, MWh, etc.

Note that $1\,J = 2.777 \times 10^{-4}\,Wh$

At its most fundamental, electrical power is the product of voltage and current.

However, for alternating current (AC) power sources, a little extra thought is required as the values of both voltage and current vary sinusoidally with time. Maximum values are unrepresentative and taking a simple average of this type of variation over a cycle would return a zero value. Consequently, root-mean-squared (RMS) values for voltage and current are in use. It can be shown that RMS values are approximately 71% of maximum cycle values.

Furthermore, depending on the type of electrical load (capacitive/inductive), the voltage and current sinusoids may be out of phase, i.e. the timing of their maxima/minima may not be contemporary. This non-synchronicity is accounted for by a parameter known as the power factor (pf) for the circuit. The power factor can be thought of as the cosine of the phase shift or difference in angle (ϕ) between voltage and current. A low value of power factor results in the power utility having to supply a higher current and more highly rated supply equipment, e.g. cables, transformers than it would for a purely resistive load.

In this occurrence, the consumer is penalised financially.

Power utilities therefore measure voltage, current, and power factor to produce three potentially chargeable power units thus:

Apparent power = voltage × current (units: Volt-amps)
Real power = voltage × current × cos(ϕ) (units: Watts)
Reactive power = voltage × current × sin(ϕ) (units: (Volt-amps)reactive)

The maximum power demand over a billing period is also recorded and charged for accordingly to encourage better demand management on the part of the consumer.

When time-based energy pricing is added in, this makes consuming electrical energy a complicated exercise. In the past, this tariff complexity was limited to large commercial/industrial consumers, but with the introduction of smart grids, smart metering, and renewable energy making a significant contribution to supply, there is likely to be some development in smaller premise and domestic tariffs along the lines of that described above.

Historically, small electricity meters were electromechanical in operation counting the number of revolutions made by an aluminium disk exposed to magnetic fields generated by the circuit voltage and current. The disk revolution was used to drive an analogue counter.

Electronic meters have no moving parts, instead sensing current by shunt resistors or current transformers. They have integrated circuits containing any necessary rectification/amplification components, processors, memory, LCD output drivers and, if smart, communication software to the outside.

11.5.2 Gas Metering

With knowledge of pipe/duct cross-sectional area, gas may be metered by restriction techniques described in Section 11.3.4.1. However, there are a range of other techniques:

(a) Positive displacement meters

These are essentially mechanical in nature relying on the movement of lobes, pistons, or diaphragms under fluid pressure. The movement is translated to a mechanical counter via a mechanical linkage. They were the meter of choice for domestic and small-scale applications until the rise of smart meters.

(b) Turbine flow meters

An axial or tangential (paddle wheel) impeller mounted in a pipe or duct rotates at a speed roughly proportional to the velocity, and hence the flow rate of the fluid in the pipe. If the blade tips are made of a magnetic material, then the number of blades passing an external pick-up will produce a pulsed output which can be processed to determine rotational speed and flow rate. See Figures 11.11 and 11.12.

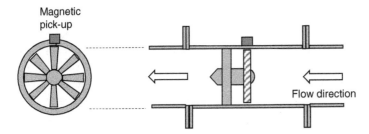

Figure 11.11 Axial turbine flow meter.

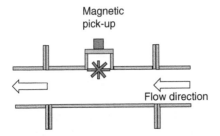

Figure 11.12 Tangential turbine flow meter.

High-quality turbine meters can have an accuracy of the order of ±0.5% of reading.

(c) Thermal mass flow meters

This type of meter is reflective of the hot-wire anemometer technique described in Section 11.3.4.2. Two probes are inserted into the flow: an upstream heater and a downstream gas flow temperature sensor. Electrical power is supplied to the heater to maintain a constant temperature difference between it and the temperature sensor. In the presence of gas flow, the energy required ($\dot{Q}$) to maintain the temperature difference (ΔT) will depend on the gas material properties such as density (ρ), thermal conductivity (k), specific heat capacity (C_p) as well as the gas velocity. With suitable calibration, the electrical energy required to maintain the temperature difference is a measure of the mass flow rate ($\dot{m}$) of the gas.

An alternative arrangement of the same principle requires a shunt 'sensing tube' in parallel with the main flow, two temperature sensors and an intervening fluid heater.

High-quality thermal mass meters can have an accuracy of the order of ±1% of reading.

(d) Coriolis or inertial mass flow meters

The Coriolis force is usually associated with systems having rotating coordinates such as, for example, weather patterns. It can also be applied to systems exhibiting angular oscillations.

Coriolis mass flow meters typically comprise a U-shaped tube containing the flow of interest. The tube is given an oscillatory motion by an electromagnetically energised leaf spring making contact at the U-extremity. As a result of the combination of the oscillation and fluid flow force, the U-tube takes up a twisting motion, whose magnitude is measured by optical pick-ups on both limbs of the U-tube. The mass flow rate is proportional to the length of the U-tube, a spring constant, and the time interval between deflections detected by the optical sensors.

High-quality Coriolis meters can have an accuracy of the order of ±0.2% of reading.

11.5.3 Water Flow Metering

Again, with the knowledge of pipe/duct cross-sectional area, water flow rate may be metered by restriction techniques described in Section 11.3.4.1. Like gas, positive displacement and turbine options are also available. However, there are a range of other techniques:

Others include:

(a) Vortex flow meters

Vortices are generated behind a 'bluff' or un-streamlined body, e.g. a tee-shaped section, when a high-speed fluid forms a boundary layer around the body then separates. For a given flow velocity, vortices shedding is regular causing measurable pressure fluctuations or vibrations on the body and in the downstream fluid. See Figure 11.13. The shedding frequency (f, Hz) can be translated into velocity ($\overline{V}$).

The relationship between flow velocity and shedding frequency is given by a dimensionless parameter known as the Strouhal number (St):

$$\text{St} = fw/\overline{V} \qquad (11.11)$$

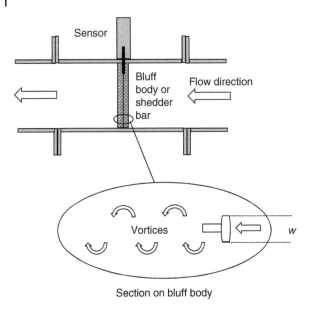

Figure 11.13 Vortex meter.

where *w* is the width of the bluff body. Their deployment for gas flow measurement is also possible.

High-quality vortex meters can have an accuracy of the order of ±1% of reading.

(b) Magnetic flow meters

The operation of an electromagnetic flowmeter is based upon Faraday's Law, which states that the voltage induced across any conductor as it moves at right angles through a magnetic field is proportional to the velocity of that conductor.

Faraday's Law indicates that signal voltage (E) is dependent on the average liquid velocity ($\overline{V}$), the magnetic field strength (B), the pipe diameter (D), and a proportionality constant or meter factor (K) to be determined by calibration.

$$E = K\overline{V}BD \tag{11.12}$$

A typical measurement arrangement is shown in Figure 11.14.

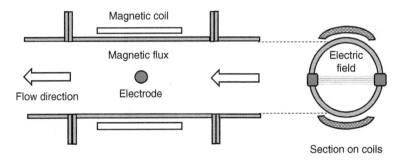

Figure 11.14 Magnetic in-line flow meter.

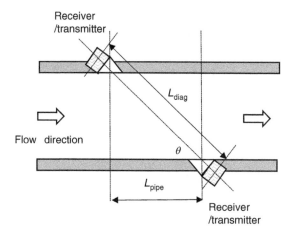

Figure 11.15 Ultrasonic flow meter schematic.

In the case of an in-line variety, the diameter is approximately equivalent to the distance between electrodes. A plastic pipe lining in the vicinity of the meter is necessary. With signal amplification, they can be suitable for low conductivity fresh water flows. Insertion variants of the above are available.

(c) Ultrasonic flow meters

The operation of an ultrasonic flowmeter is based upon the speed of propagation of a high-frequency (MHz) sound through a fluid. This type utilises the principle known as either *'transit time'* or *'time of flight'* between two sound wave transducers installed at an angle (θ) across a pipe or duct. The typical arrangement is shown in Figure 11.15. Transmitter and receiver roles are constantly switched by a control circuit to eliminate the influence of the sound wave velocity. Ultrasonic pulses travelling in the direction of the flow, travel the path between transducers in a shorter period of time (t_u), than pulses travelling in a direction counter to the fluid flow (t_d). This is because the flow speeds up pulses travelling downstream but slows down pulses travelling upstream, i.e.

Upstream pulse velocity = ultrasonic velocity − (Fluid velocity × cos θ)

Downstream pulse velocity = ultrasonic velocity + (Fluid velocity × cos θ)

The working equation for calculation of the fluid flow velocity is:

$$\overline{V} = \frac{K(L_{\text{diag}})^2}{2L_{\text{pipe}}} \left[\frac{1}{t_d} - \frac{1}{t_u} \right] \tag{11.13}$$

where K is a proportionality constant or meter factor to be determined by calibration. Receivers/transmitters may also be mounted on the same side of the pipe with ultrasonic pulses taking a 'V' or 'W' path across the pipe diameter.

Doppler shift ultrasonic devices are available for flows containing a reflective medium. High-quality ultrasonic flow meters can have an accuracy of the order of ±1% of reading.

(d) Variable area meters

One of the more common types of variable area flow meter, suitable for 'spot' checks, is the Rotameter consisting of a vertical glass tapered tube containing a 'float'. The fluid,

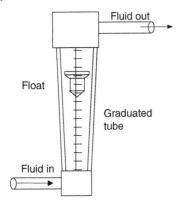

Figure 11.16 Rotameter.

which may be a liquid or gas, flows through the annular space between the float and the tube. As the flow is increased, the float moves to a greater height. The movement is roughly proportional to flow. Calibration is carried out by the supplier. Angled grooves in the rim of the float cause rotation and give the float stability. See Figure 11.16. High-quality rotameters can have an accuracy of the order of ±2% of full scale.

11.5.4 Heat Metering

Heat meters measure the consumption of thermal energy. Their use is common in district heating or community heating systems where each consumer draws heat from a central generator and a proportionate billing is required. The thermal energy (J) is calculated from a modified first law of thermodynamics and requires the mass flow, the temperature drop (ΔT) across the load and the heat capacity (C_p) of the flowing fluid, i.e.

$$Q = mC_p\Delta T \tag{11.14}$$

Flow rates can be measured by ultrasonic or turbine flow meters. Temperature measurement in flow and return pipes can be effected by platinum resistance devices integrated to provide consumption. See Figure 11.17.

Measurement of the flow rate ($\dot{m}$) will produce a value for thermal power ($\dot{Q}$, W).

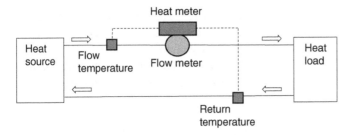

Figure 11.17 Heat meter.

11.5.5 Energy and Building Management Systems

Smart buildings are those where appliances/machines and occupants are integrated in one environment, which is controlled according to inhabitants' needs, their behaviours and variations in outdoor climate.

Building management systems provide the user with the ability to monitor, control, and budget the provision of services to a space.

Climate parameters under control of the system include:

- Air temperature, relative humidity, and air velocity
- Light intensity, sound, and the presence of impurities (particulates and gases)

Additionally, building management systems can:

- Oversee safety/security provision
- Monitor occupancy patterns
- Carry out load analysis, modelling, and forecasting of utility consumption
- Perform a preventative and breakdown maintenance function
- Allow remote connect/disconnect provision

11.6 Worked Examples

Worked Example 11.1   Resistance Thermometer Measurement

A platinum resistance thermometer detector (RTD) displays a resistance of $100\,\Omega$ at $0\,°C$ and $140\,\Omega$ at $100\,°C$. If the RTD reads $130\,\Omega$ when placed in the flow of a central heating system, what is the temperature of the water at that point?

Solution

Given: $R_0 = 100\,\Omega$, $R_{100} = 140\,\Omega$, $R_x = 130\,\Omega$

  Find: T_x

$$R_x = R_0[1 + \propto (T_x - T_0)]$$

Rewriting the relationship for the two conditions and eliminating the coefficient:

$$\frac{R_{100} - R_0}{100 - 0} = \frac{R_x - R_0}{T_x - 0}$$

$$\frac{140 - 100}{100 - 0} = \frac{130 - 100}{T_x - 0}$$

$$T_x = 75\,°C$$

Worked Example 11.2   Thermocouple Measurement

A thermocouple reads an emf value of 1 mV. Using the resistance – temperature data supplied, determine the corresponding temperature.

  Data

  At $0\,°C$ emf $= 0.18\,mV$.

  And At $100\,°C$ emf $= 4.28\,mV$.

  Assume *linear* behaviour between the two given states.

Solution

Given: $T_L = 0\,^{\circ}\text{C}$, $V_L = 0.18\,\text{mV}$, $T_H = 100\,^{\circ}\text{C}$, $V_H = 4.28\,\text{mV}$, $V_x = 1.0\,\text{mV}$

Find: T_x

Using linear interpolation:

$$T_x = T_L + \left[(V_x - V_L)\left(\frac{V_H - V_L}{T_H - T_L}\right)\right]$$

$$= 0 + \left[(1.0 - 0.18) \times \left(\frac{4.28 - 0.18}{100 - 0}\right)\right]$$

$$= 20\,^{\circ}\text{C}$$

Worked Example 11.3 Thermistor Measurement

A thermistor has the material constants shown in Table E11.3.

Approximate its temperature ($^{\circ}$C) when its resistance is 12 500 Ω.

Table E11.3 Thermistor constants for Worked Example 11.3.

Constant	Value
A	1.113×10^{-3}
B	2.353×10^{-4}
C	8.571×10^{-8}

Solution

Given: $A, B, C, R = 12\,500\,\Omega$

Find: T

$$1/T = A + B\ln R + C(\ln R)^3$$

$$= 1.113 \times 10^{-3} + [2.353 \times 10^{-4} \times \ln(12\,500)] + [8.571 \times 10^{-8} \times (\ln 12\,500)^3]$$

$$T = 293.72 = 20.72\,^{\circ}\text{C}$$

Worked Example 11.4 Estimation of Relative Humidity

The dry bulb of a hygrometer reads 20 $^{\circ}$C, whilst its wet bulb reads 14 $^{\circ}$C.

Estimate the prevailing relative humidity (%) if the atmospheric pressure is 1×10^5 Pa.

Solution

Given: $T_{\text{db}} = 20\,^{\circ}\text{C}$, $T_{\text{wb}} = 14\,^{\circ}\text{C}$, $P_{\text{atm}} = 1 \times 10^5\,\text{Pa}$

Find: ϕ

$$K = 0.000\,66(1 + 0.001\,15 T_{\text{wb}})$$

$$= 0.000\,66(1 + 0.001\,15 \times 14)$$

$$= 6.706 \times 10^{-4}$$

$$P_{sat,db} = 610.78e^{(17.2694T_{db}/(T_{db}+238.3))}$$

$$= 610.78e^{(17.2694\times20/(20+238.3))} = 2326 \text{ Pa}$$

$$P_{sat,wb} = 610.78e^{(17.2694T_{wb}/(T_{wb}+238.3))}$$

$$= 610.78e^{(17.2694\times14/(14+238.3))} = 1593 \text{ Pa}$$

$$\phi = \frac{P_{sat,wb} - (KP_{atm}(T_{db} - T_{wb}))100}{P_{sat,db}}$$

$$= \frac{1593 - (6.7 \times 10^{-4} \times 1 \times 10^5 \times (20 - 14))100}{2326}$$

$$= 51\%$$

Worked Example 11.5 Pitot-Static Tube Measurement

In order to calculate the flow rate of air entering a room, a pitot-static tube is used to measure the velocity of air in the supply duct. If the pitot-static tube is indicating a pressure difference of 10 Pa, the duct is 15 cm × 15 cm square and the density of the air is 1.2 kg/m³, determine the volume flow rate (m³/s) and mass flow rate (kg/s) of air supplied to the building.

Solution

Given: $P = 10$ Pa, $A = 0.15 \times 0.15 = 0.0225$ m², $\rho = 1.2$ kg/m³

Find: $\dot{V}$, $\dot{m}$

$$\overline{V} = \sqrt{2\frac{\Delta P}{\rho}}$$

$$= \sqrt{2 \times \left(\frac{10}{1.2}\right)}$$

$$= 4.082 \text{ m/s}$$

$$\dot{V} = \overline{V}A$$

$$= 4.082 \times 0.0225$$

$$= 0.0918 \text{ m}^3/\text{s}$$

$$\dot{m} = A\overline{V} = \rho\dot{V}$$

$$= 1.2 \times 0.0918$$

$$= 0.11 \text{ kg/s}$$

Worked Example 11.6 Average Duct Velocity Calculation

The average velocity and flow rate of air in a 20 × 20 cm duct is to be estimated using a pitot-static tube traversing in 16 equal areas of the duct. The velocity head (Pa) at the centre of each section is shown in Figure E11.6:

Determine the velocity (m/s) in each position, the average duct velocity and flow rate (m³/s) in the duct.

Solution

Given: $A_{total} = 0.2 \times 0.2 = 0.04$ m², A_i's $= 0.05 \times 0.05 = 2.5 \times 10^{-3}$ m², ΔP's

Find: $\overline{V}_{ave}$, $\dot{V}$

10	13	13	10
12	15	15	12
12	15	15	12
10	13	13	10

Figure E11.6 Velocity head data for Worked Example 11.6.

Using $\overline{V} = \sqrt{2\frac{\Delta P}{\rho}}$ at each point yields:

4.083	4.655	4.655	4.083
4.472	5	5	4.472
4.472	5	5	4.472
4.083	4.655	4.655	4.083

$$\overline{V}_{\text{duct}} = (1/A_{\text{total}}) \sum \overline{V}_i A_i$$
$$= (1/0.04) \sum \overline{V}_i \times 2.5 \times 10^{-3}$$
$$= 4.552 \text{ m/s}$$
$$\dot{V} = \overline{V}A = 0.04 \times 4.552 = 0.182 \text{ m}^3/\text{s}$$

Worked Example 11.7   Sound Meter Processing
A sound level meter records un-weighted frequency-dependent sound levels (dB) as shown in Table E11.7.

Using this data, calculate an overall sound pressure level in dB (A).

Table E11.7   Sound corrections for worked Data for Worked Example 11.7.

Frequency (Hz)	31.5	63	125	250	500	1000	2000	4000	8000	16 000
Uncorrected L_p's (dB)	50	53	55	66	69	72	69	65	60	52

Solution

Given: $L_p's$ over a range of frequencies

Find: L_{tot}

The solution is arrived at by first applying 'A' weighting corrections

f (Hz)	31.5	63	125	250	500	1000	2000	4000	8000	16 000
Uncorrected L_p (dB)	50	53	55	66	69	72	69	65	60	52
'A' weighting correction (dB)	−39.4	−26.2	−16.1	−8.6	−3.2	0	1.2	1	−1.1	−6.6
Corrected L_p dB(A)	10.6	26.8	38.9	57.4	65.8	72	70.2	66	58.9	45.4

Then using Equation (7.17):

$$L_{tot} = 10\log(10^{L_1/10} + 10^{L_2/10} + \cdots\cdots + 10^{L_n/10})$$
$$= 10\log(10^{1.06} + 10^{2.68} + \cdots\cdots + 10^{4.54})$$
$$= 10\log(355 \times 10^5)$$
$$= 76\ dB$$

Worked Example 11.8   Flow Metering

It is proposed to construct a vortex shedding meter with a bluff body width of 5 mm that operates with a Strouhal number of 0.2. When installed in a conduit, the shedding frequency is 100 Hz. Estimate the flow velocity (m/s).

Solution

Given: $w = 0.005$ m, St $= 0.2, f = 100$ Hz

Find: $\overline{V}$

$$St = fw/\overline{V}$$
$$0.2 = 100 \times 0.005/\overline{V}$$
$$\overline{V} = 2.5\ m/s$$

Worked Example 11.9   Electrical Power, Power Factor, and Energy Cost

A single phase 240 V load draws a current of 30 A with a power factor of 0.93. Determine the load's apparent (VA), real (W) and reactive power ($VA_{reacative}$).

If the load is on-line for 1500 hours/annum and the utility charge for electricity is 11 p/kWh (i.e. based on real power), calculate the annual running cost (£) for the load. All values are RMS.

Solution

Given: $V = 240$ V, $I = 30$ A, $pf = 0.93$, $t = 1500$ h/annum, unit charge $= 0.11£/kWh$

Find: Apparent power, real power, reactive power, annual running cost.

$$pf = \cos(\phi)$$
$$0.93 = \cos(\phi)$$
$$\phi = 21.565^{\circ}$$

Hence

$$\sin(\phi) = 0.367$$

Apparent power $= V \times I = 240 \times 30 = 7200$ VA

Real power $= V \times I \times \cos(\phi) = 240 \times 30 \times 0.93 = 6696$ W

Reactive power $= V \times I \times \sin(\phi) = 240 \times 30 \times 0.367 = 2646$ VA$_{\text{reactive}}$

Energy consumption $=$ Real power $\times$ time

$$= 6696 \times 1500$$
$$= 10\,044\,000 \text{ Wh} = 10\,044 \text{ kWh}$$

Energy cost $=$ Energy consumption $\times$ unit cost of electricity

$$= 10\,044 \times 0.11$$
$$= £1104.84$$

Worked Example 11.10 Heat Metering and Energy Cost

A heat meter records a cumulative hot water flow of 8800 m³ over a period of 480 hours. The flow and return temperatures of the circuit it serves are 82 and 75 °C, respectively.

Using the data supplied, determine the energy (kWh) and average thermal load (kW) indicated by the meter over the monitoring period.

Data

Specific heat capacity of water: 4200 J/kg K

Water density: 1000 kg/m³

If the utility charge for heat is £0.03/kWh, determine the total cost over this period.

Solution

Given: $V = 8800$ m³, $\Delta T = 82 - 75 = 7°C$, $C_p = 4200$ J/kgK, $\rho = 1000$ kg/m³, unit charge $= 0.03£/kWh$, monitoring period $t = 480$ h.

Find: Q (kWh), $\dot{Q}_{\text{average}}$ (kW), running cost (£)

$$Q = mC_p \Delta T$$
$$= \rho V C_p \Delta T$$
$$= 1000 \times 8800 \times 4200 \times 7$$
$$= 2.587 \times 10^{11} \text{ J}$$

Since $1\,\mathrm{J} = 2.77 \times 10^{-4}\mathrm{kWh}$

$$Q = 2.587 \times 10^{11} \times 2.77 \times 10^{-4}$$
$$= 71\,659\,900\ \mathrm{Wh}$$
$$= 71\,660\ \mathrm{kWh}$$
$$\dot{Q}_{\text{average}} = Q/t$$
$$= 71\,660/480$$
$$= 149.3\ \mathrm{kW}$$

Energy cost = Energy consumption $\times$ unit cost of heat
$$= 71\,660 \times 0.03$$
$$= £2149.9$$

11.7 Tutorial Problems

11.1 A platinum resistance thermometer has a resistance of $100\,\Omega$ at $0\,°\mathrm{C}$. At $20\,°\mathrm{C}$, its resistance has increased to $107.93\,\Omega$. Determine the temperature coefficient (/°C) for the device.
[Answer: 0.003 96/°C]

11.2 For a type K thermocouple, the variation between temperature and emf through the range between 0 and $100\,°\mathrm{C}$ can be approximated by

$$V = 39.35\,T + 2.72 \times 10^{-2}T^2$$

where V is the emf voltage in micro volts and T is the temperature in °C.
It was decided to use this type of thermocouple as an alarm when the temperature of the water heating circuit reaches $90\,°\mathrm{C}$.
Find the value of the alarm voltage (mV).
[Answer: 3.762 mV]

11.3 A negative temperature coefficient thermistor has the following expression relating the electrical resistance (R in ohms) with temperature (T in K) in the range -20 to

Table P11.3 Thermistor constants for tutorial Problem 11.3.

Constant	Value
A	1.283×10^{-3}
B	2.362×10^{-4}
C	9.285×10^{-8}

$+120\,°C$:

$$T = 1/[A + B \ln R + C(\ln R)^3]$$

where constant A, B, and C are given in Table P11.3.
Determine the temperature of the ambient when the thermistor indicates a resistance of $5\,k\Omega$.
[Answer: $25.3\,°C$]

11.4 The dry bulb of a hygrometer reads $22\,°C$, whilst its wet bulb reads $17\,°C$.
Estimate the prevailing relative humidity (%) if the atmospheric pressure is 1×10^5 Pa.
[Answer: 60.57%]

11.5 The air supply to a room in a 200 mm diameter duct is measured using a 100 mm diameter restriction measurement (orifice plate) device. The pressure drop across the device is 15 Pa. The coefficient of discharge of the orifice plate is 0.62, and the density of air is $1.2\,kg/m^3$.
Calculate the mass flow rate (kg/s) of air into the room.
[Answer: 0.029 kg/s]

11.6 A pitot-static tube is used to measure the velocity of air flowing through a duct of 30 cm diameter. The connected pressure transducer shows a difference in head of 5 Pa. If the density of air is $1.2\,kg/m^3$, determine the velocity (m/s), and the volume flow rate (m^3/s) of air.
[Answers: 2.89 m/s, $0.204\,m^3/s$]

11.7 A cross pipe ultrasonic flow meter of the type shown in Figure 11.15 is installed in a 150 mm diameter conduit. The installation angle is $30°$. The fluid velocity is 2 m/s, and the ultrasonic velocity is 1490 m/s.
Determine the time difference(s) between upstream and downstream pulses.
[Answer: 4.681×10^{-7} s]

11.8 A sound level meter records un-weighted frequency-dependent sound levels (dB) as shown in Table P11.8.

Table P11.8 Sound corrections for tutorial Problem 11.8.

f (Hz)	31.5	63	125	250	500	1000	2000	4000	8000	16 000
Uncorrected L_p (dB)	40	39.8	54.9	52.4	56.8	75	83.2	81	85.9	83.4

Using this data, calculate an overall sound pressure level dB(A).
[Answers: 89 dBA]

11.9 An electricity meter registers a real power of 11 000 W and a reactive power of 2750 VA$_{reactive}$. If the load is single phase with a voltage of 240 V, determine the power factor and the current (A) in the circuit.
[Answers: 0.97, 47.24 A]

11.10 A heat meter records a cumulative hot water flow of 5000 m^3 over a period of 650 hours. The flow and return temperatures of the circuit it serves are 82 and 71 °C, respectively.
Using the data supplied, determine the energy flow (kWh) and average thermal load (kW) indicated by the meter over the monitoring period.
Data
Specific heat capacity of water: 4200 J/kg K
Water density: 1000 kg/m^3
If the utility charge for heat is £0.035/kWh, determine the total cost (£) over this period.
[Answers: 64 172 kWh, 98.726 kW, £2246]

12

Drivers, Standards, and Methodologies

12.1 Overview

Building and building services designers do not have free range with respect to their designs. They have to comply with statutory regulations that usually set minimum standards. The regulatory environment is not static, and hence practitioners are strongly advised to consult on the regulations prevailing at the time of project conception.

Increasingly, the drivers for standards are sustainability based. Rather than being entirely prescriptive, they seek to take a holistic approach to the objective focussing on results rather than processes.

To provide transparency and judge compliance, reference must ultimately be made to nationally/internationally accepted analysis tools and standards. Some of the major methodologies used for this purpose in the United Kingdom, Europe, and around the world are described in this chapter.

The completion and handover of the building to its owner are not the end of the process. To ensure that the building achieves its sustainability intent over its lifetime, the owner/occupier should employ a further set of methodologies as part of its operational management. Some of the more common techniques are also discussed here.

All of this, of course, incurs additional cost. However, building owners need to keep in mind that achieving these high sustainability levels should not only improve building performance and lower running costs but also increase asset value or rental rates and demonstrate corporate social responsibility and sustainable business leadership.

Learning Outcomes

- To gain an understanding of the importance and influence of compliance legislation
- To gain an awareness of the role of calculation methodologies in determining the compliance of a design
- To be aware of some of the more popular voluntary sustainability benchmarks
- To be aware of some of the major operational responsibilities applicable to building owner/operators

Building Services Engineering: Smart and Sustainable Design for Health and Wellbeing, First Edition.
Tarik Al-Shemmeri and Neil Packer.
© 2021 John Wiley & Sons Ltd. Published 2021 by John Wiley & Sons Ltd.

12.2 Compliance Considerations

12.2.1 Energy Performance of Buildings (EPB) Standards and the Energy Performance of Buildings Directive (EPBD)

The Energy Performance of Buildings Directive (EPBD) is a European Union (EU) directive stretching back to 2002 and more recently revised in 2018. Its aim is 'to promote the improvement of the energy performance of buildings within the European Union, taking into account outdoor climatic and local conditions, as well as indoor climate requirements and cost-effectiveness' (*Article 1*). The Energy Performance of Buildings (EPB) standards were created to underpin the directive and achieve international harmonisation within the EU.

There is a move to amalgamate EU practices with standards formulated by the International Organisation for Standardization (ISO), and so the amended directive text (2018) states the obligation on all EU members to formulate their national calculation methodologies for energy calculations in buildings based on the following ISO standards:

- ISO 52000-1: Energy performance of buildings – Overarching EPB assessment – Part 1: General framework and procedures
- ISO 52003-1: Energy performance of buildings – Indicators, requirements, ratings, and certificates – Part 1: General aspects and application to the overall energy performance
- ISO 52010-1: Energy performance of buildings – External climatic conditions – Part 1: Conversion of climatic data for energy calculations
- ISO 52016-1: Energy performance of buildings – Energy needs for heating and cooling, internal temperatures, and sensible and latent heat loads – Part 1: Calculation procedures
- ISO 52018-1: Energy performance of buildings – Indicators for partial EPB requirements related to thermal energy balance and fabric features – Part 1: Overview of options

It is intended that further standards related to building energy performance will be added to the ISO 52000 series in the near future. As the 27 countries of Europe and the United Kingdom are currently taking a lead addressing climate change and have done much work in this area, it is likely that many of the new additions (up to 150) will be formulated by collaboration between CEN (European Committee for Standardisation) and ISO.

New standards currently in development (2018) include:

- ISO 52031, *Energy performance of buildings – Method for calculation of system energy requirements and system efficiencies – Space emission systems (heating and cooling)*
- EN ISO 52120-1, *Energy performance of buildings – Contribution of building automation and controls and building management – Part 1:*
- EN ISO 52127-1, *Energy performance of buildings – Building automation, controls, and building management – Part 1: Building management system*
- EN ISO 52016-3, *Energy performance of buildings – Energy needs for heating and cooling, internal temperatures, and sensible and latent heat loads – Part 3: Calculation procedures regarding adaptive building envelope elements*

The EPBD is a significant driver with respect to the building sustainability agenda.

12.2.2 UK Building Regulations and Approved Documents

The UK Building Regulations are concerned with the Planning and Development of buildings in the United Kingdom. *Approved documents* are a series of guidelines, published by a UK Government department, demonstrating how to meet the UK Building regulations. They are not entirely prescriptive but instead provide general statutory guidance and practical examples and solutions on how to achieve compliance.

All Approved Documents provide full reference to applicable European/British standards, legislation, health and safety requirements, and publications by other relevant industry bodies.

They cover a wide range of topics. Important elements relevant to the indoor environment include:

- Resistance to Sound: Approved Document E
  This document provides a great deal of useful guidance on material specification, installation detail, and the determination of sound insulation testing in situ.
- Ventilation: Approved Document F
  This document provides guidance on matters such as ventilation strategy, performance, and effectiveness testing for a range of new and existing buildings.
- Conservation of Fuel and Power: Approved Document L
  Approved Document L is particularly detailed being sub-divided into four parts:
  - L1A: Conservation of fuel and power in new dwellings
  - L1B: Conservation of fuel and power in existing dwellings
  - L2A: Conservation of fuel and power in new buildings other than dwellings
  - L2B: Conservation of fuel and power in existing buildings other than dwellings

As a result of review and consultation, it is likely that this arrangement will (2020) be simplified into a single document for dwellings and another for buildings other than dwellings.

Two further supplementary compliance guidance documents for building services design are also available for dwellings and non-dwellings (2019). Again, it appears likely that these will be incorporated into the approved documents in the future.

12.2.3 SAP, RdSAP, and SBEM (UK)

These are examples of government-approved calculation methodologies for predicting the energy performance of a building at the design stage. They are closely associated with complying with the building regulations mentioned above.

Underpinning the methodologies and setting the goals for improvement is a parameter known as the target emission rate (TER). This is the emission rate (kg CO_2/m^2 annum) of a '*notional*' building having the same dimensions as the proposed development. The TER is a maximum allowable emission rate or minimum standard that must be achieved by the proposed building.

In all cases, an agreed national calculation method (NCM) is employed.

SAP (Standard Assessment Procedure) is a tool for predicting the energy and environmental performance of dwellings and is relevant to Approved Document L1A (new dwellings).

RdSAP performs a similar function for existing dwellings and is relevant to Approved Document L1B.

A SAP assessment takes into account structural thermal elements, the heating and hot water system, internal lighting, heat gains, and the use of renewable energy. Standard 'notional' occupancy and owner behaviour are assumed. SAP data input and worksheet reports must be fully documented. The process will output the buildings energy consumption (kWh/m^2), a CO_2-based *Environmental Impact rating* and a fuel cost-based *SAP rating* (1–100). A rating of 100 signifies a building requiring no energy input. SAP ratings in excess of 100 are possible if the building is a net exporter of energy. SAP can also output the risk of summer overheating and the need for cooling. SAP calculates a Dwelling Emission Rate (DER) for TER comparison as well as an energy equivalent (kWh/m^2 annum) Fabric Energy Efficiency (DFEE) for comparison with a maximum Total Fabric Energy Efficiency (TFEE) target. At the time of writing (2020), replacing these fabric factors with a *Primary Energy Rate* factor (kWh$_{PE}$/m^2 annum) is being considered.

SBEM (Simplified Building Energy Model) is a software tool for predicting the energy and environmental performance of non-dwellings and is therefore relevant to Approved L2.

Its required building data inputs are in many ways similar to SAP. SBEM then models the actual CO_2/m^2 annum building emission rate (BER) and, again, compares it to a '*notional*' TER based on the proposed building size and shape. Again, the BER cannot exceed the TER. Additionally, occupancy and weather patterns are normalised. Some occupant-related activities and patterns are recognised within the software.

12.2.4 Energy Performance Certificates (EPCs)

An Energy Performance Certificate (EPC) makes transparent the energy use of a building. These certificates are a requirement of the aforementioned EPBD.

EPCs are required for new constructions and the letting or sale of existing buildings and are valid for 10 years.

For dwellings, the certificate provides:

- colour-coded, alphabetised energy and environmental asset ratings A–G (A excellent, G poor) similar to that already in use for white goods and automobiles
- a three-year estimated energy cost broken down into heating, lighting, and hot water provision
- suggestions for energy reduction measures and the resultant energy cost savings

For non-domestic buildings, an energy-based asset energy A–G rating is, again, used. If the building is not new but, instead, refurbished, then the certificate compares the resulting rating with that of a new built equivalent and the rating typical of existing stock in the same sector.

Display Energy Certificates (DECs) are EPCs applicable to the operation of existing buildings and will be returned to in Section 12.4.3.

For smaller buildings having a total floor area between 500 and 1000 m^2, DEC's are renewed every 10 years. For buildings having a floor area over 1000 m^2, a new DEC must be produced annually.

The creation of EPCs must be carried out by an accredited assessor using accredited software.

12.2.5 Ecodesign and Energy Related Products (ErP)

Correctly sizing plant for a given duty is insufficient. It is incumbent on designers to specify equipment having an acceptable energy conversion efficiency and hence low environmental impact. This can be fulfilled by only purchasing products with a recognised environmental quality standard.

The EU Ecodesign Directive 2009/125/EC seeks to set out a framework providing threshold standards for fossil fuel and electric boilers, space and water heaters, air-conditioning plant, ventilation units, electric motors, pumps and lamps as well as a number of domestic appliances. Under the arrangements, product testing is open, transparent, and equitable.

Directive 2009/125/EC is supported by the EU Energy Labelling Regulation 2017/1369. The system uses a colour-coded, alphabetic (A, excellent – G, poor) ranking to categorise and make public the energy efficiency of products.

In the United States, a similar system of certification is provided by the Government – backed (EPA, DOE) ENERGY STAR scheme.

12.2.6 Lamp and Lighting Standards

A number of lighting standards are set by the International Commission on Illumination (CIE) which is an International Organisation for Standardisation (ISO) recognised body. The CIE produces standards on topics such as photometric performance, test methods, lighting of workplaces, and the modelling of daylight. Examples include:

- CIE S 017/E:2011 International Lighting Vocabulary
- ISO/CIE 20086:2019(E) Light and Lighting – Energy Performance of Lighting in Buildings
- CIE S 025/E:2015 Test Method for LED Lamps, LED Luminaires, and LED Modules
- ISO 8995-1:2002 Lighting of work places – part 1: indoor

12.2.7 Noise Standards

With respect to noise, the following calculation methodologies are produced by the International Organisation for Standardisation (ISO):

- ISO 12354-2:2017 Building acoustics – Estimation of acoustic performance of buildings from the performance of elements
  Part 1: Airborne sound insulation between rooms
  Part 2: Impact sound insulation between rooms
  Part 3: Airborne sound insulation against outdoor sound

  Some examples of permissible sound levels are provided by:

- Control of Noise at Work Regulations 2005 (UK).
  This statutory instrument is concerned with occupational noise and provides exposure action values, exposure limit values, and guidance on risk assessment and the elimination and/or control of exposure in the workplace.

- World Health Organisation (WHO 2018): Environmental Noise Guidelines for the European Region
  This guideline document provides exposure action levels for environmental noise associated with road traffic, railway, aircraft, wind turbines, and leisure. The exposure levels are justified with quantitative risk assessments correlating the prevailing sound level with a range of different health effects.

The International Electrotechnical Commission (IEC) publishes international standards for a range of electrical and electronic technologies. The following IEC standards are applicable to sound measurement:

- IEC 61672:2013 Sound level meters and electroacoustics
- IEC 61252:1993 Personal exposure meters (Noise dosimeters)

12.2.8 Indoor Environmental Design Parameters

The European standard EN 16798-1:2019 Energy Performance of Buildings provides up to date indoor environmental input criteria, including occupancy schedules, for design and energy prediction purposes while addressing the issues associated with poor indoor air quality, lighting, and acoustics.

The standard is applicable to buildings with a low process contribution to its prevailing operating conditions.

With respect to the thermal environment, it also advises on the effects of draught, floor surface temperature, and vertical temperature gradients. Its thermal criteria are based on ISO 7730:2005 Ergonomics of the thermal environment – Analytical determination and interpretation of thermal comfort using calculation of the PMV and PPD indices and local thermal comfort criteria.

It does not specify design methodologies, but its inputs are suitable for use in NCMs as discussed in Sections 12.2.3 and 12.2.4.

12.3 External Certification and Recognition

12.3.1 BREEAM

BREEAM (Building Research Establishment Environmental Assessment Method) is a certificated, standard setting, sustainability assessment method for communities, infrastructure, non-domestic new build, residential new build and refurbishment and fit-out projects.

A successful certification is dependent on a sustainability analysis of the planning, procurement, design, specification, construction, and operation of a scheme.

The scheme provides a series of weighted benchmarks against which to grade any proposal.

Typically, the areas covered include:

- Energy consumption
- Health and well-being
- Land use

- Management
- Material resource procurement, use, and waste minimisation
- Pollution generation
- Transport
- Water
- Innovation

Under the scheme, a number of credits are available to each category. As part of the assessment, a number of these credits are allocated by the assessor and adjusted for that particular category weighting (0–1). The weighted credits are summed to produce the final rating. Additional 'innovation' credits can be added for the introduction of particularly novel solutions.

Certification is a two-stage (pre/post construction) process, carried out by independent licensed assessors. The performance of a project is, essentially, rated against current practice in the sector. Table 12.1 relates the broad statistical relevance of new UK building projects in each BREEAM category.

Table 12.1 BREEAM rating system (2018).

Rating	Description of practice	Achieved by
Outstanding > 85%	Innovative	<1% of UK building projects
Excellent > 70%	Best practice	Top 10% of UK buildings projects
Very good > 55%	Advanced good practice	Top 25% of UK building projects
Good > 45%	Intermediate good practice	Top 50% of UK building projects
Pass > 30%	Standard good practice	Top 75% of UK new non-domestic buildings

Source: Adapted from Scoring and Rating BREEAM assessed buildings, www.breeam.com.

Although not stipulated in the EPBD, successful planning permissions for building developments in the United Kingdom and Europe are much facilitated by having carried out environmental assessments such BREEAM or an equivalent.

12.3.2 Passivhaus

A voluntary scheme for dwellings that seeks to exceed the prevailing building regulations on energy efficiency and thereby minimise the building's energy demand related to heating the internal environment. Passivhaus designs rely on such measures as the specification of super-insulating wall elements, multi-pane glazing, super-air tightness, optimal use of available solar gain, knowledge of the structure thermal storage properties, and heat recovery from exhaust air in mechanical ventilation systems.

It is claimed that the appropriate application of these measures can reduce a building's heating/cooling requirement to a maximum $15\,kWh/m^2$ annum and the primary energy demand to below $120\,kWh/m^2$ annum.

Certification is made by independent assessors. Designers, tradespeople, and components can also be Passivhaus certified. The Scheme originated in Germany and has over $65\,000$ examples worldwide.

12.3.3 WELL Standards

The International WELL Building Institute standards provide a series of voluntary guidelines and rating systems focussing on the creation of internal environments having high values of health and wellness for their occupants.

WELL certification addresses concepts and features. The major concepts are:

- Air
- Water
- Nourishment
- Light
- Movement
- Thermal comfort
- Sound
- Materials
- Mind
- Community
- Innovations

Each concept applicable to a project is allocated to a number of *pre-conditions* that must be achieved and a number of sub-divisions or *optimisations* contributing to an overall rating.

For example, the *thermal comfort* category has one pre-condition (T01 Thermal performance) and six sub-divisions (T02 Enhanced thermal performance, T03 Thermal zoning, T04 Individual thermal control, T05 Radiant thermal comfort, T06 Thermal comfort monitoring, and T07 Humidity control).

The final certification (Silver/Gold/Platinum) is dependent on the achievement of pre-conditions and the overall percentage achievement of potential optimisations. The highest certification (Platinum), typically, requires 80% of optimisations to be achieved.

12.3.4 LEED Leadership in Energy and Environmental Design

A green building certification scheme originating in the United States. Again, the scheme requires an independent assessment of the building design and construction, interior design and construction, building operations and maintenance, and neighbourhood development from an environmental perspective. The scheme features a credits acquisition, points-based assessment and certification (Silver/Gold/Platinum) similar to those described above.

Other such schemes can be found across the globe, e.g. HQE (France), CASBEE (Japan), and Green star (Australia).

12.4 Operational Considerations

12.4.1 Post Occupancy Evaluation

Building services designers may adhere to the best design principles, installation contractors may use proscribed installation techniques, and the building tested and commissioned to the accepted format, meeting all its specifications.

Yet, the building may still fail in its objectives if the building users, upon occupation, are not satisfied with the result. Post occupancy evaluations and surveys seek to elicit feedback from building users as a means of gauging the effectiveness of the building as a working/living space. Many pro formatted surveys are available for use.

For example, the international WELL Building Institute suggests a survey based on indoor environmental air quality factors such as acoustics, cleanliness, odours and emissions, thermal comfort and workplace light levels and quality amongst other parameters. The initial evaluation carried out within 12 months of the building handover will provide invaluable information for owners to clear up what may be teething problems, within contractor liability.

WELL recommend that, to be meaningful and act as a tool in maintaining an acceptable environment, the survey should be an on-going, annual event and be completed by at least 30% of the occupants of the space. Importantly, the results of the survey should be made available to the building owners/operators with 30 days of its completion.

12.4.2 Building Owner Manuals and Building Logbooks

A *building owner's manual* is an extensive document supplied to the building owner at project practical completion. The manual will typically contain a building asset register, construction elements, 'as-built' drawings and specifications, commissioning and test results, and instructions for operation and maintenance. This list is not exhaustive. The responsibility for the manual's creation lies principally with the project's contractors, although input from designers and suppliers will, inevitably, be required.

A *building logbook* is an important user guide, prepared by the project's designer, for a building's facilities manager. It is essential for new projects and existing buildings where extensive modifications have been carried out. The logbook is a descriptive document covering such topics as layout, occupancy strategy, building services plant, and maintenance. More specifically, from an energy perspective, it should contain data on the building's original design energy targets and EPC. These can be used to gauge actual operational energy use. The logbook is not a static document and should be updated annually by the building facility manager.

12.4.3 Display Energy Certificates (DECs)

DECs are applicable to the operation of existing buildings. Using operational data on a building's size and energy consumption, the certificates output a comparative energy use metric or operational rating which is illustrated on a colour-coded, alphabetised (A, excellent – G, poor) basis.

In addition to the current year, the certificate also illustrates two previous years' worth of operational ratings and carbon dioxide emissions (kg CO_2/m^2 annum) for the building. Further, a breakdown of energy consumption (kWh/m^2 annum) between heating and electricity as well as any renewable energy generated is also provided.

DECs are required by all owned and occupied public authority buildings having a total floor area of over 500 m^2 and frequently visited by the public. DECs must be prominently displayed within the building, often in a foyer.

For smaller buildings having a total floor area between 500 and 1000 m², DECs are renewed every 10 years. For buildings having a floor area over 1000 m², a new DEC must be produced annually.

The creation of DECs must be carried out by an accredited assessor using accredited software.

12.4.4 Air-Conditioning Reports

The EPBD recognises the widespread use of air systems and the importance of their correct operation by requiring the production of air-conditioning reports on all systems having an effective rated output, as stated by the manufacturer under continuous operation, of >12 kW.

The report is essentially the end point of an energy-related practical assessment/inspection carried out by a certified assessor that covers topics such as general maintenance, state of cleanliness, refrigerant replacement and substitution, escape of refrigerant gases.

The inspection also includes air movement equipment and controls.

Although the inspection is not specifically related to public health, the assessor may highlight any issues associated with Legionella.

The owner/operator of the plant has no legal requirement to act on the report recommendations. Inspection reports can be no more than five years old, and there are cumulative financial penalties not having an up to date report.

12.4.5 F-Gas Regulations

The EU F-Gas regulations (517/2014) are concerned with putting a check on and reversing the use of refrigerant gases having climate modifying properties within the EU. The policy is one of emission containment and prevention. They seek to limit the deployment of some gases to 1/5th of 2014 levels by 2030. They also lay out a number of responsibilities that apply to owner/operators of cooling equipment insisting on:

- Use of trained and certified personnel associated with refrigeration equipment maintenance
- Equipment leak detection and checks
- Recovery and the appropriate disposal of refrigerant gases at the end of equipment life
- A verifiable system of record keeping and reporting with regard to refrigerant gases

The EU is also committed to maintaining an electronic registry for the placement of refrigerants in the marketplace within the bloc.

12.4.6 Indoor Air Pollutants

Consideration of indoor air pollutant quality is not straightforward. 'Ambient' air quality *standards*, i.e. an enforceable maximum value with a risk of penalty for failure, are common, but they do not generally distinguish between indoor and outdoor conditions. Instead, they simply say what time-related level of exposure can be tolerated without exposing individuals to significant risk. Many air pollution *guidelines,* i.e. an obligation to reduce health effects but not at disproportionate costs, are also common.

World Health Organisation (WHO) guidelines attempt to set out international limits and are adopted or used as a foundation for many national air pollution standards. Table 12.2 provides some examples of EU standards under directive 2008/50/EC.

However, the application of 'ambient' or existing occupational exposure standards for hazardous substances to all indoor environments is questionable. There are several issues to consider. For example, perhaps a lower *sensory comfort/odour detection* limit is more appropriate than a health limit in some cases? Are the averaging times appropriate to a particular application? Occupational exposure levels consider occupants to be healthy individuals. With medical and care facilities particularly in mind, this is not always the case.

To try to address these questions some countries, e.g. Germany, Norway, and Finland, are setting specific internal air pollutant standards, but, in relative terms, this subject is still in its infancy.

The performance of air filters with respect to particulate removal efficiency and particle size is addressed in ISO 16890. Typically, filter test results are classified with reference to 1, 2.5, and 10 µm particles, and the results coded as ePM1, ePM2.5, and ePM10, respectively, plus the actual collection efficiency at that particle size. For example, a filter having

Table 12.2   European Union air quality standards.

Pollutant	Concentration	Averaging period	Permitted exceedances each year
Fine particles (PM2.5)	25 µg/m^3	1 yr	n/a
Sulphur dioxide (SO$_2$)	350 µg/m^3	1 h	24 3
	125 µg/m^3	24 h	
Nitrogen dioxide (NO$_2$)	200 µg/m^3	1 h	18 n/a
	40 µg/m^3	1 yr	
Larger particles (PM10)	50 µg/m^3	24 h	35 n/a
	40 µg/m^3	1 yr	
Lead (Pb)	0.5 µg/m^3	1 yr	n/a
Carbon monoxide (CO)	10 mg/m^3	Max daily 8-h mean	n/a
Benzene (C$_6$H$_6$)	5 µg/m^3	1 yr	n/a
Ozone (O$_3$)	120 mg/m^3	Max daily 8-h mean	25 d averaged over 3 yr
Arsenic (As)	6 ng/m^3	1 yr	n/a
Cadmium (Cd)	5 ng/m^3	1 yr	n/a
Nickel (Ni)	20 ng/m^3	1 yr	n/a
Polycyclic aromatic hydrocarbons (PAH)	1 ng/m^3	1 yr	n/a

Source: Adapted from Air Quality Standards, European Commission: Environment https://ec.europa.eu/environment/air/quality/standards.htm used under CC BY 4.0. Last accessed 24/04/2020.

a classification of ePM2.5 65% will collect 2.5 µm sized particles at an efficiency of 65%. All categories are required to have a minimum collection efficiency of 50%.

Use of abbreviations to describe particulate filters such as HEPA (high efficiency particulate air filters) and ULPA (ultra low penetration air filters) is in common use.

For the best of these (ULPA), collection efficiencies of the order of 99.999% for contaminants of diameter 0.1 µm and above is expected. HEPA filters typically have collection efficiencies of 99.995 for particulate sizes of 0.3 µm and above.

12.4.7 Prevention of Legionellosis

The possibility of the occurrence of the legionella bacterium, as mentioned in Chapter 4, places a responsibility on building owner/operators to provide risk identification and control. Requirements will vary with installed system but will likely, as a minimum, include six-month checks in line with manufacturer's instructions and/or as detailed in a risk assessment exercise. Actions may include cleaning, disinfection, descaling, corrosion prevention, and sampling for the bacterium itself. Some excellent advice on policy and practice can be found on the European Agency for Health and Safety at Work website.

Locally, from a legal perspective, things can become a little complicated. For example, in the United Kingdom duties of care regarding a risk of exposure to Legionella falls under the Health and Safety at Work Act (HWSA) 1974. This is supported by the Control of Substances Hazardous to Health Regulations (COSHH) 2002 which covers assessment, prevention, and control strategies. The UK Health and Safety Executive (HSE) publishes detailed help in its document: *Legionnaires' disease. The control of legionella bacteria in water systems: Approved Code of Practice and guidance.*

13

Emerging Technologies

13.1 Overview

Among its many aims, the Energy Performance of Buildings Directive (2018), as discussed in Chapter 12, is concerned with the establishment of a '*smart readiness indicator*' (SRI) for buildings. This will require a harmonised methodology that recognises the extent to which a building is able to adapt its performance to changes in the occupiers' needs and the needs of electricity grid smart, interconnected devices.

A range of technologies will need to be recognised including renewable energy generation, energy storage, smart metering, and building management control systems.

The SRI will help identify the presence of a healthy internal environment and a high energy efficiency and low energy consumption through the use of sensors, self-regulating devices, and automation. In parallel, the presence of localised or remote occupant initiated controls will also be recognised as playing a part.

The indicator will also display electrical load flexibility in terms of managing maximum demand and load shifting patterns without distressing building occupants. This will include addressing the issue of e-mobility and e-charging points.

The methodology will be constructed as to not result in any deleterious effects with respect to data protection, privacy, and security of individuals or owners.

A consultation on its further development is in progress (2019).

It is quite clear that new standards, regulations, and legislation such as the above will be driving building environmental systems design and engineers will have to be prepared to consider developing and deploying new technologies to keep up.

This chapter details some emerging technologies that may, perhaps, be common place in the not too distant future. Many are smart. Some make use of existing science and technology in a smarter manner.

13.2 Smart Ventilation

Varying the flow rate in a '*dumb*' ventilation system might be limited to time control, i.e. either on or off.

'*Smart*' or demand controlled ventilation (DCV) systems vary the flow rate according to any prevailing set of circumstances. In such an arrangement, a wide range of controlled

Building Services Engineering: Smart and Sustainable Design for Health and Wellbeing, First Edition.
Tarik Al-Shemmeri and Neil Packer.

space parameters are monitored including carbon dioxide, relative humidity, total volatile organic compounds (TVOCs), and occupancy, perhaps requiring a cluster of sensors. Outside temperature may also be added to the collection.

Allowing the flow rate to vary according to demand will require a new approach. Rather than maintaining a constant set condition, the control system will seek to reduce the air flow providing that the overall equivalent dose or exposure to any given pollutant is not exceeded in any given set time period. Thus, comparison with a standard based on a continuously operated system with constant pollutant generation will be required.

The control system might then reduce flow over one time period providing that the space is over-ventilated at another time period to compensate.

In addition to, or as an alternative to, internal air quality considerations, the system might seek to minimise energy consumption by either increasing ventilation when the internal–external temperature difference is small (winter use) or perhaps decrease the flow rate to minimise fan electricity consumption during peak electrical grid charge times.

The choice of sensor will depend on the use of the space to be controlled. For example, the ventilation of a 'wet' area such as a water closet (WC) or shower room would be dominated by occupancy and humidity considerations. Alternatively for offices, meeting rooms, and classrooms, control might be dominated by CO_2 and volatile organic compound (VOC) considerations.

13.3 Smart Active Glazing

Traditionally, glazing is a passive, transparent barrier allowing in daylight, facilitating a connection with the exterior and, as rooms without glazing are felt to be oppressive, providing a wellness factor. They also, when openable, provide the opportunity for fresh air. However, glazing is non-discriminatory, at times, allowing in too much daylight (glare) and facilitating unwanted energy exchange (heat loss/gain).

Multi-pane arrangements, inert gas filled cavities, and selective reflective coatings/layers are now commonplace, but twenty-first century glazing is moving towards having a more active role as window technologies become '*smart*'. Smart windows aim to actively control the chromogenic properties of a building's glazing removing the need for blinds/shades and helping with cooling resulting from solar heat gain.

Passive chromogenic glazing systems rely on a coating having optical properties that vary with changes in temperature. The coating is engineered such that at a set activation temperature its transmissivity–wavelength relationship is switched from having a positive to a negative characteristic essentially reflecting infrared wavelengths. The challenge is to engineer materials that behave in this manner at commonly encountered building temperatures. Current research focusses on doped vanadium dioxide (VnO_2) as the thermo-chromic material.

Active smart glazing systems are controllable and divided into three varieties: electrochromic (EC), liquid crystal (LC), and suspended particle (SP) types.

- *Electrochromic (EC)*: Electrochromic glazing uses a small DC electric field to facilitate (lithium or hydrogen) ion passage resulting in a change in molecular colour in an electrochromic film. When energised, the opacity (or '*colour*') of the glazing is increased due

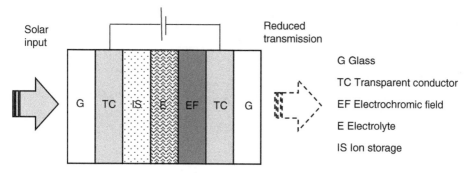

Figure 13.1   Generalised layer arrangement of (energised) EC glazing.

to a reduction in visible and near infrared wavelength transmission. On removal of the field, the transparency returns. A generalised cross-section is shown in Figure 13.1.

Both inorganic (metal oxides) and organic (e.g. polymers) electrochromic compounds are available. The degree of transmissivity and switching time are controllable by electric potential. Current research is focussing on selectable wavelength tuning to, say, transmit visible light only, reducing solar gain.

- *Suspended particles (SPs)*: In these materials, the orientation state of suspended particles determines its opacity. When unenergised, the particles (typically, polyhalogens in an organic fluid) are randomised and unaligned and the material has a low light transmission rate. However, upon the application of a small AC voltage, particle alignment is achieved and transmission rate increases. See Figure 13.2a.
- *Liquid crystals (LCs)*: This technology relies on a similar energising principle to that described previously with the difference that when unenergised the predominant mechanism is scattering and so the glazing appears translucent. Upon the application of a voltage, crystals in a polymer matrix become single plane-aligned resulting in an increased light transmission. See Figure 13.2b.

Other 'smart' technologies are also becoming more commonplace such as self-cleaning (or hydrophobic) glass and glazing integrated photovoltaics (PVs).

13.4   Cooling Technologies

Caloric effect technologies bring about a change in temperature in a solid-state material by a change in the application of an external field (electric, magnetic, pressure, or stress). One of the attractions of this approach is the non-requirement for the presence of a potentially climate modifying, toxic, or flammable refrigeration fluid.

13.4.1   Elasto-Caloric Refrigeration

At the time of writing, a great deal of research is focussing on the thermo-elastic or *elasto-caloric* effect of materials, a phenomenon related to stress-induced phase transformations.

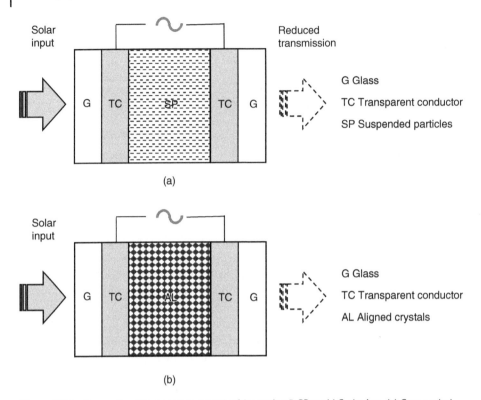

Solar input

Reduced transmission

G Glass
TC Transparent conductor
SP Suspended particles

(a)

Solar input

G Glass
TC Transparent conductor
AL Aligned crystals

(b)

Figure 13.2 Generalised layer arrangement of (energised) SP and LC glazing. (a) Suspended particle glazing and (b) liquid crystal glazing.

Most students of science and engineering will be familiar with force-related elasticity phenomena for ductile materials. This typically results in a linear relationship between a supplied stress (force/area, Pa) and strain (change in dimension/original dimension, %) in the elastic region, typically followed by a plastic region and fracture.

Force-related elastic moduli (stress/strain, Pa) tend to decrease with increasing temperature.

Less well known is the phenomenon of *entropy-related* elasticity, for some materials, which is dependent on atomic lattice (Austenitic – Martensitic) transformations in structure. The stress–strain relationship is far from linear in this case. Moreover, as a result of the phase transformation, the removal of the applied stress results in a hysteresis or irreversibility effect.

Entropy-related elastic moduli (stress/strain, Pa) tend to increase with increasing temperature.

Loading results in an increase in a material's Martensitic (low symmetry) fraction and an increase in material temperature, under adiabatic conditions. If the process is conducted under isothermal conditions, a release of heat will result. Unloading produces a return transformation to an Austenitic (high symmetry) lattice and a cooling effect.

Research is focussing on materials that exhibit a large entropy and temperature change over the cycle. Required applied stresses are in the range of 100s of MPa.

Typical measured strains are up to 5%. Current likely candidates for exploitation include non-magnetic (e.g. nickel-titanium) and magnetic (e.g. nickel-manganese) shape memory alloys (SMA).

Under adiabatic conditions, temperature changes of up to 30 K have been reported but 15 K or less is more common. More work is required on a bulk material configuration providing useful heat capacity characteristics and the loading/unloading mechanism. Fatigue, durability, and mechanical failure after a limited number of cycles are also issues, requiring attention be paid to material microstructure.

13.4.2 Magneto-Caloric Refrigeration

A technology relying on the relationship between magnetism and heat.

Entropy is a state of atomic disorder. For most engineering materials, thermal entropy, i.e. associated with temperature dominates its presence. However, for some materials, magnetic entropy, and the condition of its inherent magnetic domains are important. Here, total entropy is a balance between its thermal and magnetic forms. For these materials, the magnetic entropy contribution can be altered by the presence of an applied magnetic field.

If the magnetic domains are randomised, the application of a magnetic field will line them up and thus reduce the magnetic entropy of the material. Providing the magnetic field is applied adiabatically (i.e. in a thermodynamically reversible manner), then the total entropy of the material will not change and its thermal entropy must increase to compensate. Thus, its temperature will rise. On removal of the magnetic field, the reverse will take place and the temperature of the material fall.

Magnetic refrigeration requires the use of rare earth metals or their alloys (known as *Magneto-Calorics*), an environment of variable magnetic field strength and cooling and heating fluid circuits.

To utilise the magneto-caloric (MC) phenomenon to facilitate refrigeration, four processes are required.

1. The unmagnetised material is introduced to a strong magnetic field causing any magnetic dipoles in the material to line up. If the process is adiabatic, then this increase in order (i.e. decrease in entropy) will result in a temperature increase and the material becomes heated.
2. The magnetised/heated material gives up its heat externally to circulating fluid.
3. The magnetised material is adiabatically demagnetised by altering the operating magnetic field conditions from that in process 1 and cools.
4. The demagnetised material absorbs external heat from a circulating fluid.

If processes 1 and 3 occur under conditions of a constant magnetic field strength, then the cycle diagram is analogous to that of the reversed Brayton case.

A number of ingenious, compact, rotary designs are available that either move the MC material through a stationary magnetic field or, alternatively, rotate the magnetic field relative to a number of stationary magnetic-caloric materials. Depending on layout, a number of diverting valves/actuators and a small pump will be required to facilitate the cooling/heating fluid paths.

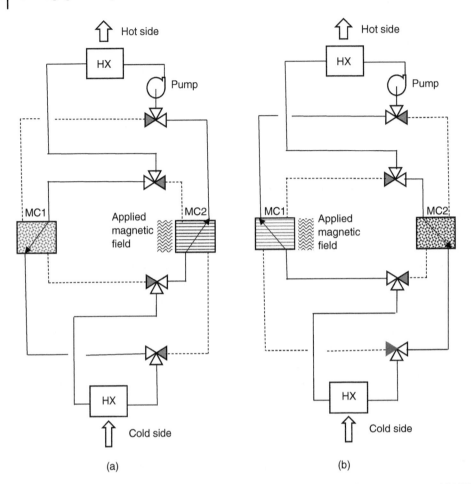

Figure 13.3 Example of stationary (two) bed magneto-caloric refrigeration arrangement. (a) MC2 core energised. (b) MC1 core energised.

A typical schematic is shown in Figure 13.3. Here, cold side-to-hot side heat extraction is maintained by alternating the magnetic field between two stationary MC materials.

Materials exhibiting a useful temperature change for practical purposes include gadolinium, lanthanum, manganese, praseodymium, and their compounds. These tend to be expensive, however, and research is focussed on finding cheaper materials that will function at lower magnetic field strengths (<5 T).

13.4.3 Electro-Caloric Refrigeration

A technology relying on the relationship between electrical charge and heat.

The degree of polarisation (*lining up*) of electrical charges (+/−) in a material is another measure of entropy. Dielectric materials are essentially electrical insulators lacking enough free electrons to conduct a significant current. Instead, the application of an electric field (+/−) across such a material slightly displaces its charges resulting in an increase in orderliness and hence a decrease in entropy which induces a temperature change (rise) under adiabatic conditions. Removal of the electric field results in material depolarisation bringing about the return of disorder and a cooling effect. Therefore, after polarisation, heat may be removed from the material whilst after depolarisation the absorption of heat by the material is possible, both effected by suitable heat transfer fluids. Suitable materials are ceramics and thin-film polymers.

Typical electric field strengths required are, currently, high being of the order of 10–100 MV/m. Current thin film-based devices suffer from low thermal mass.

13.4.4 Baro-Caloric Refrigeration

This technology relies on the application of pressure to a molecularly disordered solid to bring about entropic change. Historically, bringing about a small entropy change has required quite large pressure resulting in disappointing temperature effects. However, some recent research (2019) using an orientationally disordered plastic material (NPG, neopentylglycol) operating at lower pressures indicates quite significant (order of magnitude) improvements in the effect and a better future for the technology.

13.4.5 New and Re-Emerging Refrigerants

As a result of the disastrous effect of chlorine-loaded refrigerants on the planet's ozono-sphere, non-ozone depleting hydrofluorocarbons (HFCs) now dominate the refrigerant market. However, HFCs do not solve the second issue associated with refrigerants, i.e. their role as climate-modifying gases.

This unwelcome refrigerant characteristic is quantified by its Global Warming Potential (GWP). This factor is a (relative) per molecule measure, with a carbon dioxide molecule arbitrarily given a GWP value of unity.

Typically, GWP values of 150 are regarded as 'low'. GWPs of 150–2500 are 'medium' and GWPs of over 2500 are 'high'. Approximately one-third of HFCs lie in the 'high' band with the remainder having a rating of 'medium'.

At the time of writing, the solution being proposed by the refrigeration industry is the introduction of hydrofluoroolefins (HFOs). These are a structural variant of HFCs. The variation lays in a double bond between its carbon constituent atoms making it, essentially, an alkene and can be simply thought of as halogenated propene with fluorine replacing hydrogen in its structure.

Examples include 2,3,3,3-tetrafluoropropene (HFO-1234yf) and 1,3,3,3-tetrafluoro-propene (HFO – 1234ze).

Single component HFOs have GWP values in single figures and so do not attract any controls (<150) under the European Union (EU) F – Gas regulations, at the time or writing.

They are chemically stable but more reactive than HFCs; hence, they have short atmo-spheric lifetimes, typically in the range of one to three weeks.

HFOs have been used in automotive air conditioning systems for some time, and there has been some concern about marginal flammability at elevated temperatures in some isolated cases. Nevertheless, their development continues to expand into areas such as supermarkets and other chilling applications.

It appears their introduction into the larger market will take place via blending with HFCs to accommodate existing systems. Low (<10) GWPs are difficult to achieve via this route, but reductions in GWP of more than 50% are possible.

In the case of using a blend, care must be taken with operation as a result of the inherent variable phase change temperature range or '*glide*'.

Meanwhile, some early refrigerant types are making a partial comeback. As a result of their flammability issues, the industry may have called time on hydrocarbon refrigerants, e.g. R290 (propane), R600a (iso-butane), etc., regardless of their very low GWP. However, R717 (ammonia, GWP 0) and R744 (carbon dioxide, GWP 1) are still in favour for chilled water and low temperature applications, respectively.

13.5 Smart Tuneable Acoustic Insulation

Traditional sound absorption techniques rely on pressure changes, thermal effects, or the mass/deflection of the intervening material. Generally, these are passive, non-tuneable sound control methods. An alternative, tuneable approach is suggested by the use of non-Newtonian fluids as the sound absorbing medium.

In the study of fluid mechanics, Newtonian fluids are materials that display a constant stiffness or resistance to flow when subjected to force. For example, water is regarded as Newtonian. On the other hand, the stiffness exhibited by non-Newtonian fluids will depend on the rate at which it is being stressed. There is a category of non-Newtonian EMR (electro-magneto-rheological) fluids whose properties are altered by the presence of an electric or magnetic field. Upon application of the field, the material changes its nature from one of a free flowing liquid to that of a quasi-solid.

In the case of electro-rheology, the material comprises an electrically insulating fluid containing non-conducting particles, e.g. silica in silicone oil. As a result of attractive particle clustering, changes in fluid viscosity of several orders of magnitude are reported for electric field strengths of the order of 1 kV/mm.

In magneto-rheology, the material comprises a non-magnetic fluid containing magnetisable (e.g. iron) particles. The clustering effect in this case is the result of attractive dipolar forces between magnetised particles forming chains and structures.

If such a material is incorporated as a sandwich or mid-layer in a composite panel, it should be possible, with suitable sound monitoring, to actively and reversibly tune its response to any incoming sound by the application of a suitable electric or magnetic field across its volume. Research appears to indicate that the effect is, however, far from simple as the reduction in sound transmitted through any such material will depend on panel geometry, particle geometry, and incoming sound frequency as well as applied field magnitude and application points. Theoretical studies show that EMR-energised fluids are able to out-perform air and other non-energised solid layers of equivalent thickness at the lower frequency end (<5000 Hz) of the range of hearing. Work shows that this type of acoustic insulation may also have an application in the infra-sound range. The challenge is to manufacture suitable materials that can provide attenuation, cheaply, at low applied energy field levels, over the full range of human hearing.

13.6   Smart (Human Centric) Lighting Design

The existence of light sensitive rod and cone cells in the human eye and their respective roles in helping the brain making sense of the local environment has been known for some time. However, more recently, a third type of retinal light receptor, the *intrinsically photo-sensitive retinal ganglion cell* (ipRGC), and associated photo-pigment has been discovered that appears to perform a different function.

Human beings have a biologically programmed sleep–wake cycle over a 24-hours period. This daily (or *circadian*) rhythm is controlled by a part of the brain (the *suprachiasmatic nuclei*) receiving input concerning ambient light levels. This input is provided by the ipRGC cells. Low light levels tend to facilitate the release of the 'sleep' hormone *melatonin* promoting low activity. In the morning, higher levels of blue wavelength light tend to reduce its levels and therefore facilitate the effects of *cortisol* – the 'awake' hormone.

An understanding of the light intensity – wavelength relationship can allow designers to optimise the visual environment in a space to enhance alertness, task performance, and well-being.

The simplest lighting systems are designed to provide a set lighting level (lux) at a working plane in the absence of daylight. If sufficient daylighting is available and the luminaire dimmable, a photodetector could be used to reduce the lighting system power input. Occupancy detection providing an on/off function is also common.

However, a smart lighting system seeks to support the circadian rhythm and provides the optimum light intensity and colour, over the day, through the use of daylighting in combination with a Wi-Fi controlled, tuneable LED lighting design. Further, a smart lighting system could take into account the occupant activity and the optimum light level/colour to help reduce eye strain and headaches resulting from high contrast environments. These considerations are leading engineers to increasingly make use of the term '*Human-centric lighting design*'.

13.7   Active Botanical Air Filtration

Providing there is no release of pollen or spores, it is generally recognised that the presence of plants in an indoor space is psychologically positive. However, there is now some reason to believe that plants may be able play a role in cleaning indoor air, as indeed they already do in soil and water remediation schemes.

Pot plants and green roofs already play their part but they are, however, passive devices. *Green walls* seek to go some way to purification by having the air mechanically pulled through their volume.

A section through a green wall would typically comprise a substrate or packing medium (e.g. coconut coir – fibre from husks) for root growth into which plants would be inserted. The substrate is trickle fed with nutrients. The substrate section is also punctuated by a perforated air conduit which itself is attached to a supporting wall.

The conduits converge in a manifold which supplies a duct to the rear of the wall.

Air is pulled by a fan through the foliage and substrate into the conduits and ductwork before being supplied to a space. See Figure 13.4.

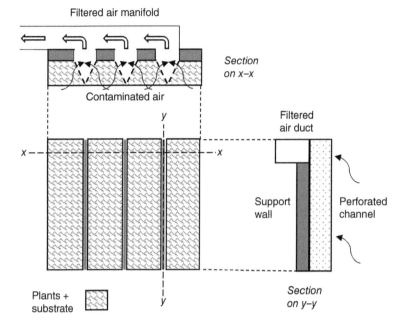

Figure 13.4 Schematic layout of wall mounted botanical air filter.

Central to the purification process is biodegradation of pollutants in the *rhizosphere* or root environment where pollutant gases are absorbed and diffused by the liquid film before processing by substrate micro-organisms.

Gases may also be absorbed directly via leaves (stomatal uptake).

Leaves also provide surface area for particulate collection (foliar interception) and from this perspective the rougher the leaf the better.

In addition to ambient CO_2/O_2 levels, there is a growing body of evidence to show that plants can have a significant effect on VOC level and, to a lesser extent, particulates.

There are some challenges to be overcome, however. It is necessary to ensure that the wall does not become a secondary source of pollutant. Additionally, there is some evidence to suggest that Green walls can significantly increase the local relative humidity and so air purity gains may have to be balanced by the requirement for dehumidification.

Over time the plants may suffer from the build-up of toxins as a result of their pollutant load. The pollutant removal and resistance will vary with species. If there is to be a scale adoption of this technology, there may be a case for the genetic engineering of pollutant-hardy species.

13.8 Peak Lopping Thermal Mass

Thick, dense walls of high thermal mass can be useful in reducing temperature fluctuations due to external variation. However, solar energy may be captured passively in the elements of a building's structure using phase change materials (PCMs). Closed cell PCM wall boards, ceiling tiles, and composite PCM/concrete blocks are available. See Figure 13.5. Here the

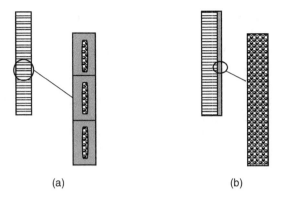

(a)   (b)

Figure 13.5   Encapsulated PCM wall components. (a) PCM cavity filled blocks and (b) PCM filled wall lining board.

PCM serves to moderate daytime cooling loads and, if required, release its heat overnight to an occupied space to keep inside temperatures at a suitable level. Microencapsulation and even the incorporation of PCM's into a gypsum matrix are under consideration.

With careful design, reductions in peak loads and hence plant sizing is possible.

Of course, in all cases, the structural integrity of the building must not be compromised by their inclusion.

13.9   Smart Batteries

When reliance on electrical energy storage becomes common place in buildings, its performance/condition monitoring will become critical and all battery systems will have to become '*smart*'. Systems will then have to able to accurately report on factors such as state of charge/state of health, remaining run time, and deliverable power output.

Charge management systems will have to feature flexibility over and above the commonplace '*bulk, absorption, maintenance*' strategy, be able to react to the battery charge/recharge state and temperature as well as the power availability from a renewable supply during the charge cycle.

Such systems will also be able to log charge cycles, usage patterns, state of self-discharge, as well as schedule maintenance. Clear and timely data communication with user and charger will be the key.

It is likely that some efficiency and capacity battery developments in the automotive industry will migrate across to the building sector.

13.10   Smart Sensors and Meters

Traditional, '*dumb*' building sensors simply convert the condition of a local physical variable, e.g. temperature, humidity, light level into an electrical signal for action by a remote controller/control element. They tend towards little or no on-board intelligence or memory and communicate via a hardwire arrangement.

'*Smart*' sensors are the antithesis of this approach instead utilising the Internet of Things (IoT) philosophy. They are Wi-Fi connected, can be linked to a cloud computing resource enabling the output of actionable information, and able to receive central commands.

In addition to basic sensing, signal processing, and communication (send/receive) technology, they should have sufficient on board processing power (*edge computing*) to carry out functions such as self-condition monitoring and calibration. When physical variables are grouped in sensor combination, e.g. room temperature, air flow rate, occupancy, light levels, and solar irradiation, there should be sufficient processing power to calculate derived values locally prior to transmission. Combinations such as these provide an opportunity for improved energy management and sustainability.

Smart sensors should also be low cost, unobtrusive, and low power. Long life batteries are an option but sustainability could be optimised by the sensor being able to harvest locally available energy. In addition to controlling internal environment condition, smart sensors can also play an important role in plant condition monitoring in devices such as pumps and filters providing a predictive role with respect maintenance and failure. Technical concerns remain regarding compatibility and retrofitting, communication protocols, security, and future proofing. From a social perspective, personal privacy may also be an issue if a sensor is focussed on human movement factors such as occupancy, location, and status.

At their most simple '*dumb*' meters measure and record cumulative consumption parameters for local access. In some larger consumer units, they may also record the maximum rate demanded over any period. Crude analogue counter mechanisms are common. Manual reading is required and for many years this was deemed acceptable.

However, with changes in the energy supply/demand market, their inflexibility renders this no longer the case. Drivers for change include the growth in renewable energy generation, the commitments to electric vehicle conversion, and tariff complexity. With on-going developments like this, all meters must become '*Smart*' and this change is already under way.

It is possible to differentiate 'smart' from 'dumb' meters by the presence of significant on-board processing power, memory, and the ability to receive/transmit data with the outside world. From the perspective of a utility supplier, a smart meter can receive software updates including tariff changes, provide remote reading, and record/supply detailed consumption patterns and rates. Estimated billing is eliminated.

Real time consumption/rate data can also be transferred to consumers. This, of itself, does not reduce consumption, but it is hoped that with this information, consumers will be willing to alter their behaviour/usage patterns to minimise costs.

Traditionally, conventional building services control systems can be thought of as dedicated and hard wired to a specific building. The IoT philosophy will change this approach. As described earlier, smart, sustainable sensors/controllers will contain embedded memory and data exchange capabilities able to report their location, status, and use in a wireless network. Currently, Wi-Fi is the standard communication 'language' devices use to talk to each other because of its range. However, it is relatively power hungry. With improved range, lower power variants may supersede it.

13.11   Smart Microgrids

National (or Mega) electricity grids traditionally comprise a relatively small number of large-scale energy generators linked by a network of cables/substations to each other and consumers. If the power is generated by steam- or gas-based cycles using fossil fuel sources, then low generation efficiency (30–35%) results in unacceptable emission rates and high costs.

Furthermore, this traditional power supply strategy has problems maintaining quality (e.g. voltage, frequency) when confronted with the intermittent nature of renewable energy sources. The matching of energy supply with use can be exacerbated by a rapidly fluctuating or 'peaky' energy demand.

One proposed solution to these problems is to change the nature of the national grid from a centralised format to one of distributed, localised microgrids.

A microgrid is an energy-flexible entity having a clearly defined boundary with respect to its inherent energy sources and loads. The boundary might, for example, encompass a business park, industrial estate, or university campus. It is designed to be able to operate either synchronised with the national grid via *a single connection* or as a stand-alone system or electrical energy *'island'*.

It is envisaged the energy sources within the microgrid will be diverse and resilient to accommodate changes in supply/demand patterns.

Inside the microgrid, on the supply side, this might entail the deployment and inter-connection of renewables, e.g. photovoltaics (PVs) and wind turbines, combined heat and power (CHP) plant (e.g. fuel cells and reciprocating engines), as well as energy storage devices such as batteries and PCMs.

On the load side, flexibility and diversity also play an important role.

Demand management is key with loads classified as *sensitive* (i.e. data, security, and safety systems that must be maintained), *adjustable* (i.e. flexible with respect to timing), and *sheddable* (i.e. can be cut-off from supply without penalty or risk).

Without the intervention of the national grid, the microgrid has the potential for higher energy usage efficiencies, reduced resource consumption, and reduced emissions at the point of use. From the perspective of the microgrid community, the attractiveness of the scheme lies in the avoidance of national grid utility energy charges, enhanced reliability, and the possibility of bi-directional power flow.

Problems with the operation of microgrids include levels of circuit protection, the injection of unwanted frequency harmonics, and unbalanced loads affecting the voltage.

Energy governance (i.e. power, voltage, and frequency) inside the microgrid is carried out by a smart microgrid controller. This controller must be able to communicate with the microgrid community as well as linking the microgrid with the national grid when not in 'island' mode. See Figure 13.6.

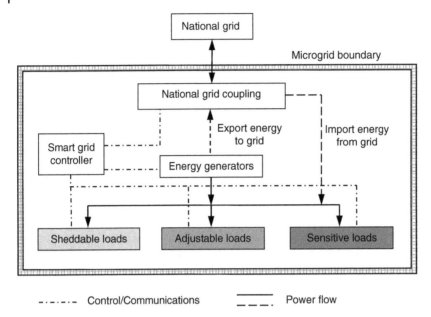

Figure 13.6 Microgrid interactions.

13.12 Hydrogen

13.12.1 Methane Combustion Chemistry

For much of the Northern hemisphere, the heating fuel of choice is natural gas. Considering natural gas to be primarily CH_4, the stoichiometric (or chemically correct) combustion of the gas with air (21% O_2 + 79% N_2) can be described in molar terms by

$$CH_4 + 2(O_2 + 3.76N_2) \rightarrow CO_2 + 2H_2O + 7.52N_2$$

Making the simplification that the nitrogen is unchanged, the products are carbon dioxide and water vapour.

In mass terms, every kilogram of methane results in the production of 2.75 kg of CO_2.

On a gross calorific value basis, an emission factor of approximately 0.18 kg of CO_2 per kWh of natural gas energy is often used.

Heating buildings make a significant contribution to a country's carbon footprint and the mid-century (2050) carbon reduction targets agreed by many nations seem unlikely to be met with a continuance of the current fossil fuel strategy.

Of course, demand reduction is the priority but after that, a new direction is required on the supply side.

13.12.2 Hydrogen Combustion Chemistry

Using hydrogen (H_2) to, initially supplement and later, replace natural gas is a zero (*at the point of use*) carbon solution.

The stoichiometric (or chemically correct) combustion of hydrogen with air can be described in molar terms by

$$2H_2 + (O_2 + 3.76N_2) \rightarrow 2H_2O + 3.76N_2$$

13.12.3 Fuel Property Comparison

Although hydrogen is a component of methane, the former has some significantly different characteristics. For example, the energy content or *Gross calorific value* of hydrogen (~142 MJ/kg) is approximately 2.5× times that of methane, while its flame speed (~3 m/s) can be an order of magnitude greater than methane. On the other hand, some characteristics are similar, e.g. both fuels produce flames having a temperature of around 1900–2100 °C.

A comparison of some of the thermophysical and transport properties of hydrogen and methane at ambient pressure is provided in Table 13.1

13.12.4 Fuel Substitution

In any suggested fuel substitution to hydrogen, there are, perhaps, three aspects to consider:

- *Gas production*: Global hydrogen requirements could be sourced by the splitting of methane itself with steam in the presence of a catalyst. In simple terms:

$$CH_4 + 2H_2O \rightarrow 4H_2 + CO_2$$

  However, the carbon footprint is not attractive as the process itself would result in CO_2 requiring capture.
  A more sustainable solution is to electrolyse water with a source of renewable electricity. Again, this requires the presence of a catalyst but is capable of supplying very pure hydrogen.
- *Gas distribution and storage*: Existing natural gas grids, typically, restrict the presence of hydrogen to less than 1%. Completely replacing the grid with new pipelines would

Table 13.1 Approximate properties of methane and hydrogen.

Property	Methane (CH$_4$)	Hydrogen (H$_2$)
Molar mass (kg/kmol)	16.04	2.016
Density (kg/m³) at 0 °C, 1 atm	0.72	0.0899
Ratio of specific heats at 1 atm	1.307 at 16 °C	1.4 at 0–200 °C
Dynamic viscosity (kg/ms) at 1 atm	8.7×10^{-6} at 15 °C	10.6×10^{-6} at 4.4 °C
Boiling point (°C) at 1 atm	−161.5	−252.9
Stoichiometric air–fuel ratio	17	34
Flammability limits (% fuel by volume)	5.3–14	4–75
Auto-ignition temperature (°C)	540	585
Gross calorific value (kJ/kg) at 25 °C, 1 atm	55 510	141 790

Source: Adapted from Speight, J.G. (2017) *Rules of thumb for Petroleum Engineers*, John Wiley and Sons.

be a massive financial undertaking but may not be necessary. Research is already being carried out on the effect of increasing the grid supply hydrogen content to 20% with positive results. A complete conversion to hydrogen would require some significant number crunching to determine the effect of hydrogen transport properties on the pressures in existing storage vessels, compressors, and ancillary line equipment.

From the above, it can be seen that methane has a much higher density than hydrogen although their dynamic viscosities are comparable in magnitude.

In terms of safety, hydrogen flames are difficult to detect visually. Further, methane flame detection systems are unsuitable for hydrogen. Like methane, hydrogen is odourless and so odourising additives will be required.

- *Point of use*: In terms of combustion, the spontaneous ignition temperatures and flame temperatures of the two gases are of similar order. However, differences in gross calorific value and flame speed will necessitate a re-think in burner design. The safety points made above also apply to the point of use.

A move towards hydrogen is now beginning to figure strongly in many international economic, industrial, and energy strategies as part of a desire to provide energy security and resilience in a low or zero-carbon world. For example, the European Union (EU) plans to become the first climate neutral continent on the planet by 2050 and, in 2020, it announced its plan to set up a '*clean hydrogen alliance*' across the bloc with a view to maintaining their leadership in the technological development and deployment of hydrogen.

14

Closing Remarks

Chapter 13 provides a view on what might be. It seems that technology is able to solve most problems (as well as, sometimes, creating unintended consequences!).

Whether all, some, or none of the 'emerging' systems described become commonplace will likely be determined by factors other than science and technological development. Ultimately, the resulting technological demands placed on building services engineers are a consequence of planning, building form, and building envelope design. Therefore, it is at this early stage that decisions regarding health and well-being in the indoor environment must be taken. It is worth remembering that a simple, well-designed building often requires only simple, low-impact service solutions. Hence, the input of the building services engineer at a project's inception will be crucial. There are many cases where it has been discovered, too late, that a glass palace is best suited to plants and not people, even after the expenditure of large sums on service equipment and energy bills.

Looking deeper, clearly, building planning is driven by government regulation and policy which is, in part, informed by industry advisory bodies. Perhaps, it is by sitting on such bodies and leading the discussion that experienced building services engineers can have their greatest effect on the creation of low emission, habitable, indoor spaces.

In 2020, the surface and airborne Covid-19 virus swept across the planet. The internal built environment with its movement and mixing of people provided an ample opportunity for the virus to thrive. Building services engineers have played their part in retarding the spread as some of the standard practices described in this book, e.g. mixing of air streams in air conditioning systems, use of thermal wheels, etc., have been deemed by the relevant engineering institutions to be inappropriate or questionable under current circumstances. These developments, again, underline the important role played by building services engineers in providing a safe and healthy internal environment.

Building Services Engineering: Smart and Sustainable Design for Health and Wellbeing, First Edition.
Tarik Al-Shemmeri and Neil Packer.
© 2021 John Wiley & Sons Ltd. Published 2021 by John Wiley & Sons Ltd.

Appendix A

The Psychrometric Chart

Building Services Engineering: Smart and Sustainable Design for Health and Wellbeing, First Edition.
Tarik Al-Shemmeri and Neil Packer.
© 2021 John Wiley & Sons Ltd. Published 2021 by John Wiley & Sons Ltd.

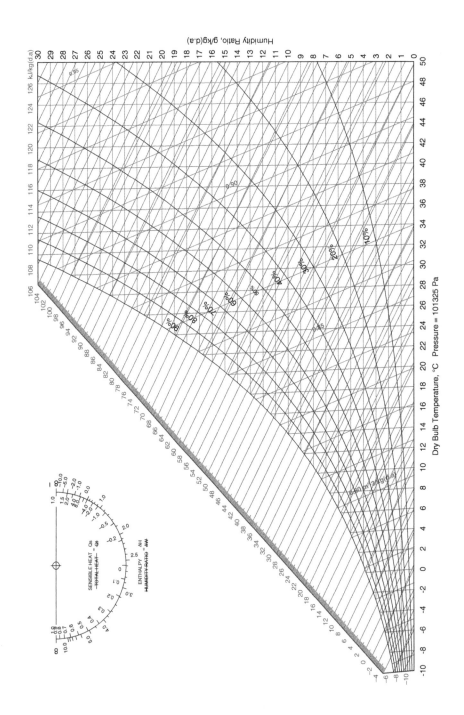

Source: Courtesy of Flycarpet Inc, NY, USA.

Appendix B

Refrigerant Thermodynamic Properties

Building Services Engineering: Smart and Sustainable Design for Health and Wellbeing, First Edition.
Tarik Al-Shemmeri and Neil Packer.
© 2021 John Wiley & Sons Ltd. Published 2021 by John Wiley & Sons Ltd.

Table Appendix B.1 1,1,1,2 Tetrafluoroethene (R134a).

Enthalpy

Pressure	Saturation temperature	Saturated conditions (kJ/kg)		Enthalpy (h, kJ/kg) at stated superheat temperature (°C)							
(bar)	(°C)	Saturated liquid (h_f)	Saturated vapour (h_g)	−10	0	10	20	30	40	50	60
2.0	−10.09	36.84	241.3	241.38	250.1	258.89	267.78	276.77	285.88	295.12	304.5
4.0	8.93	62.0	252.32	—	—	253.35	262.96	272.54	282.14	291.79	301.51
6.0	21.58	79.48	259.19	—	—	—	—	267.89	278.09	288.23	298.35
8.0	31.33	93.42	264.15	—	—	—	—	—	273.66	284.39	294.98
10.0	39.39	105.29	267.97	—	—	—	—	—	268.68	280.19	291.36
12.0	46.32	115.76	270.99	—	—	—	—	—	—	275.52	287.44

Entropy

Pressure	Saturation temperature	Saturated conditions (kJ/kg K)		Entropy (s, kJ/kg K) at stated superheat temperature (°C)							
(bar)	(°C)	Saturated liquid (s_f)	Saturated vapour (s_g)	−10	0	10	20	30	40	50	60
2.0	−10.09	0.1481	0.9253	0.9256	0.9582	0.9898	1.0206	1.0508	1.0804	1.1094	1.1380
4.0	8.93	0.2399	0.9145	—	—	0.9182	0.9515	0.9837	1.0148	1.0452	1.0748
6.0	21.58	0.2999	0.9097	—	—	—	—	0.9388	0.9719	1.0037	1.0346
8.0	31.33	0.3459	0.9066	—	—	—	—	—	0.9374	0.9711	1.0034
10.0	39.39	0.3838	0.9043	—	—	—	—	—	0.9066	0.9428	0.9768
12.0	46.32	—	—	—	—	—	—	—	—	0.9164	0.9527

Source: Adapted from Moran et al. (2018).

Table Appendix B.2 Ammonia (R717).

Enthalpy

Pressure	Saturation temperature	Saturated conditions (kJ/kg)		Enthalpy (s, kJ/kg K) at stated superheat temperature (°C)							
(bar)	(°C)	Saturated liquid (h_f)	Saturated vapour (h_g)	−10	0	20	40	60	80	100	120
2.0	−18.86	93.8	1419.31	1440.31	1463.59	1509.18	—	—	—	—	—
4.0	−1.9	171.18	1439.89	—	1444.86	1495.33	1543.38	1590.17	1636.41	1682.58	1728.97
6.0	9.27	223.32	1451.12	—	—	1480.55	1532.23	1581.38	1629.22	1676.52	1723.76
8.0	17.84	263.95	1458.3	—	—	1464.70	1520.53	1572.28	1621.86	1670.37	1718.48
10.0	24.89	297.76	1463.18	—	—	—	1508.2	1562.86	1614.31	1664.10	1713.13
12.0	30.94	327.01	1466.53	—	—	—	1495.18	1553.07	1606.56	1657.71	1707.71

Entropy

Pressure	Saturation temperature	Saturated conditions (kJ/kg K)		Entropy (s, kJ/kg K) at stated superheat temperature (°C)							
(bar)	(°C)	Saturated liquid (s_f)	Saturated vapour (s_g)	−10	0	20	40	60	80	100	120
2.0	−18.86	0.3843	5.5969	5.6781	5.7649	5.9260	—	—	—	—	—
4.0	−1.9	0.6776	5.3548	—	5.3731	5.5515	5.7101	5.8549	5.9897	6.1169	6.2380
6.0	9.27	0.8649	5.2122	—	—	5.3145	5.4851	5.6373	5.7768	5.9071	6.0304
8.0	17.84	1.0054	5.1099	—	—	5.1318	5.3161	5.4763	5.6209	5.7545	5.8801
10.0	24.89	1.1191	5.0294	—	—	—	5.1768	5.3460	5.4960	5.6332	5.7612
12.0	30.94	1.2152	4.9625	—	—	—	5.0553	5.2347	5.3906	5.4433	5.6620

Source: Adapted from Moran et al. (2018).

Bibliography

Al-Shemmeri, T.T. (2011). *Energy Audits: A Workbook for Energy Management in Buildings*. Wiley.

Billington, M.J., Barnshaw, S.P., Bright, K.T. et al. (2017). *The Building Regulations: Explained and Illustrated*. Wiley Blackwell.

Borgnakke, C. and Sonntag, R.E. (2016). *Fundamentals of Thermodynamics*, 9e. Wiley.

CIBSE (2015). *Guide A: Environmental Design*. London, UK: The Chartered Institution of Building Services Engineers.

CIBSE (2020). *TM40: Health and Wellbeing in Building Services*. UK: The Chartered Institution of Building Services Engineers.

CORGI (2008). *Ground Source Heat Pumps*, 1e. Livewire Intelligent Media Ltd.

Davies, M.G. (2004). *Building Heat Transfer*. Wiley.

Dincer, I. (2017). *Refrigeration Systems and Applications*, 3e. Wiley.

Dincer, I. and Rosen, M.A. (2010). *Thermal Energy Storage: Systems and Applications*, 2e. Wiley.

Duffie, J.A. and Bechman, W.A. (2013). *Solar Engineering of Thermal Processes*, 4e. Wiley.

Engineering ToolBox (2004a). Clo – clothing and thermal insulation [online]. https://www.engineeringtoolbox.com/clo-clothing-thermal-insulation-d_732.html (accessed 01 January 2020).

Engineering ToolBox (2004b). Met – metabolic rate [online]. https://www.engineeringtoolbox.com/met-metabolic-rate-d_733.html (accessed 01 January 2020).

Engineering ToolBox (2011). Building materials – vapour resistance [online]. https://www.engineeringtoolbox.com/vapour-resistance-d_1807.html (accessed 31 December 2019).

Etheridge, D. (2011). *Natural Ventilation of Buildings: Theory, Measurement and Design*. Wiley.

Fox, R.W., McDonald, A.T., Pritchard, P.J., and Mitchell, J.W. (2015). *Introduction to Fluid Mechanics*, 9e. Wiley.

Harper, P., Wilday, J., and Bilio, M. (2011). *Assessment of the Major Hazard Potential of Carbon Dioxide (CO_2)*. UK: HSE.

Health and Safety Executive (HSE) (2013). *Legionnaires' Disease. The Control of Legionella Bacteria in Water Systems*: Approved Code of Practice and Guidance (HSG274), 4e. HSE Books.

Hens, H. (2007). *Building Physics – Heat, Air and Moisture Fundamentals and Engineering Methods with Examples and Exercises*. Ernst & Sohn – A Wiley company.

Building Services Engineering: Smart and Sustainable Design for Health and Wellbeing, First Edition.
Tarik Al-Shemmeri and Neil Packer.
© 2021 John Wiley & Sons Ltd. Published 2021 by John Wiley & Sons Ltd.

HM Government (2015). *Ventilation: Approved Document F*. London, UK: The Building Regulations, NBS, RIBA Enterprises Ltd.

HM Government (2016). *Conservation of Fuel and Power: Approved Document L*. UK: The Building Regulations, NBS, RIBA Enterprises Ltd.

Hodge, B.K. (2017). *Alternative Energy Systems and Applications*, 2e. Wiley.

Kaminski, D.A. and Jensen, M.K. (2017). *Introduction to Thermal and Fluids Engineering*. Wiley.

McQuiston, F.C., Parker, J.D., and Spitier, J.D. (2004). *Heating, Ventilation, and Air Conditioning: Analysis and Design*, 6e. Wiley.

Mitchell, J.W. and Braun, J.E. (2012). *Principles of Heating, Ventilation, and Air Conditioning in Buildings*. Wiley.

Moran, M.J., Shapiro, H.N., Boettner, D.D., and Bailey, M.B. (2018). *Fundamentals of Engineering Thermodynamics*, 9e. Wiley.

Munson, B.R., Rothmayer, A.P., Okiishi, T.H., and Huebsch, W.W. (2017). *Fundamentals of Fluid Mechanics*, 8e. Wiley.

National Refrigerants Inc (2016). *Refrigerant Reference Guide*, 6e.

Packer, N. and Al-Shemmeri, T.T. (2018). *Conventional and Alternative Power Generation – Thermodynamics, Mitigation and Sustainability*. Wiley.

Pita, E.G. (2002). *Air Conditioning, Principles and Systems – An Energy Approach*, 4e. Wiley.

Rhodes, M. (2008). *Introduction to Particle Technology*, 2e. Wiley.

Salvendy, G. (2012). *Handbook of Human Factors and Ergonomics*. Wiley.

Speight, J.G. (2017). *Rules of Thumb for Petroleum Engineers*. Wiley.

Welty, J.R., Wicks, C.E., Wilson, R.E., and Rorrer, G.L. (2008). *Fundamentals of Momentum, Heat and Mass Transfer*, 5e. Wiley.

Williams, I. (2001). *Environmental Chemistry*. Wiley.

Websites

BREEAM, https://www.breeam.com

Chartered Institution of Building Services Engineers, https://www.cibse.org

Engineering toolbox, www.engineeringtoolbox.com

EPBD, https://ec.europa.eu

European Agency for Health and Safety at Work, https://osha.europa.eu

European Committee for Standardisation (CEN), https://standards.cen.eu

Flycarpet Inc., www.flycarpet.net

H_2 Hydrogen Tools, https://h2tools.org

Health & Safety Executive, UK, www.hse.gov.uk

International Commission on Illumination (CIE), www.cie.co.at

International Organisation for Standardisation, https://www.iso.org

International WELL Building Institute, https://www.wellcertified.com

LEED, http://leed.usgbc.org

National Refrigerants Ltd., https://nationalref.com

Passivhaus, www.passivhaustrust.org.uk

Planning and Building Regulations, UK, www.planningportal.co.uk

World Health Organisation (WHO), https://www.who.int

Index

Building Services Engineering: Smart and Sustainable Design for Health and Wellbeing, First Edition.
Tarik Al-Shemmeri and Neil Packer.
© 2021 John Wiley & Sons Ltd. Published 2021 by John Wiley & Sons Ltd.